48th Vibration Institute Annual Training Conference 2025 (VIATC 2025)

Newport News, Virginia, USA
6-8 August 2025

ISBN: 979-8-3313-2579-4

Printed from e-media with permission by:

Curran Associates, Inc.
57 Morehouse Lane
Red Hook, NY 12571

Some format issues inherent in the e-media version may also appear in this print version.

Vibration Institute
2625 Butterfield Road
Suite 128N
Oak Brook, IL 60523-3415
USA

Phone: (630) 654-2254
Fax: (630) 654-2271

information@vi-institute.org

Additional copies of this publication are available from:

Curran Associates, Inc.
57 Morehouse Lane
Red Hook, NY 12571 USA
Phone: 845-758-0400
Fax: 845-758-2633
Email: curran@proceedings.com
Web: www.proceedings.com

VIATC

TRAINING
CONFERENCE
& EXPO

NEWPORT NEWS, VIRGINIA | AUGUST 6-8, 2025

CONFERENCE PROCEEDINGS

Marriott Newport News at City Center

Newport News, VA

www.vi-institute.org | 630-654-2254

TABLE OF CONTENTS

TABLE OF CONTENTS

CENTRIFUGAL FANS AND VARIABLE FREQUENCY DRIVES: RISKS AND MITIGATIONS

Author: David Sheffler, CMRP, EIT

VIATC TRAINING CONFERENCE & EXPO
NEWPORT NEWS, VIRGINIA | AUGUST 6-8, 2025

Centrifugal Fans and Variable Frequency Drives: Risks and Mitigations

David Sheffler, CMRP, EIT Nuclear Facilities Engineer, Los Alamos National Laboratory

David Sheffler is a facilities engineer leading the predictive maintenance program for the nation's premier nuclear research facility at Los Alamos National Laboratory, implementing PdM technologies including vibration monitoring and ultrasonic compressed gas leak detection. David has over 10 years of maintenance and reliability engineering experience in automotive, paper, steel, and nuclear industries. He is a Certified Maintenance and Reliability Professional as well as a Vibration Analyst Category III.

Abstract

Variable frequency drives (VFDs) have become a preferred method of process control for electric motor driven equipment ranging from vertical water pumps to centrifugal air fans. Their use allows for a more energy efficient operation than controlling process flows with valves and dampers. However, the improper implementation of VFDs can lead to a number of other problems and increased maintenance costs. The following is a basic discussion of risks of using a VFD with direct driven centrifugal fans and methods of mitigating those risks to maintain equipment reliability.

Key Words: Variable Frequency Drive, VFD, Natural Frequency, Resonance, Fluting, Grounding, Centrifugal Fan

Risk - Natural Frequencies

All objects have frequencies at which they tend to vibrate, called "natural frequencies." If forced to vibrate at a natural frequency, an object will amplify the vibration to greater amplitudes. This forcing, or "excitation," can be from the speed, or frequency, that a motor or fan rotates. An amplified vibration as a result of excitation of a natural frequency can cause catastrophic equipment failure. The Tacoma Narrows Bridge disaster is a notable example, where the bridge shook violently until it collapsed because the wind passing over it excited the bridge's natural frequency.

Complex objects such as an electric motor driven fan have a number of different natural frequencies that need to be considered [1][2][3]. Fans that run at a single speed without a VFD are typically designed to run at a speed that does not excite any of the natural frequencies present in the fan assembly. When one of these fans is then run using a VFD, which operates at a wide range of running speeds, there is a risk that one of the possible new running speeds of the fan will excite a natural frequency [4].

Fan Wheels

Fan wheels have their own set of unique natural frequencies that can cause issues separate from the rest of the machine. The warping and bending of the side rings and fan blades at natural frequencies can cause failure of the fan wheel, resulting in imbalance and loss of process efficiency. Before the fan wheel fails, excessive vibration caused by operating at a natural frequency can often be detected in the fan bearings [1].

Equipment Base and Foundation

The foundation and base that the fan assembly rests on also has a unique set of natural frequencies. Similar to fan wheels, operating the equipment at one of these natural frequencies can cause failure of the base and foundation. However, because everything is directly mounted onto this part of the machine, vibration at a foundation's natural frequency can more easily cause excessive vibration in the fan bearings as well as in the motor [1][2].

System Ducting – Acoustic Natural Frequencies

When the vane pass frequency, or the running speed of the fan multiplied by the number of blades on the fan, matches the acoustic natural frequency of the ducting it feeds, the ducting can fail, resulting in process losses [3].

VFD Electrical Risks

The most common VFDs use Pulse Width Modulation (PWM) to modify the utility voltage wave. This method of frequency modification is achieved by essentially turning switches on and off very quickly to approximate an output voltage wave with a different frequency than the utility voltage wave. The higher the switching speed, the more closely the output wave approximates a pure sine wave, and the smoother the equipment runs as a result [5]. Low VFD switching speeds can generate unwanted vibration in the driven equipment. However, the switching action within a VFD generates voltage spikes, and the faster the switching speed, the larger the voltage spike [3][4][5]. Without proper insulation on the magnet coils in the motor, these voltage spikes can cause unexpected winding failure [5].

Another risk of stray voltage generated by VFDs is the presence of a motor shaft voltage. This will arc across the motor bearings, creating thousands of microscopic defects in the rolling elements and races which eventually results in a failed bearing with a pattern known as electrical fluting visible in the bearing races, shown in Figure 1 [6][7].

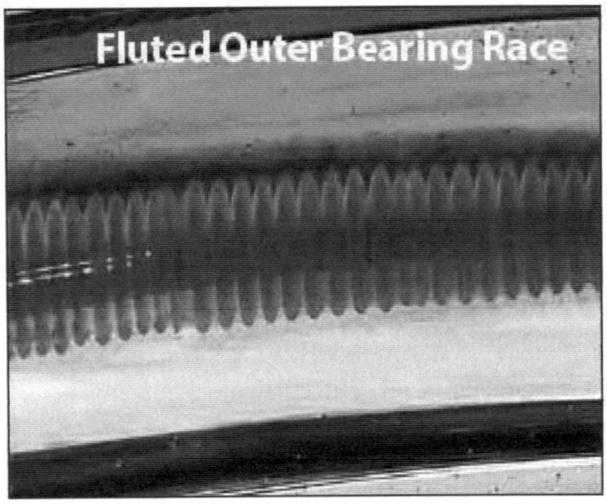

Figure 1: Electrical Fluting in a Bearing Race

Small issues in the design and manufacture of VFD circuitry create variations, or harmonics, in the voltage waveform signal generated by the drive [4][5]. These harmonics have unwanted effects on the torque of the motor, making its running speed fluctuate if the effect is severe enough, creating noise. The voltage harmonics are also influenced by the total number of rectifiers in the drive circuit, which generates characteristic signatures in the voltage signal depending on that number. Typically, more rectifiers means higher frequency harmonics which have smaller effects on noise generated by the motor. The effects are exacerbated by the length of cable between the VFD and the motor. If there are two identical motors with identical drives but one has problems and the other does not, and the problematic motor has a longer cable, the length of cable is likely the source of the problems [5].

Operation Risk

If a VFD driven fan is connected with a common header to constant speed fans, there is a risk that the fan using a VFD would run slow enough to either be starved or be unable to generate enough air velocity to overcome reverse flow through its outlet produced by the other fans, effectively dead-heading the fan. In this situation, the VFD driven fan is running without contributing to the process flow.

Mitigation – Determination of Natural Frequencies

Avoiding exciting the natural frequencies of a centrifugal fan first requires determination of those natural frequencies. This information is often available, especially for the fan wheels, from the manufacturer of the fan. If this is not available, the first step in determining possible natural frequencies is typically some type of impact testing. The most basic type of test to get a rough estimate of the natural frequencies is referred to as a "bump" test. A bump test is done by simply impacting the fan using a block of wood or dead blow hammer, which in turn causes the system to ring and the resulting vibration response is measured. Where the fan rings or vibrates the most will be the Natural frequencies, so this is a quick and fairly simple rough evaluation. If bump testing does suggest that there is a natural frequency that could be problematic, then Impact or Modal Testing would be the next step. This involves placing accelerometers on the equipment at key locations to measure the equipment's response, and striking it with an instrumented force hammer. This allows an accurate measurement of the resulting force/response vibration signatures and to identify the natural frequencies of the equipment. The terms Impact testing and Modal testing are often interchanged, however Modal testing is more involved since it includes not only the natural frequencies but also the vibration's associated deformation of "shape" typically shown as a 3D model. It is important to note that any type of impact testing should only be performed when the equipment is not running.

If the equipment in question is already running using a VFD, other tests such as Start-up and Coast-down and negative averaging can be used to determine the critical speed and other natural frequencies of the fan wheel. Start-up and coast-down testing is best performed with a VFD to control the rate of speed change during equipment start-up and coast-down. Otherwise, without a reduced rate of speed change, fans often start up and coast down too quickly to collect useful data. This test involves using two, preferably 4, sensors on the fan shaft bearings separated by 90 degrees and a tachometer signal. Proximity probes can be used to generate a plot of the shaft's orbit. Sudden changes in the shape of the orbit at a given running speed would indicate a natural frequency. Velocity and acceleration sensors can be used to generate a Bode Plot which displays vibration amplitude and phase versus RPM. A significant increase in amplitude with a corresponding phase change indicates a natural frequency.

Mitigation – Equipment Design and Modification

If an output setting of a VFD driven fan makes the fan turn at or close to a natural frequency, there are a number of possible solutions. The first, and easiest, is to consider a programming change to keep the fan from operating at that speed for any significant period of time or to shift the speed setting away from the natural frequency. If, however, the process the fan is feeding requires the fan to run at a natural frequency, physical modifications would be required to shift the equipment's natural frequency away from the fan speed. Examples of typical modifications to change natural frequencies include adding mass and/or installing stiffeners [1][2][3]. Typically, it is more cost effective to stiffen equipment and supporting structures than to add mass. Figure 2 is an example where mass was added to a vertical pump motor to lower the natural frequency of the machine. Below are some general rules of thumb to help avoid equipment structures that would be prone to excessive vibration and natural frequency excitation [1][2].

re 2: Mass Added to a Motor to Lower Natural Frequency

1. Total weight of the foundation should be greater than 5x the weight of all the equipment supported on the foundation, including housings, ducts, and other structures. The foundation weight should never be less than 3x the equipment weight.
2. Total weight of the foundation should be at least 10x the weight of the rotating parts of the machine. The foundation weight should never be less than 6x the weight of rotating parts.

3. The width of the foundation should be at least 2x the distance measured from the base of the foundation to the center of gravity of the whole machine including the foundation.
4. The thickness of the foundation should be at least 33% of the clear distance between the inboard and outboard bearing pedestals.
5. The height to thickness ratio of the outboard bearing pedestal should not exceed 3. The flexural stiffness of the outboard bearing pedestal, assuming rigid constraint at the foundation, should be at least 5,000,000 lb/in in both the axial and horizontal directions.
6. The span/depth ratio of any beam that supports equipment should not be more than 10. For example, the minimum depth for a beam spanning 20 feet would be 24 inches (2 feet).
7. Web stiffeners should be provided in steel wide flange beams at all locations where equipment is bolted. Web stiffeners connect the upper and lower flanges of the beam. The web stiffeners increase torsional stiffness of the beam and minimize deformation of the flanges.
8. Lateral supports and bracing should be provided to yield a slenderness ratio no larger than 100 for any beam or column that supports a machine. Concrete floor slabs should not be considered as lateral support.
9. A deeper beam should never be framed into a shallower beam.

The above rules of thumb are not meant to replace engineering calculations, but only to serve as guidelines for identification of robust and reliable machine supports.

While faster switching speeds within a VFD smooth out inherent motor vibrations, the risk of electrical damage to the motor is increased. Therefore, an acceptable compromise must be found between elevated vibration due to slower switching speed and increased risk of electrical damage. However, the risks of electrical damage to motors and bearings can be reduced by doing the following [5][8]:

1. Ensure a motor to be driven by VFD has insulation rated for VFD applications.
2. Keep the cable length between the motor and the VFD to a minimum.
3. Use a shaft grounding device such as an Aegis Pro grounding ring.
4. Insulate motor bearings.

Mitigation - Operation

As mentioned above, one of the easier methods of avoiding exciting natural frequencies with a VFD is to prevent the VFD from running at a natural frequency for any significant amount of time. It is acceptable to allow the rotating speed of the equipment pass through natural frequencies to change output, but not to linger there.

To prevent inadvertently starving or dead-heading, at least one of the following should be done. First, it is ideal that all fans connected with common headers be driven the same way, either all fans using single-speed drives or all fans using VFDs [3]. Second, there should be a minimum speed set for VFD driven fans such that the minimum speed still allows the fan to avoid starvation or a dead-head condition.

Acknowledgements

This paper was originally written by David Sheffler for and published by IVC Technologies in 2020, and was submitted to Vibration Institute with prior approval from IVC Technologies.

References

[1] Sayer, Robert J. "Structural Dynamics of Centrifugal Fans." Vibrations, pp. 91–103.

[2] Sayer, Robert J. "The Effect of Structural Support Conditions on the Vibration Characteristics of Machinery." Vibrations, pp. 23–33.

[3] Baxter, Nelson L. "The Perils of Installing VFD Drives on Vertical Pumps." Vibrations.

[4] Eshleman, Ronald L. "Vibration Considerations in VFD Driven Systems." Vibrations.

[5] Hall, Harlow C. "Variable Frequency Drives: Are They Heroes or Villains?" Vibrations, vol. 17, no. 4, Dec. 2001, pp. 3–6.

[6] Willwerth, Adam. "Detecting and Preventing Electrical Bearing Damage in VFD Driven Motors Through Shaft Voltage Measurement and Analysis - Part 1: Basic Principles." Vibrations, vol. 29, no. 4, Dec. 2012, pp. 14–17.

[7] Willwerth, Adam. "Detecting and Preventing Electrical Bearing Damage in VFD Driven Motors Through Shaft Voltage Measurement and Analysis - Part 2: Test Procedures." Vibrations, vol. 30, no. 1, Mar. 2013, pp. 12–18.

[8] Protecting Large Motors and Generators from Damaging Bearing Currents with AEGIS® PRO Series Rings, Electro Static Technology, 2017.

HOW TO COMPUTE AND USE THE FULL SPECTRUM PLOT

Author: Michael J. Titone, PE

VIATC TRAINING CONFERENCE & EXPO

NEWPORT NEWS, VIRGINIA | AUGUST 6-8, 2025

HOW TO COMPUTE AND USE THE FULL SPECTRUM PLOT

Michael J. Titone, PE
MJDynamics, LLC

Michael J. Titone, PE, is the semi-retired owner of MJDynamics, LLC. He sits on the Board of Directors for the Vibration Institute and is the Chair of the Annual Training Conference Committee. Prior to his retirement, he was a Machinery Diagnostic Services Technical Leader for Bently Nevada. In other job assignments, he worked for a power generation company and a vibration consulting firm in various roles. Since 1989, Michael has been a practicing engineer in the areas of vibration analysis, balancing, condition monitoring, mechanical system design/analysis, and I&C system design.

Michael earned a BS in mechanical engineering from Louisiana Tech University and an MBA from Mississippi College. He also earned 18 hours of graduate credit in mechanical engineering classes. He is a licensed professional engineer in the state of Mississippi and certified as a Category III Vibration Analyst through the Vibration Institute.

Abstract

The full spectrum plot was introduced by Bently Nevada in an Orbit Magazine article from 1993 as the "Plus and Minus Spectrum." [1] This plot takes advantage of the fact that the FFT algorithm is based on complex numbers as inputs and outputs. Over time, other software and instrument companies have included the full spectrum as a standard plot type. It is the author's perception, however, that the plot remains confusing to many analysts, and that most of the documentation on using the plot comes from Bently Nevada sources. The current paper reviews the methodology and algorithm used to compute the plots and provides discussions on how to use the plots. Detailed examples are shown using Python as the programming language since it is open source and popular.

Key Words: Fast Fourier Transform, FFT, Full Spectrum, Signal Processing, Python

Introduction

The classic spectrum plot used in vibration analysis is constructed from the signal from a single transducer. The well-known plot shows the frequency content and the amplitude of the frequencies from zero to the maximum configured frequency range (F_{max}) without displaying any negative frequencies.

The full spectrum plot was introduced by Bently Nevada in an Orbit Magazine article from 1993 as the "Plus and Minus Spectrum" since the plot displayed the full content of the calculated spectrum including the negative frequencies. [1] Shortly thereafter, Bently Nevada began referring to the plot as the full spectrum plot and to the classic spectrum plot as the half spectrum. The full spectrum plot takes advantage of the fact that the commonly available Discrete Fourier Transform (DFT) and Fast Fourier Transform (FFT) algorithms are based on complex numbers as inputs and outputs and that these algorithms calculate the full range of frequencies from $-F_{max}$ to $+F_{max}$. Over time, other vibration software and instrument companies have included the full spectrum as a standard plot type.

The full spectrum plot is constructed from the signals of two orthogonal or approximately orthogonal transducers sampled simultaneously. The signal from one measurement is input as the real component, and the signal from the other measurement is input as the imaginary component into the Fourier Transform algorithm. The resulting full spectrum plot provides the same information as the classic spectrum plus information about the precession and shape of the response at the various frequencies.

It is the author's perception that the details of computing the full spectrum are not well understood by potential users of the plot. Therefore, the current paper provides details and examples to illustrate how the data points for this plot type are computed. For this paper, the author developed examples using a simple Python code. The code was developed in the Spyder environment. Some examples of interpreting full spectrum plots are provided.

For additional information regarding Fourier Transform concepts, the reader should consider consulting an informative YouTube Channel, *Mark Newman Abstract concepts made clear*. The videos provided on his channel are highly informative. They provide both mathematical and intuitive content. Additionally, his web site, howthefouriertransformworks.com provides a wealth of information. [2] [3]

A Review of Some Principles of the Fourier Transform

The Fourier Transform is a well-known mathematical process that transforms an input in the time domain to the frequency domain. Wikipedia and other sources such as Brandt [4], Smith [5], Newman [6] [7], and practically any signal processing text, discuss the details and history of the Fourier Transform.

The Fourier Transform is used in vibration signal analysis to determine the frequencies and amplitudes of interest in a sampled signal. Generally, the signal is a two-dimensional signal over time. Typically, complex numbers are used to represent the signal, and the coordinates are the real parts of the numbers, the imaginary parts of the numbers, and the corresponding time. The resulting frequencies and amplitudes in the spectrum range from a negative frequency to a positive frequency.

A basic understanding of the history and characteristics of Fourier analysis should help the reader relate to the concepts needed to calculate and interpret the full spectrum. After describing the history of the Fourier Transform in Chapter 8 of his referenced book [5], Smith lists the following four categories of Fourier Transforms as they relate to signals:

- **Aperiodic – Continuous** – Signals are not periodic, but they extend from negative infinity to positive infinity without repeating in a periodic pattern. Fourier analysis is performed via the Fourier Transform.
- **Periodic – Continuous** – Signals are periodic (such as sine waves) and extend from negative to positive infinity. The applicable Fourier Transform is the Fourier Series.
- **Aperiodic – Discrete** – Signals are not periodic and are defined only at specific points. The Discrete Time Fourier Transform (DTFT) would be used for Fourier analysis.
- **Periodic – Discrete** – The signals are periodic and defined at specific points (samples). The Discrete Fourier Transform (DFT) is used to analyze these signals.

In Chapter 29, Smith covers the complex Fourier Transform. He notes that each of the four transforms listed above has a real and a complex version. The complex versions of the transforms have a complex signal in the time domain and produce useful positive and negative frequencies in the frequency domain. [5]

Since computers work with discrete signals only, the only two candidates for use in digital signal process would be the DTFT and the DFT. However, an infinite number of sine waves is required to decompose an aperiodic signal. Therefore, the only practical transform for use in a computer system is the Discrete Fourier Transform.

In the realm of the vibration analyst, the Fourier transform is often referenced as the FFT. FFT is the abbreviation for Fast Fourier Transform. Cooley and Tukey publicized [8] the FFT algorithm in 1965 in their paper titled, "An Algorithm for the Machine Calculation of Complex Fourier Series." The algorithm was discovered many years prior to their paper, but it was not utilized much until their paper since the computer was a new technology at the time of their paper. The FFT algorithm is a method of performing a DFT that substantially reduces the calculations required. As noted in the title, the FFT is an algorithm that transforms a signal made up of complex numbers into a frequency domain with positive and negative frequencies.

Complex Numbers

The DFT (and the other transforms) can be described in terms of real or complex numbers. Since many readers may not be familiar with complex numbers, a short introduction is in order. While there are many complicated reasons that complex numbers are used, a simple view is a good starting point. Consider that a complex number is simply a convenient way of expressing a two-dimensional quantity with a single two-part number.

Real numbers are the numbers that are used for typical everyday life. Real numbers are manipulated using simple arithmetic and algebra. Complex numbers are used in higher math, engineering, and other scientific applications when it is useful to define a two-dimensional quantity with a single number. Even though a single complex number contains two parts, real and imaginary, the fact that it is a single number can simplify some mathematical concepts and processes, especially when arrays are required. Python, for example, can perform calculations directly using complex numbers, and this feature simplifies coding.

For some applications, X-Y coordinates are used. For these cases, the two-dimensional quantity is defined by two separate numbers. Similarly, a polar coordinate system can also be used. In this case, the two-dimensional quantity is defined by a radius and an angle.

Figure 1 shows a graphical example of the relationship between X-Y, polar, and complex notation for a vector. In this example, the vector in polar coordinates would be (r, ϴ). The complex number would be in the format of $z = a + bj$, where $j^2 = -1$. Note that "i" is used instead of "j" in a mathematical context, and "j" is preferred in the electrical engineering field where "i" is used for current.

Note that phase angles are shown using math convention. Phase angles using vibration convention depend on rotation, probe location, etc. Therefore, phase, when computed using typical computer algorithms must be converted to vibration convention for plotting and analysis.

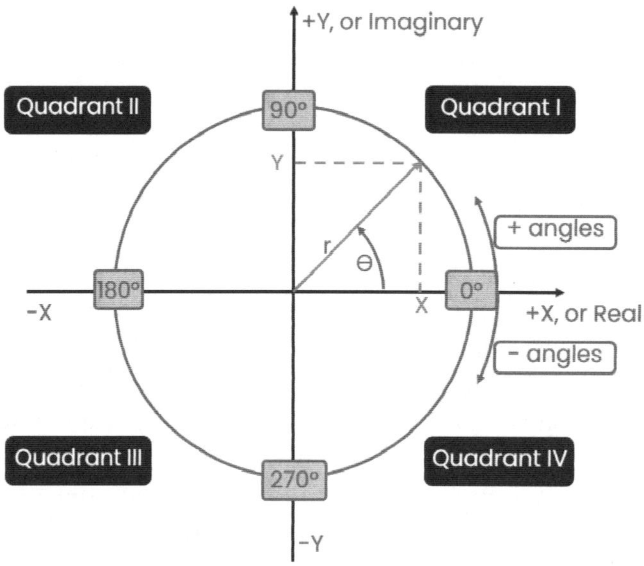

Figure 1, Polar, x-y, and real-imaginary coordinates

The following equations convert the vector from polar coordinates to rectangular components.

$$X = r * cos(\Theta) \tag{1}$$
$$Y = r * sin(\Theta) \tag{2}$$

Recognizing that the X component is equivalent to the real part of the complex number (a) and that the Y component is equivalent to the imaginary part (b), the complex number representing the vector would be as shown.

$$z = r * cos(\Theta) + r * \sin sin(\Theta)\, i \qquad (3)$$

Although not shown in Figure 1, the Imaginary axis is sometimes referred to as the Quadrature axis, and the Real axis is sometimes referred to as the Direct axis.

Positive and Negative Frequencies

The concept of positive and negative frequencies is likely one idea that confuses vibration analysts. Most analysts learned from the beginning of their training and experience to use the classic spectrum generated from a single measurement and displaying only the positive half of the spectrum. However, the negative frequencies have always been present, and the negative frequencies are real, physical things.

While it is often stated that the FFT decomposes a signal into its component sine waves, the better statement would be to say that the Fourier Transform describes a frequency in terms of sine and cosine components. In other words, it is inherently a two-dimensional process. Each sinusoid at each frequency that results from the decomposition contains a sine and a cosine component. Using complex numbers, the cosine components represent the real part of the signal, and the sine components represent the imaginary part of the signal.

Bently Nevada uses the concept of rotating vectors to illustrate positive and negative frequencies. A positive frequency rotates in the same direction as rotor rotation, signifying forward precession. A negative frequency rotates opposite the direction of rotor rotation, signifying negative precession. This conceptual explanation appears in many Orbit Magazine articles [9], Bently Nevada training material, and in Appendix 5 of Bently's book [10].

The use of the positive and negative frequencies in the full spectrum plot involves the following principles:

- Peaks are compared at a given +/- frequency
- If there is a large amplitude difference in the size of the peaks, the filtered orbit at that frequency is circular
- If the peaks are equal in size, then the filtered orbit at that frequency will be highly elliptical
- If the positive peak is larger than the negative peak, the precession of the filtered orbit at that frequency will be positive (in the direction of shaft rotation)
- If the negative peak is larger than the positive peak, the precession of a filtered orbit at that frequency will be negative (opposite shaft rotation)

To summarize, negative frequencies should not be viewed as an artifact of the math behind the spectrum. They are simply a natural output of the FFT/DFT, and they are useful for interpreting the plot when the input to the FFT consists of two simultaneously sampled orthogonal signals.

Signals in the Time Domain

Mathematically, the vibration signal from a single probe is considered to be a "real" signal, and the imaginary part of the signal is zero. Using the routine developed for this paper and setting the amplitude of all frequencies to zero except for a single frequency on the X probe signal, the plots shown in Figure 2 - Figure 4 were produced.

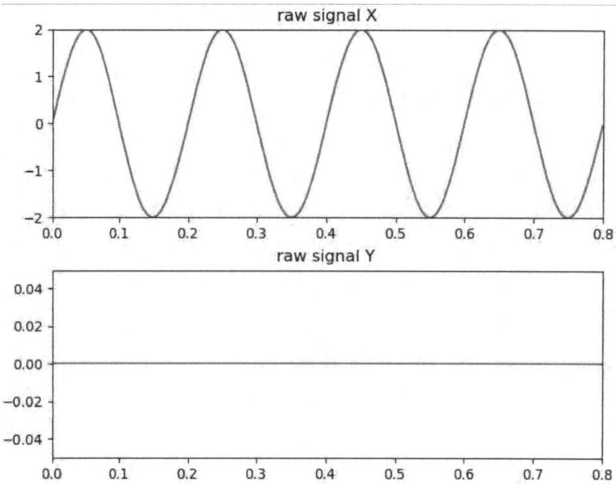

Figure 2, Signal vs. time for both probes

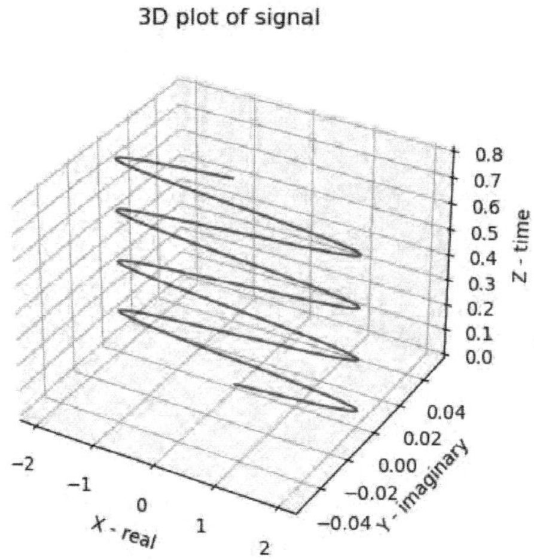

Figure 3, Three-dimensional plot with one frequency on one measurement

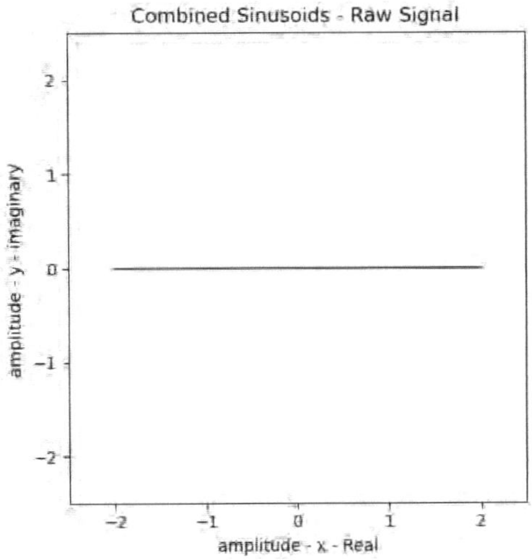

Figure 4, Two-dimensional view (Orbit plot) of real and imaginary signals with the imaginary component set to zero corresponding to the previous figures

Viewing a vibration signal as a one-channel measurement is useful for many applications. However, analyzing the path of the shaft dynamic motion (shaft orbits) is important for machines with fluid film bearings, especially turbomachinery. An orbit plot of two orthogonal simultaneously sampled acceleration signals or of two velocity signals shows the motion of the non-rotating structure. This use of an orbit plot is not as well-established as the shaft orbit plots, but it is equally valid. The orbit plot is simply a plot of the combined signals from two orthogonal measurements. Note that for shaft orbits, the plot shows the dynamic path of the motion of the shaft centerline plus any noise and runout in the measurements. The two signals that comprise the orbit plot are an example of a two-dimensional signal over time. Therefore, the use of the complex Fourier Transform is a natural application to produce a full spectrum plot.

To further explain the concept of a complex (two-dimensional) signal over time, a three-dimensional plot of the signal from two probes can be generated with the simple code developed for this paper. For this plot, it is assumed that both probes see the same frequency and amplitude of the signal with a ninety-degree phase lag and that the signal only includes a sine wave, generating a circular orbit. The plots of the individual signals are shown in Figure 5. The three-dimensional plot is shown in Figure 6.

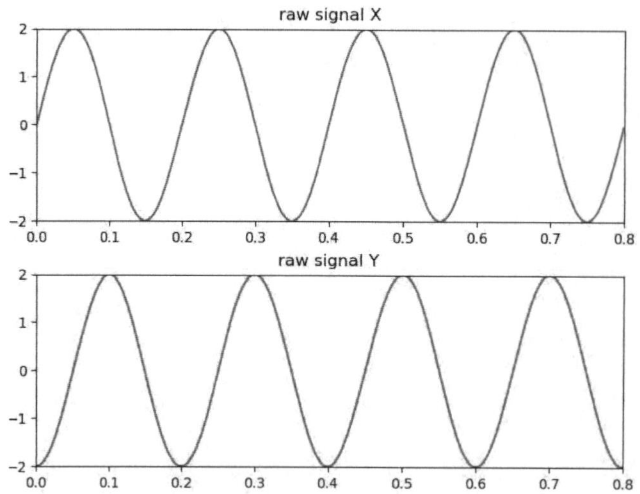

Figure 5, Plots of individual probe signals

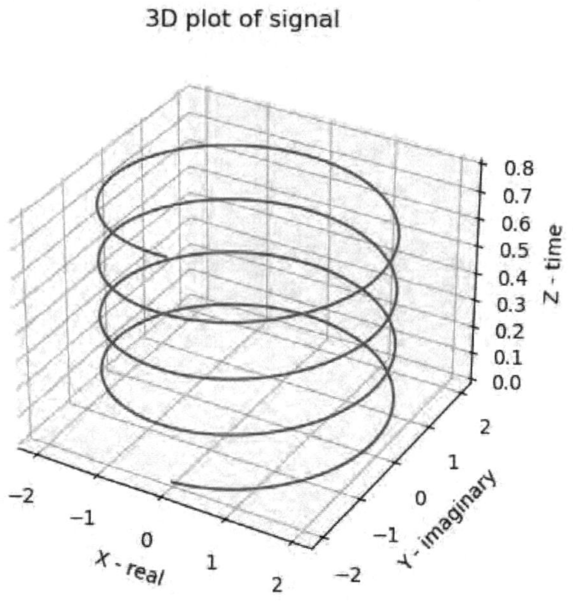

Figure 6, Three-dimensional plot of two probe signals over time

The plot can be rotated to show the signals together on the XY plane, resulting in the orbit plot shown in Figure 7. The plot can also be rotated to show the individual probe signals vs. time. Figure 8 shows the plot rotated to show the Y signal over time. Note that standard tools provided as part of Spyder (Python environment used by the author – details provided in a subsequent section) are difficult to use to precisely rotate the plots.

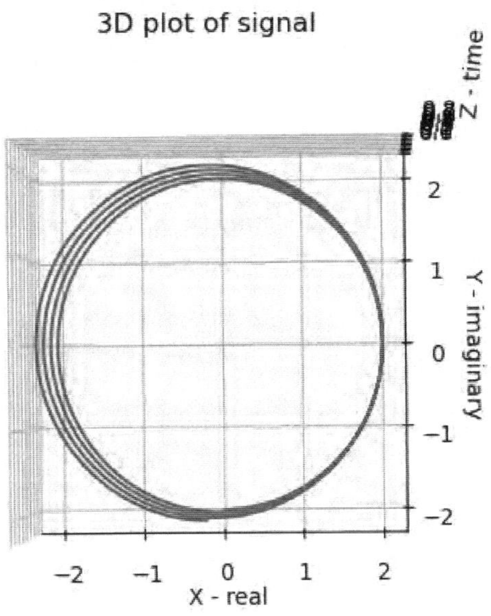

Figure 7, Plot rotated to show the X-Y plane

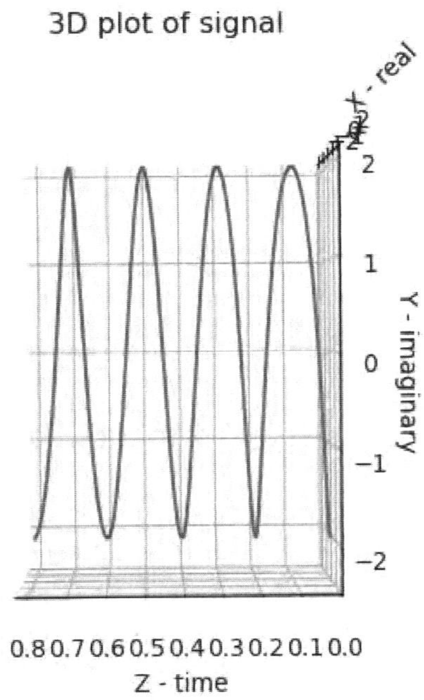

Figure 8, Plot rotated to show the Y signal vs. time

Principles of Computing the DFT/FFT Using a Standard Algorithm

As previously mentioned, using the two signals as input to an FFT algorithm results in a spectrum having both positive and negative frequencies. Even if the signal is one-dimensional (from a single measurement), the resulting frequencies include both positive and negative frequencies. The traditional process for plotting the half spectrum (or the classic spectrum) simply hides the negative frequencies. More accurately, the plotting process combines the negative frequencies with the positive frequencies and displays only the modified positive frequencies.

The reason that the negative and positive frequencies can simply be combined for a classic spectrum plot is that the negative frequencies mirror the positive frequencies when the signal is one-dimensional. Using the code written for this paper, the inputs are configured with only one sine wave for the signal from the X probe, and the Y probe signal is set to zero amplitude. Figure 9 and Figure 10 show the two signals that are used to produce the full spectrum plot provided in Figure 11. The amplitude of the X signal is 4.0 mils pk-pk, and the amplitude of the signal from the Y probe is zero. The resulting plot shows a positive and negative peak of half the amplitude at -100 Hz and 100 Hz. To convert this to a classic half spectrum plot, the two peaks would be added resulting in a single peak of 4.0 mils pk-pk at 100 Hz.

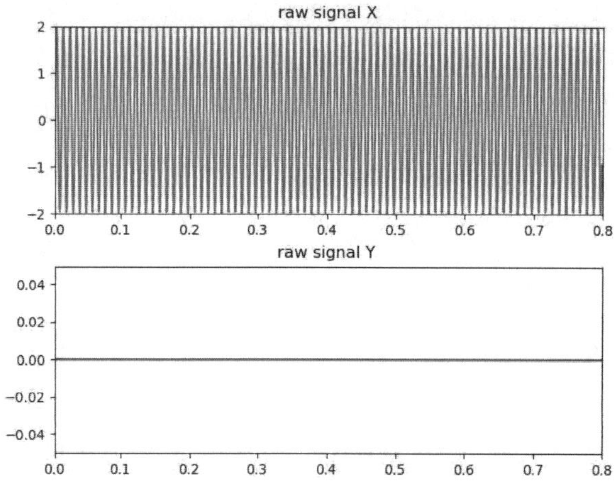

Figure 9, Input used to produce a full spectrum with only one signal

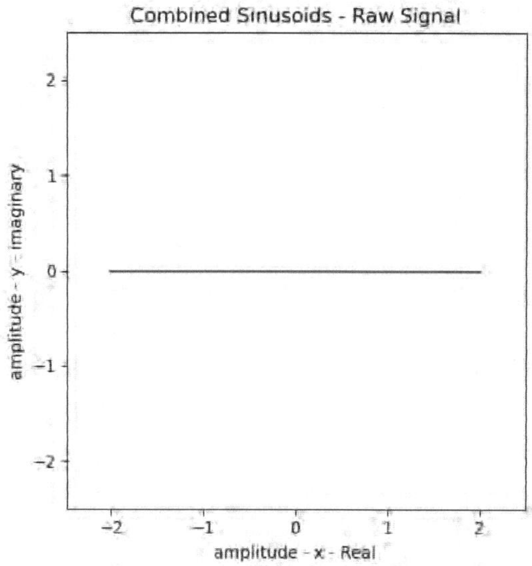

Figure 10, Orbit plot of the two signals with Y set to zero amplitude

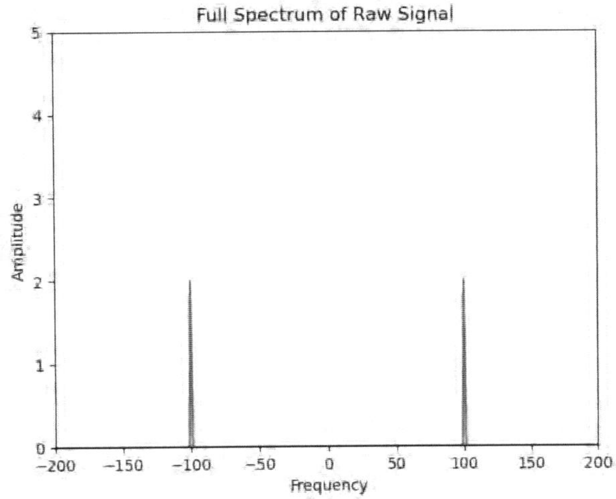

Figure 11, Full spectrum plot with one signal

Using the code developed for this paper, the signal can be modified to include the Y probe component. For example, a Y signal of the same frequency and amplitude with a 90-degree phase lag can be input resulting in the plots shown in Figure 12 - Figure 14. Note the amplitude of the peak in the spectrum is 4.0 mils pk-pk reflecting the fact that the signal has the same amplitude in both probe's signals.

If the input signals are changed such that the amplitude are different, say 2.0 mils pk-pk for the X probe and 1.0 mils pk-pk for the Y probe, without changing any other input, the plots shown in Figure 15 and Figure 16 are produced.

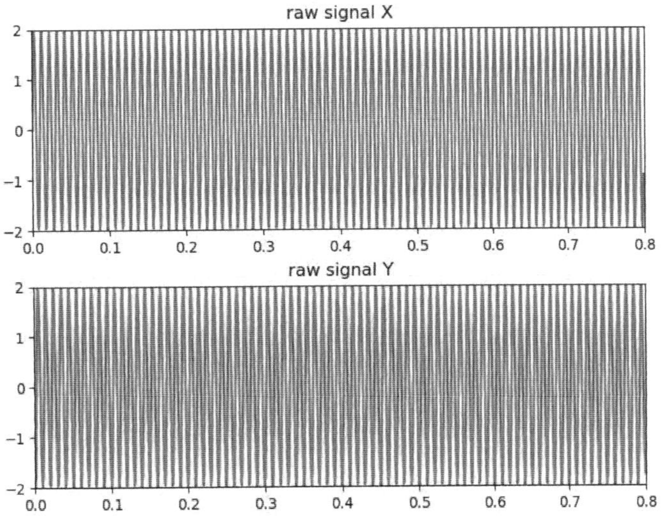

Figure 12, Input used to produce a full spectrum with two equal amplitude signals at the same frequency with a 90-degree phase difference

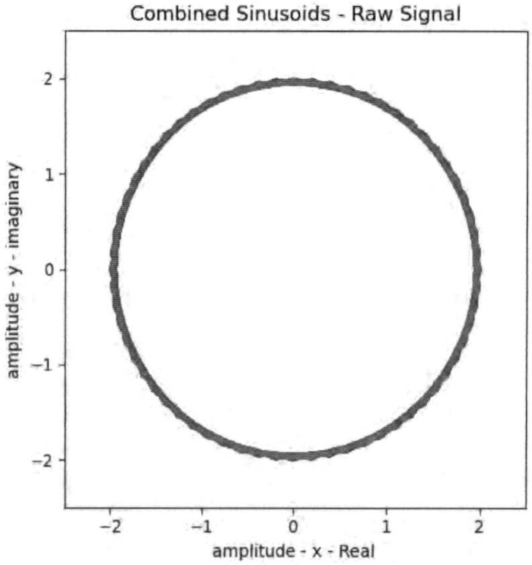

Figure 13, Orbit plot of two signals with equal amplitude and a 90-degree phase difference

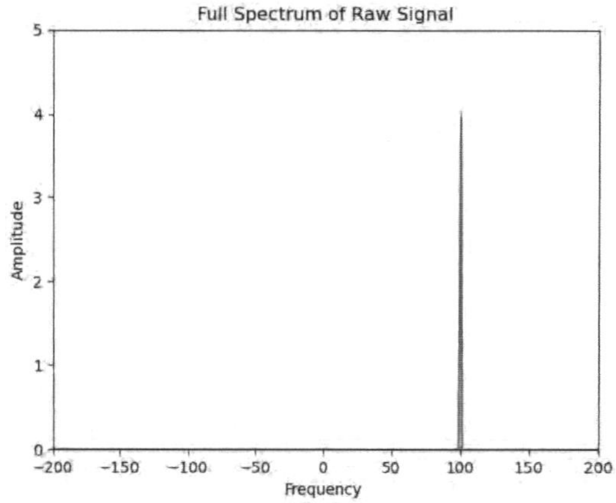

Figure 14, Full spectrum resulting from the two equal signals

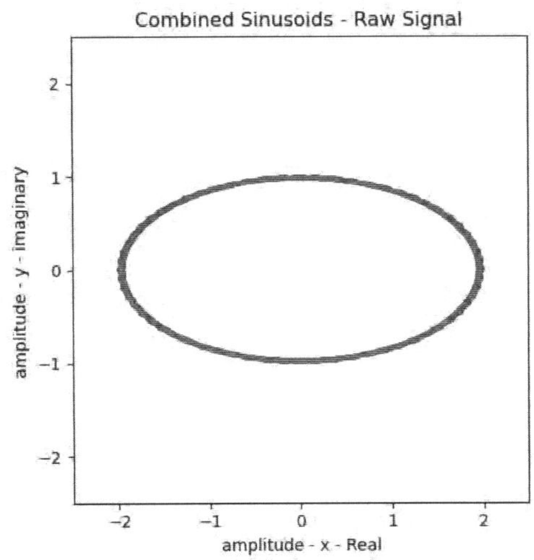

Figure 15, Orbit plot with two signals of unequal amplitude

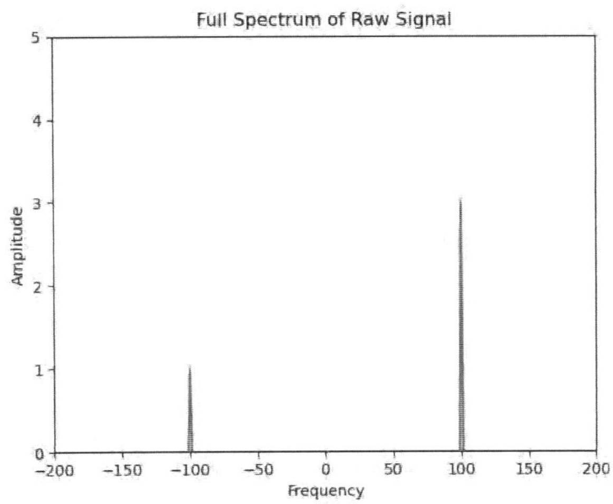

Figure 16, Full spectrum resulting from two unequal amplitude signals

Implementation of the FFT and/or DFT in Python

Python was chosen as the tool to simulate signals and produce plots for this paper. Python is open source and popular. The author used Spyder for the development environment since it caters to scientific programming and since the author had used it for previous work. Version details are provided in Figure 17. A complete listing of the computer code is provided in the Appendix.

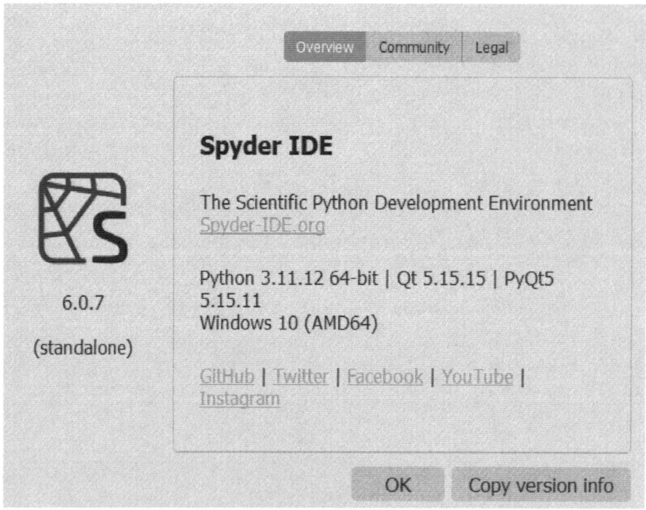

Figure 17, Spyder IDE version details

Computer codes such as Python are often popular because of the large number of libraries that are available. These libraries contain pre-defined functions that extend the usability of the code to rank amateurs such as the author. The libraries used in the code presented in this paper are as follows:

- Scipy – used for performing the FFT (scipy.org)
- Numpy – used for general math functions (numpy.org)
- Matplotlib – used for plotting (matplotlib.org)
- Os – used for calls to the operating system (lib/os.py)

As noted above, the typical or standard DFT/FFT algorithm takes a complex signal input regardless of whether the signal contains both complex components. The full spectrum plot assumes that the Y probe signal is the imaginary part of the complex signal and that the X probe signal is the real part of the complex signal. Southwick clearly states this in his article from 1993. [1] This point is further expressed in an article by Goldman and Muszynska from Orbit in 1999. [9]

An additional step is required to properly implement the full spectrum plot. The plot convention is that the positive side of the spectrum contains the components that precess in the same direction as shaft rotation. Using a popular convention, viewing the machine from driver to driven, the shaft rotates clockwise for some machines and counterclockwise for some machines. Therefore, an algorithm is required to incorporate a rotation definition in order to sort out which half of the full spectrum is positive (forward), and which is negative (reverse). This part of the process is not covered by the simple code written for this paper.

To demonstrate how the DFT is implemented in Python, the author wrote a simple program. Some code snippets are shown in the body of the paper, and the full code is provided in the Appendix. Note that the program is not interactive. The user must edit the file to change the results. The user inputs include the following (Figure 18):

- F_{max}. This is the maximum frequency displayed in the spectrum.
- Number of lines of resolution (LOR, or Nlines in the code).
- Four input signals that are sine waves. The purpose of this is to allow two input frequencies per vibration probe, assuming a typical system with XY displacement measurements. The sine waves are defined by the following:
    - Frequency – xfreq1, yfreq1, xfreq2, yfreq2
    - Amplitude – xamp1, yamp1, xamp2, yamp2
    - Phase – xphase1, yphase1, xphase2, yphase2

```
# Assume a 400 line spectrum with Fmax set to 200 Hz
# set variables for user input
# user input is Fmax and number of lines for the spectrum
# and the frequencies for the sine waves the make up the signal

Fmax = 500.0        # set typical maximum displayed frequency to 200 Hz
Nlines = 400        # number of spectral lines is set to 400, actual lines = 512

xfreq1 = 5.0        # user input for frequency of first sine wave - x probe
xamp1 = 2.0         # user input for amplitude of first sine wave - x probe
xphase1 = 0.0       # user input for phase of first sine wave - x probe

yfreq1 = 5.0        # user input for frequency of first sine wave - y probe
yamp1 = 2.0         # user input for amplitude of first sine wave - y probe
yphase1 = -20.0  # user input for phase of first sine wave - y probe

xfreq2 = 100.0        # user input for frequency of second sine wave - x probe
xamp2 = 1.0         # user input for amplitude of second sine wave - x probe
xphase2 = 0.0       # user input for the phase of the second sine wave - x probe

yfreq2 = 100.0        # user input for frequency of second sine wave - y probe
yamp2 = 1.0         # user input for amplitude of second sine wave - y probe
yphase2 = -45.0       # user input for the phase of the second sine wave - y probe
```

Figure 18, Screenshot of user inputs

The program does not enforce any limits on the inputs. For example, it would be logical to assume that the first and second frequencies seen by the two probes were the same for each probe. In other words, xfreq1 should equal yfreq1, and xfreq2 should equal yfreq2. For this simple program, the user has free reign to choose illogical inputs. For the screenshots provided in this section, the inputs were edited to make the figures clearer. Therefore, the figures were not all based on the same inputs.

The program then generates the sine waves for each set of inputs and plots these sine waves as shown in Figure 19.

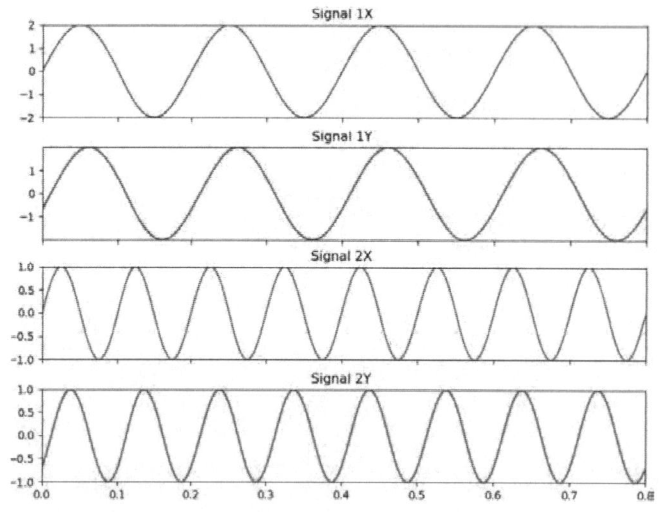

Figure 19, Plots of input sine waves

Next, the program adds the two sine waves for each measurement and plots the "raw" signals that are produced. See Figure 20.

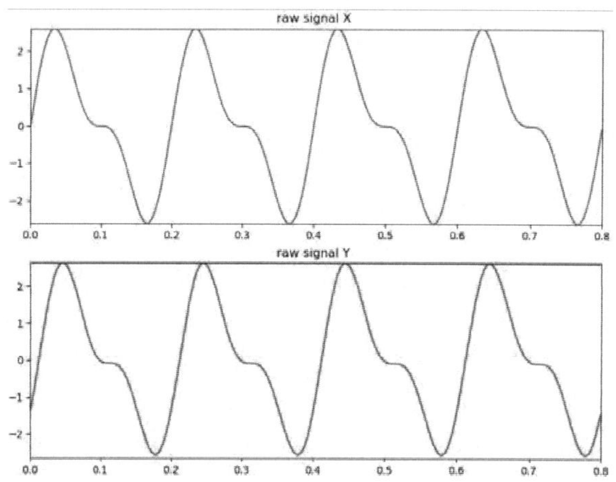

Figure 20, Plots of the raw signals for each probe

The program then plots the raw signals in an X-Y (orbit) plot format as shown in Figure 21 and in a three-dimensional orbit plot format with time as the third dimension as shown in Figure 22.

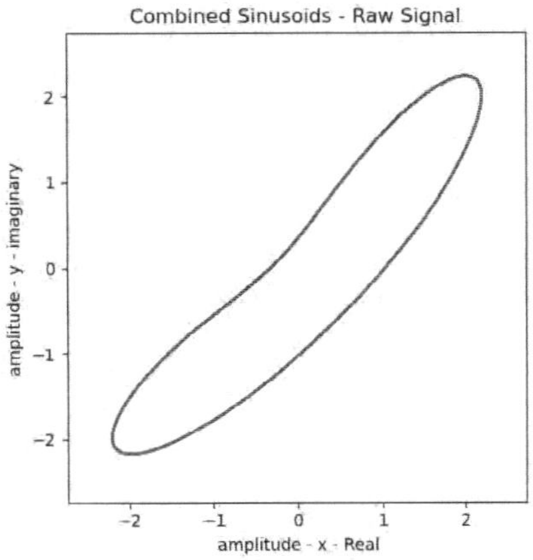

Figure 21, Orbit plot of raw signals

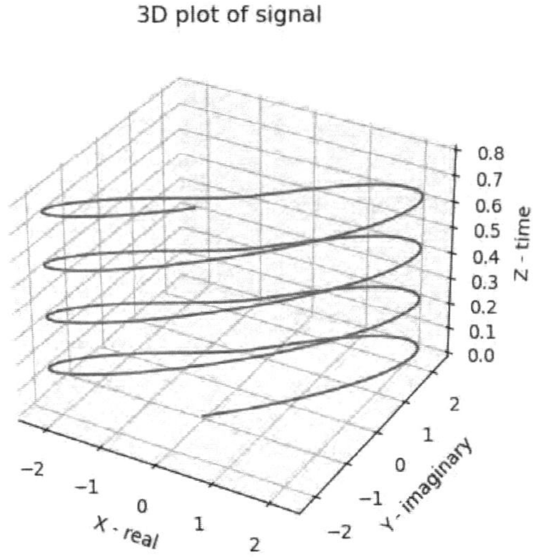

Figure 22, Three-dimensional orbit plot

Finally, the program calculates and plots the full spectrum. The code snippet used for calculating and plotting the spectrum is provided in Figure 23 and the resulting plot is provided in Figure 24. The code shows that generating the spectrum using Scipy is a simple process. The tools provided allow the user to bypass the complexities of formatting the output.

```
# begin FFT part

spec_amp = (2.0/ns)*fft(sig_raw)     # calculate the complex spectral amplitudes
spec_freq = fftfreq(ns,1/Fs)         # calculate the frequencies - array of float

fig = plt.figure()
plt.title('Full Spectrum of Raw Signal')
plt.xlabel('Frequency')
plt.ylabel('Amplitude')
# plot on a frequency axis from -200 to +200 (hard coded for now)
plt.axis([-200, 200, 0, 2.0*max_signal])  # multiply signal by 2.0 to account for pk-pk
# fftshift rearranges the amplitudes and frequencies to the normal order
plt.plot(fftshift(spec_freq), fftshift(np.abs(spec_amp)))
plt.show()
```

Figure 23, Code snippet used to calculate and plot the full spectrum

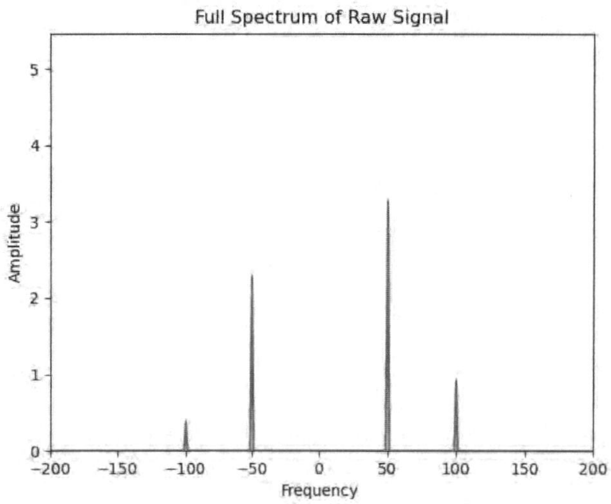

Figure 24, Full spectrum plot

Using the Full Spectrum Plot

Based on the author's experience, the orbit plot is well-established and well-understood in the machinery diagnostics field. While the full spectrum plot is simply a plot of the orbit in the frequency domain, the plot does not seem to have reached the same level of acceptance. This paper has attempted to emphasize the point that the plot is just as naturally useful as the classic half spectrum.

Some of the principles of using the full spectrum plot are stated above in the Positive and Negative Frequencies section. Some additional considerations include the following:

- The peaks in the full spectrum represent the frequency content of the two-dimensional signal.
- Since the full spectrum is based on the two-dimensional signal, the spectral content is not dependent on the probe locations, although the probes should be orthogonal.
- The full spectrum provides insight into the precession and shape of a filtered orbit at various frequencies. This means that the user does not need to configure filters at an unknown frequency to evaluate the shape and precession at that frequency.

When should the full spectrum plot be used? In the author's opinion, the plot should be reviewed as a normal part of analyzing the machinery response whenever it is available, especially for proximity probe measurements.

Some specific situations where the plot could be highly useful include the following:

- When the data is very noisy and orbit plots are obscured by the noise, the full spectrum can provide insight into the frequencies of interest.
- When the user wants to present a lot of information in a single plot, the full spectrum supports this. For example, cascade plots during transient conditions work well with full spectra providing more information than the classic half spectra. Similarly, waterfall

plots collected during transient conditions and/or steady state conditions often lend themselves to full spectra. See Figure 25 and Figure 26 for examples.

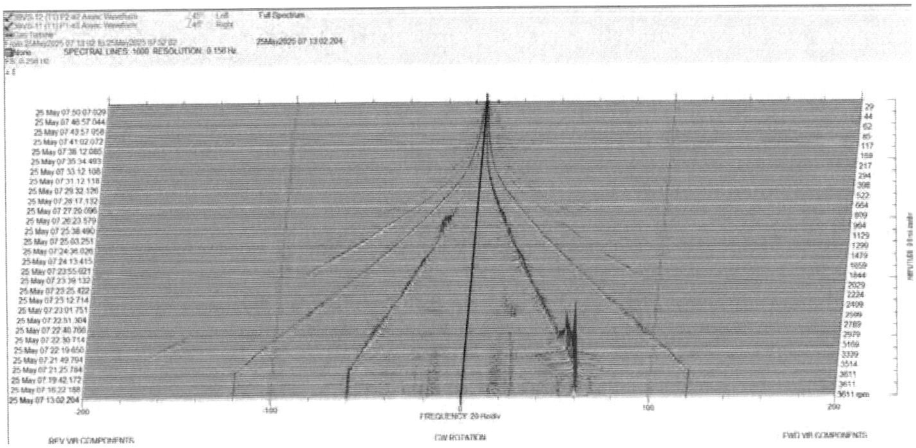

Figure 25, Example full spectrum waterfall plot during a machine shut down transient

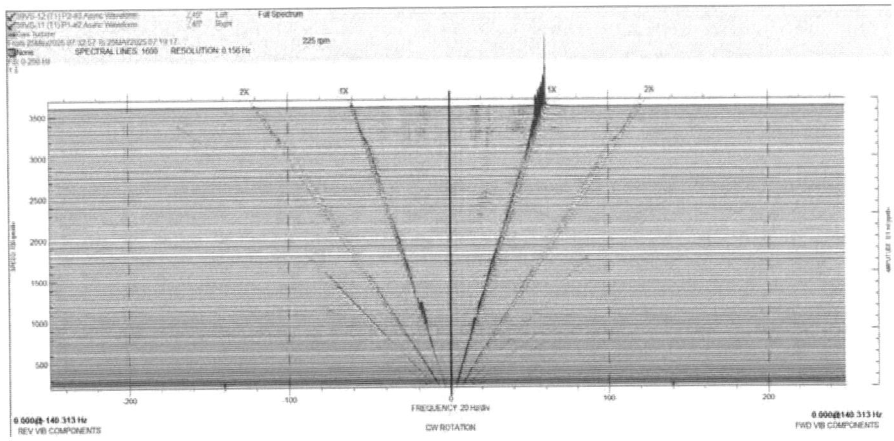

Figure 26, Example full spectrum cascade plot during a transient startup or shutdown

When should the full spectrum plot not be used? In some cases, depending on the audience, the user might prefer to present the traditional half spectrum in reports.

Conclusions and Future Work

The work performed for this paper demonstrates the simplicity of computing the full spectrum plot using Python. The explanations provided herein should provide prospective users with a comfort level for using the plots.

The code developed for this paper is simple, and some refinements would improve its usefulness for demonstrating Fourier Transform concepts. Some desirable improvements would be to create a user interface, to improve the plots, and to add some animations.

References

[1] D. Southwick, "Plus and Minus Spectrum, Enhancement to Transient Data Manager 2 Software," *Orbit,* pp. 18-20, June 1993.

[2] M. Newman, "Mark Newman Abstract concepts made clear," [Online]. Available: howthefouriertransformworks.com.

[3] M. Newman, "Mark Newman Youtube Channel," [Online]. Available: https://www.youtube.com/@MarkNewmanEducation.

[4] A. Brandt, Noise and Vibration Analysis: Signal Analysis and Experimental Procedures, Second ed., Hoboken, NJ: John Wiley and Sons, 2023.

[5] S. W. Smith, The Scientist and Engineer's Guide to Digital Signal Processing, San Diego: California Technical Publishing, 1997.

[6] M. Newman, How the Fourier Series Works, Independently Published, 2022.

[7] M. Newman, How the Fourier Transform Works, Independently published, 2023.

[8] J. W. Cooley and J. W. Tukey, "An Algorithm for the Machine jCalculation of Complex Fourier Series," *Mathematics of Computation,* vol. 19, no. 90, pp. 297-301, 1965.

[9] P. Goldman and A. Muszynska, "Application of Full Spectrum to Rotating Machinery Diagnostics," *Orbit,* pp. 17 - 21, April 1999.

[10] D. E. Bently and C. T. Hatch, Fundamentals of Rotating Machinery Diagnostics, Minden: Bently Pressurized Bearing Press, 2002.

Appendix – Python code

```
# -*- coding: utf-8 -*-
"""
Created on Wed Aug  2 15:46:52 2023

@author: Michael J. Titone
"""
# This program is an attempt to:
# 1. generate a signal with four sine waves
# 2. Assuming two probes, each probe has two sine waves
# 3. plot the sine waves and the combined signal
# 4. take the fft of the combined signal and plot the spectrum

# from scipy import signal
from scipy.fft import fft, fftfreq, fftshift
import numpy as np
import matplotlib.pyplot as plt
#from mpl_toolkits import mplot3d
# import warnings
import os
os.system('cls')

# Assume a 400 line spectrum with Fmax set to 200 Hz
# set variables for user input
# user input is Fmax and number of lines for the spectrum
# and the frequencies for the sine waves the make up the signal

Fmax = 500.0        # set typical maximum displayed frequency to 200 Hz
Nlines = 400        # number of spectral lines is set to 400, actual lines = 512

xfreq1 = 50.0        # user input for frequency of first sine wave - x probe
xamp1 = 2.0         # user input for amplitude of first sine wave - x probe
xphase1 = 0.0       # user input for phase of first sine wave - x probe

yfreq1 = 50.0        # user input for frequency of first sine wave - y probe
yamp1 = 2.0         # user input for amplitude of first sine wave - y probe
yphase1 = -20.0    # user input for phase of first sine wave - y probe

xfreq2 = 100.0        # user input for frequency of second sine wave - x probe
xamp2 = 0.5         # user input for amplitude of second sine wave - x probe
xphase2 = 0.0       # user input for the phase of the second sine wave - x probe

yfreq2 = 100.0        # user input for frequency of second sine wave - y probe
yamp2 = 0.5         # user input for amplitude of second sine wave - y probe
yphase2 = -45.0       # user input for the phase of the second sine wave - y probe

# calculate sample variables based on user input
Fs = 2.56*Fmax              # set sample rate to typical oversample rate
NLFFT = Nlines*1.28         # calculate actual number of lines based on user setting
ns = int(round(2*NLFFT))    # number of samples to be generated
tsamp = ns/Fs               # sample time, samples start at 0.0 and end at tsamp

t_stamp = np.linspace(0, tsamp, ns, False)  # generate array of time stamps

# Generate sine waves with x as real and y as imaginary
sig1x = np.sin(2*np.pi*xfreq1*t_stamp+np.deg2rad(xphase1))*xamp1
sig2x = np.sin(2*np.pi*xfreq2*t_stamp+np.deg2rad(xphase2))*xamp2
sig1y = np.sin(2*np.pi*yfreq1*t_stamp+np.deg2rad(yphase1))*yamp1
sig2y = np.sin(2*np.pi*yfreq2*t_stamp+np.deg2rad(yphase2))*yamp2

# plot the individual sine waves created above
# Create a figure with 4 subplots.  4 rows and 1 column
fig, (ax1, ax2, ax3, ax4) = plt.subplots(4, 1, layout = "constrained", sharex=True)

#plot the x probe signal in the first subplot using the default color line
ax1.plot(t_stamp, sig1x)
ax1.set_title('Signal 1X')
ax1.axis([0, 0.8, -np.max(np.abs(sig1x)), np.max(np.abs(sig1x))])
```

```
# plot the y probe signal in the second subplot using red colored line
ax2.plot(t_stamp, sig1y, 'r')
ax2.set_title('Signal 1Y')
ax2.axis([0, 0.8, -np.max(np.abs(sig1y)), np.max(np.abs(sig1y))])

#plot the x probe signal in the first subplot using the default color line
ax3.plot(t_stamp, sig2x)
ax3.set_title('Signal 2X')
ax3.axis([0, 0.8, -np.max(np.abs(sig2x)), np.max(np.abs(sig2x))])

# plot the y probe signal in the second subplot using red colored line
ax4.plot(t_stamp, sig2y, 'r')
ax4.set_title('Signal 2Y')
ax4.axis([0, 0.8, -np.max(np.abs(sig2y)), np.max(np.abs(sig2y))])

sig1 = sig1x+sig1y*1.j  # create the complex version of signal 1
sig2 = sig2x+sig2y*1.j  # create the complex version of signal 1
sig_raw = sig1+sig2     # combine the signals to create the complex signal

#combine the signals to create the raw signal components
sig_rawx = sig1x+sig2x  # create the raw x signal
sig_rawy = sig1y+sig2y  # create the raw y signal

# Create a figure with 2 subplots.  2 rows and 1 column
fig, (ax1, ax2) = plt.subplots(2, 1, layout = "constrained", sharex=False)

#plot the x probe signal in the first subplot using the default color line
ax1.plot(t_stamp, sig_rawx)
ax1.set_title('raw signal X')
ax1.axis([0, 0.8, -np.max(np.abs(sig_rawx)), np.max(np.abs(sig_rawx))])

# plot the y probe signal in the second subplot using red colored line
ax2.plot(t_stamp, sig_rawy, 'r')
ax2.set_title('raw signal Y')
ax2.axis([0, 0.8, -np.max(np.abs(sig_rawy)), np.max(np.abs(sig_rawy))])

# plot the combined signal (orbit plot) in a separate plot using green colored line
fig = plt.figure()
plt.title('Combined Sinusoids - Raw Signal')
plt.xlabel('amplitude - x - Real')
plt.ylabel('amplitude - y - imaginary')

# force orbit plot to be square
# first find maximum absolute value for raw signal from x and y probes
# and add 0.5 mils to make the plot look better.
max_signal = max(np.max(np.abs(sig_rawx)), np.max(np.abs(sig_rawy)))+0.5
plt.axis([-max_signal, max_signal, -max_signal, max_signal])
axes = plt.gca()
axes.set_aspect('equal', adjustable = 'box')
plt.tight_layout()

plt.plot(sig_rawx, sig_rawy, 'g')

plt.show()

#plot a 3D version of the signal
fig = plt.figure()
ax = fig.add_subplot(111, projection='3d')
#fig.set_size_inches(10,10, forward = False)
z = t_stamp
x = sig_rawx
y = sig_rawy
ax.plot3D(x, y, z, 'green')
ax.set_title('3D plot of signal')
ax.set_xlabel('X - real')
ax.set_ylabel('Y - imaginary')
ax.set_zlabel('Z - time')
plt.show()

# begin FFT part
```

```
spec_amp = (2.0/ns)*fft(sig_raw)     # calculate the complex spectral amplitudes
spec_freq = fftfreq(ns,1/Fs)         # calculate the frequencies - array of float

fig = plt.figure()
plt.title('Full Spectrum of Raw Signal')
plt.xlabel('Frequency')
plt.ylabel('Amplitude')
# plot on a frequency axis from -200 to +200 (hard coded for now)
plt.axis([-200, 200, 0, 2.0*max_signal])  # multiply signal by 2.0 to account for pk-pk
# fftshift rearranges the amplitudes and frequencies to the normal order
plt.plot(fftshift(spec_freq), fftshift(np.abs(spec_amp)))
plt.show()
```

USING 32-BIT FLOATING POINT DATA ACQUISITION OF AN AUDIO RECORDER TO COLLECT VIBRATION MEASUREMENTS

Author: Dr. George Zusman and Ed Spence

VIATC | TRAINING CONFERENCE & EXPO

NEWPORT NEWS, VIRGINIA | AUGUST 6-8, 2025

USING 32-BIT FLOATING POINT DATA ACQUISITION OF AN AUDIO RECORDER TO COLLECT VIBRATION MEASUREMENTS

Dr. George Zusman
President, Vibration Measurement Solutions Inc.
1710 South Dairy Ashford Road, Houston TX 77077
zusger@gmail.com

Ed Spence
Managing Director, The Machine Instrumentation Group
535 Washington St, Dedham MA 02026
ed.spence@machineinstrumentation.com

George Zusman holds a Ph.D. in Electrical Engineering, a D.Sc. in Mechanical Engineering, and is currently the president of Vibration Measurement Solutions, Inc. (Houston, TX).

Prior to that, he was Director of Product Development at PCB Piezotronics, Inc. (Buffalo, NY), Director of Engineering and later President of Metrix Instruments (Houston, TX), and is the founder of ViCont, Ltd.

George has designed numerous sensors and systems, holds 30 patents, has over 150 technical publications, and is author of two chapters in the Handbook of Noise and Vibration Control. He is Vibration Specialist Level III and Ultrasonic Specialist Level II in accordance with DIN EN 473, a member of editorial board of Control/Diagnostics magazine, and a member of International Institute of Acoustic and Vibration.

Ed Spence is the Founder and Managing Director of The Machine Instrumentation Group, representing a network of contract engineering

service providers helping clients define and develop new sensor solutions for machine health applications.

Previously, Ed was the Marketing Manager for Analog Devices MEMS Sensor Technology Group, where he defined the MEMS accelerometer roadmap for vibration based Condition Monitoring. Ed has been engaged in the development of Condition Monitoring sensor solutions since 2008. Ed has published or presented on the subject of sensors for Condition Monitoring at Vibration Institute, Mobius Institute, SMRP, MFPT and Reliable Plant.

Abstract

Handheld digital data collectors have been used by vibration analysts since the 1980s. Over the decades, several models have been introduced by various companies, with most contributing to a rise in the average cost to the user. Much of the increased cost is due to increased performance but also the expansion of features, such as the color display and the embedded signal processing. Higher performance requires more processor power, memory, and battery capacity – all adding to the costs.

Lower cost means of capturing vibration data have come on the market over the years in the form of accelerometers with embedded data acquisition (A/D converter), but also requiring the device to be tethered to a laptop, limiting their utility in the field.

But it turns out that there are devices readily available on the market today that are not only very low cost and highly portable but have more than enough data acquisition fidelity to satisfy most vibration analysts.

In this paper we describe how several audio recorders on the market today might serve as vibration data collection devices in some applications. Designed to capture high fidelity audio signals from an attached microphone, these devices can be repurposed to capture vibration signals with both high resolution (32-bit floating point[1] provided by two 24-bit A/D converters) and high sample rates (48ksps) from an accelerometer with modified output circuitry. With both embedded memory and use of external MicroSD cards (512GB to 1TB) these devices can collect hours of vibration data.

Key Words: Accelerometers; Condition Monitoring; Data Collector; Vibration;

What is an Audio Recorder?

Low-cost audio recorders have been developed as a way to capture high quality audio to accompany video without depending on the camera's built in microphone. Filming the wedding

[1] See IEEE 754 for details

couple from a distance of more than a few feet in the presence of background noise results in poor quality audio. Battery powered audio recorder devices can be clipped to the subject much like a portable microphone is used for public speakers who like to move as they talk.

A typical audio recorder weighs about an ounce and can store hours of audio on a combination of internal memory and auxiliary MicroSD card. Because audio signals in the presence of noise exhibit such a high dynamic range, the digitizing electronics are typically very high quality. Recorders today use two 24-bit A/D converters with autoranging to capture high dynamic range signals without clipping (saturating) as well as whisper quiet audio. The data capture rates are very high, with 44 or 48ksps being standards. This performance rivals many commercially available vibration data loggers or wireless sensor systems.

Figure 1: Commercially available audio recorders, from LEFT to RIGHT: Tascam DR-10L, Zoom F2 and Diety PR-2.

Audio recorders are designed to be very easy to use, with minimal manual controls and start / stop button operation. Parameters are configurable with a phone app. WAV files are date/time stamped then downloaded for analysis to a PC via USB cable.

A survey of audio recorders available on the market today turns up the following:

Table 1: Commercially available audio recorders.

	Tascam DR-10L	Zoom F2	Diety PR-2
Weight (excl. batteries)	43 g	32 g	40 g
Dimensions (mm)	53 x 50.7 x 21.4	57.5 × 46.4 × 22.4	57.9 x 55.1 x 18
List Price	$ 149	$ 180	$ 239
Memory (MicroSDXC)	512 GB	1 TB	128 GB
Sample Rates (ksps)	44 / 48	44 / 48	48
Maximum Recording time (hours)[1]	24.5	15	30

Note 1: Recording time for two AAA batteries.

Maximizing Measurement Fidelity

Audio signals can have very high dynamic range, particularly in a crowd of people (such as a wedding) where the primary subject is some distance away but other audio sources can be quite close. The bane of videographers can be saturation or distortion of sound close by that is too loud when applying gain to the signal for clarity when the desired object is further away. Today's audio recording minimizes this problem with intelligent digitizing. Two separate 24-bit A/D Converters are used to auto range the signal to prevent distortion due to clipping or saturating of the input signal. To maximize the captured fidelity, the data is formatted as 32-bit floating point for optimal processing of the signal to fit the A/D Converter full scale range. This is accomplished without any user intervention, DAQ settings or other configuration.

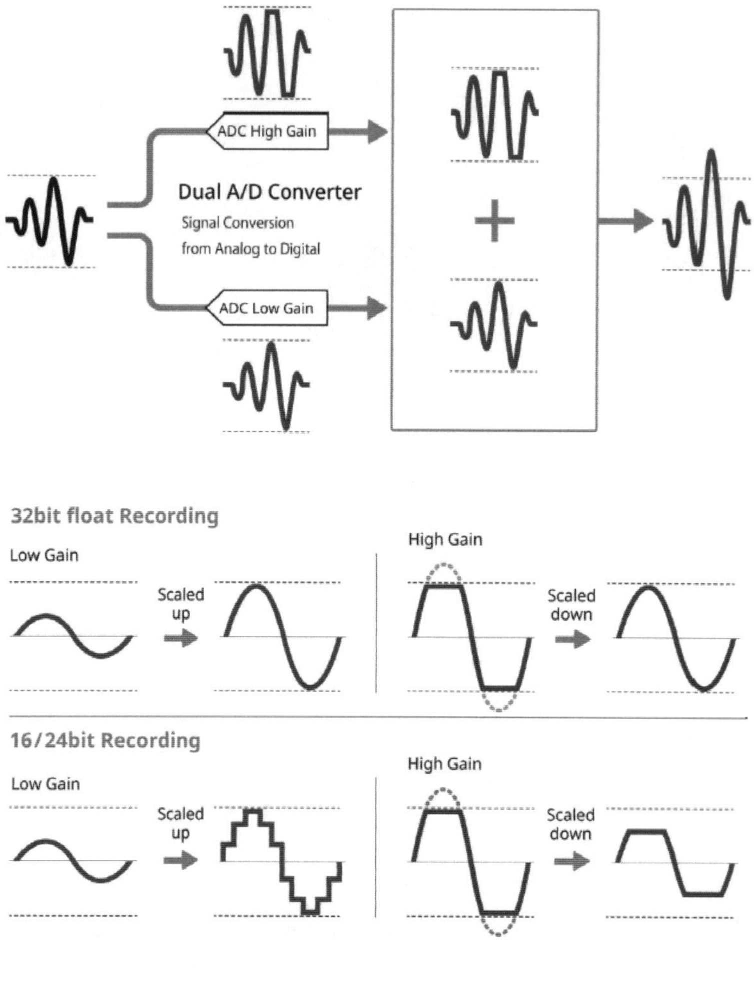

Figure 2: Graphical demonstration of the benefits of the dual A/D Converter approach to data acquisition, coupled with 32-bit floating point data processing. Source: Tascam DR-10L Pro datasheet.

Vibration Measurement Innovation

Although audio recorders typically use an electret microphone to provide the input signal, the a stud mounted accelerometer was designed with an interface that provides an analog vibration signal to the same audio interface instead. Vibration measurements can now be made with an accelerometer with a familiar look and feel, but with a data collector that fits in a shirt pocket.

Much like the innovation of a USB digital oscilloscope, the physical size and cost of data acquisition was greatly reduced by moving the interface knobs, touchscreen and integrated processing features typically found on a bench top oscilloscope to software running on a laptop where all those capabilities can be accessed through familiar laptop controls. Bidirectional control and data viewing is enabled via the USB connection. Although physical start/stop record buttons reside on the recorder, these and other controls can be viewed and accessed via a mobile app supplied by the vendor.

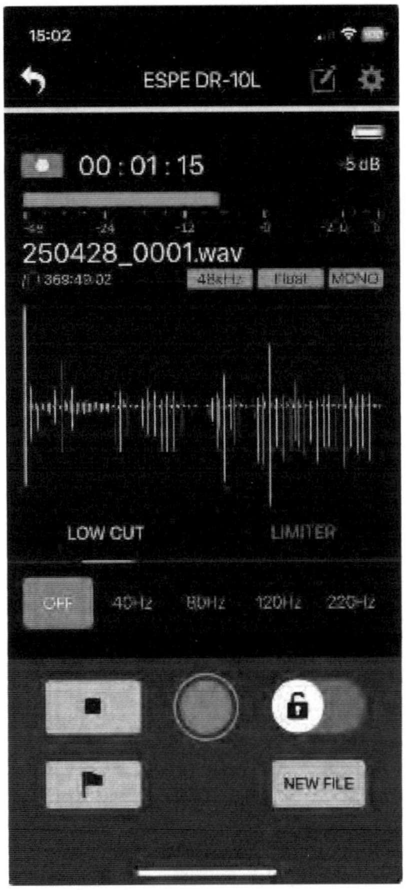

Figure 3: Audio recorder vendors offer a mobile app for configuration and control of the recorder. Source: Tascam DR-10L Pro mobile app.

Accelerometer Performance

A sensor with a standard IEPE interface cannot be used with the recorder input designed for an Electret microphone. The Junction Field-Effect Transistor typically used as the microphone part is not suitable for vibration measurement- the gain is too low, and the temperature dependence is high.

An accelerometer has been specially designed with a low voltage and low current charge amplifier which allows direct connection to the recorder via the microphone jack. The sensor has an output of 2 microA/g on standard recorder input resistor 2.4 kOhm, which enables maximum sensor measurement range of 100 G peak with a frequency range of 10 Hz - 20 kHz. The recorder input noise spectral density is 5 μg/√Hz, and the accelerometer noise density is 80 μg/√Hz, which enables a collective dynamic range of about 90 dB. This is quite respectable when compared to a leading data collector costing $20-30k.

The photo of the accelerometer designed for use with the audio recorder is shown on Fig. 4.

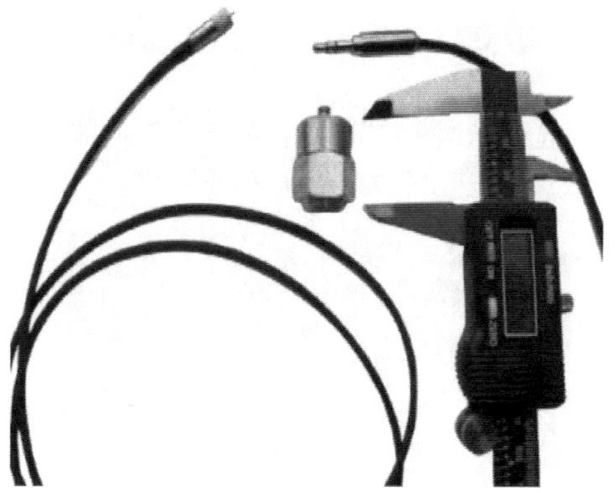

Figure 4: A photo of an accelerometer designed for the audio recorder interface, with cable. The sensor is 1" in height. An adapter for the accelerometer 10-32 connector to provide the standard IEPE 100 mV/g output for calibration purposes is available.

The sample of recorded vibration data with spectrum taken from a small electrical motor running at 3600 RPM is shown in the plots in Figure 5.

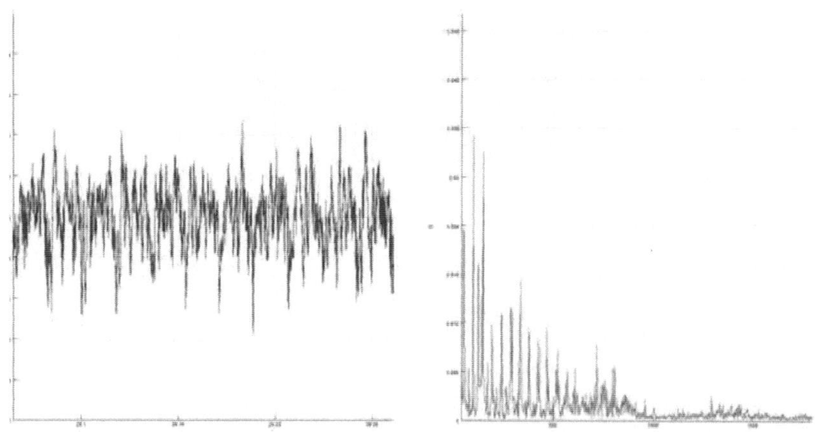

Figure 5. A time waveform with associated spectrum is shown, with the signal acquired by the accelerometer interface directly to the audio recorder, with some looseness observed.

Application

Although the audio recorders provision for modifying the names of the files, the changes are restricted in character length and not particularly convenient to make if one is collecting data on a route in the field. Here again, low cost 3rd party applications come to the rescue. There are a number of low cost mobile applications available that are optimized for collecting and tracking information, and annotating accompanying photos or videos.

Figure 6: Timemark time stamp camera mobile application.

For example, *Timemark* is an application that automatically adds accurate data/time stamps and GPS coordinates to photos and videos, which can then be matched to the audio recorder data files for renaming during analysis on your laptop. Custom text or notes can also be embedded, to include asset nomenclature from the CMMS hierarchy. The app will even include temperature and humidity information from the environment. This application is free (perhaps with Ads). Other mobile apps available include Timestamp Camera or Solocator.

Summary

An ultra portable, low-cost recording system using 32-bit float point data has been presented for potential use as a minimalist vibration data collector. A custom accelerometer with specialized interface to the recorder was constructed and characterized with large amplitude signals and high frequency vibration measurements. Signals were collected and stored in WAV files in the integrated memory of the device, then downloaded for analysis. This configuration demonstrates measurement fidelity on par with data collectors that are much more expensive, the cost savings achieved by moving the controls and interface to a mobile app and the analysis to a laptop.

TROUBLESHOOTING HIGH VIBRATION ON A CIRCULATING WATER PUMP IN A POWER PLANT USING VIDEO VIBRATION AMPLIFICATION AND VIDEO-BASED MOTION TRACKING METHODS

Author: Andrew Lerche, and Maki Onari

VIATC
TRAINING
CONFERENCE
& EXPO
NEWPORT NEWS, VIRGINIA | AUGUST 6-8, 2025

TROUBLESHOOTING HIGH VIBRATION ON A CIRCULATING WATER PUMP IN A POWER PLANT USING VIDEO VIBRATION AMPLIFICATION AND VIDEO-BASED MOTION TRACKING METHODS

Andrew Lerche, Senior Staff Engineer & Business Development
Maki Onari, President
Mechanical Solutions, Inc.
11 Apollo Drive
Whippany, NJ 07981
(001) 973-326-9920
Andy.Lerche@mechsol.com

Mr. Lerche is a mechanical engineer, and over the past 25 years has held roles in both technical and business development capacities. His technical experience has been focused on the areas of vibration and rotating machinery. He has a diverse background of mechanical engineering experience and skill sets, including research and development, product design, finite element analysis, computational fluid dynamics, production and manufacturing engineering, laboratory and field testing, equipment troubleshooting, engineering sales, and business development. Mr. Lerche has authored and presented more than a dozen technical papers in the area of vibration and turbomachinery, and holds two patents related to industrial compressors and valves. He holds a B.S. and M.S. in Mechanical Engineering from the University of Texas at Austin and the University of Texas at San Antonio, respectively. Mr. Lerche currently works at Mechanical Solutions, Inc., helping customers to resolve difficult problems involving vibration, as well as developing new business.

Mr. Onari is the President and Director of Turbomachinery Testing at Mechanical Solutions, Inc. (MSI), in Whippany, New Jersey. He is responsible for field vibration testing involving Operating Deflection Shape (ODS) and Experimental Modal Analysis (EMA) testing. His career spans more than 25 years, primarily working with rotating equipment analysis and troubleshooting in the petrochemical, refinery, and power generation industries. Before joining MSI, Mr. Onari was a Rotating Equipment Engineer in PDVSA-Venezuela responsible for the predictive maintenance of one of

the largest petrochemical complexes in Latin America. Mr. Onari received his B.S. degree (Mechanical Engineering, 1996) from Zulia University in Venezuela. Mr. Onari is co-author of the pump vibration chapter, McGraw-Hill Pump Handbook. He is a member of ASME and the ISO TC108/S2 Standards Committee for Machinery Vibration as well as the Hydraulic Institute Vibration Committee.

ABSTRACT:

In recent years, Video Vibration Amplification (VVA) has emerged as a valuable tool for troubleshooting vibration issues in rotating equipment. This paper presents a case study where VVA, combined with advanced video-based motion tracking and measurement techniques, was used to troubleshoot high vibration levels in a Circulating Water Pump (CWP) at a power plant. These video-based methods were applied alongside traditional diagnostic techniques to isolate and identify the root cause of excessive vibration. The findings indicate that integrating video-based analysis with conventional troubleshooting enhances diagnostic accuracy, providing insights that traditional methods alone may not reveal. This study details the methodologies applied, compares the effectiveness of video-based and conventional approaches, and offers insights into the root causes of vibration in large, high-value rotating equipment systems.

KEY WORDS: Field Testing; Machinery Diagnostics; Motion Magnification; Pump Vibration; Rotating Machinery; Video Motion Tracking; Video Vibration Amplification

INTRODUCTION:

Video-based diagnostic tools, such as VVA, are highly effective in identifying the root causes of excessive vibration in machinery by measuring, visualizing, and analyzing dynamic motion. Recently, similar video-based technologies known as Video Motion Tracking (VMT) tools have emerged to capture positional changes over time. While closely related to VVA, VMT specifically focuses on non-cyclic movements occurring over extended periods, making it particularly suitable for monitoring static rather than dynamic responses.

In this paper, we collectively refer to these video-based tools as VMT. Piotrowski, Marscher, and Lerche [1] previously discussed VMT technology, emphasizing its application in capturing alignment changes between offline and operational equipment states [1]. Alignment shifts due to thermal expansion or contraction of equipment represent just one of many potential applications for VMT.

Additionally, VMT technology can effectively measure and quantify positional changes in machinery subjected to varying loads. For instance, pumps and associated piping systems often experience positional shifts after startup, primarily due to hydraulic thrust loads. Tracking these movements allows engineers to assess resultant stresses within piping systems, verifying whether these stresses remain within acceptable limits. For pumps specifically, VMT can be utilized to estimate nozzle loads after systems reach hydraulic and thermal equilibrium, accounting for thermal growth in connected piping.

Although traditional measurement methods such as laser-based tools, string potentiometers, calipers, and tape measures remain viable, these approaches typically demand frequent manual intervention and are limited to discrete measurement points. Conversely, VMT captures an entire visual scene, enabling detailed, high-resolution analysis of positional changes down to thousandths of an inch. Consequently, VMT provides significantly greater measurement coverage than traditional methods, while substantially reducing manual effort since data collection occurs passively via video recording.

In the following case study, both VVA and VMT were employed alongside conventional diagnostic methods to investigate the cause of high vibration levels in a circulating water pump at a power plant. The integration of VMT and VVA methodologies enhanced diagnostic precision, facilitated comprehensive data collection, and reduced overall effort, ultimately contributing to an improved understanding of the factors influencing excessive vibration.

BACKGROUND:

A 1,500 horsepower and 166,000 gallon per minute circulating water pump used in a power plant application began exhibiting elevated vibration levels. Following a scheduled refurbishment, the vibration increased more rapidly than expected based on historical trends. The measured vibration exceeded the allowable limits defined by ANSI/Hydraulic Institute Standard 9.6.4 for this pump type, size, and speed [2]. Two identical pumps were installed at the site (Figure 1); however, the alpha pump was the only unit displaying significantly elevated vibration levels, prompting further investigation.

Figure 1: Circulating water pumps at power plant.

VIBRATION TROUBLESHOOTING

To identify the root cause of the elevated vibration, a series of diagnostic tests were conducted on the pumps, support structure, and associated piping. The primary objective was to determine the source(s) of the excessive vibration so appropriate corrective actions could be implemented to reduce it to acceptable levels.

The following tests were performed:

1. Experimental modal analysis (EMA)

2. Continuous monitoring of vibration
3. VVA
4. VMT
5. Operating Deflection Shape testing (ODS)

These tests were used to assess whether system dynamics, such as structural resonances or natural frequencies, were contributing to the abnormal vibration. They also helped identify potential secondary contributors, including excessive pump movement under load (due to thrust forces reaching tens of thousands of pounds), foundation integrity issues, or rotor-related problems.

General Instrumentation

Proximity probe data and accelerometer readings were integral to the troubleshooting process for characterizing rotor vibration and overall system movement.

- Radial X and Y probes were installed at the inboard bearing of the motor shaft to monitor rotor vibration and displacement behavior.
- In addition to the proximity probes, tri-axial accelerometers were used at the base of the discharge head, the top of the discharge head, near the bottom motor bearing, near the top motor bearing, and discharge piping flanges (on both sides of the flexible pipe coupling) to monitor vibration over time.

These point measurements provided critical insight into the dynamic response at specific locations within the system over time. However, to understand how the entire system was behaving as a whole, both dynamically and under static loading, additional full-field motion analysis techniques were employed. These methods included EMA, ODS, motion magnified video (MMV) testing (interchangeably used with VVA), and video motion tracking. ODS and VVA/MMV are well-established in prior literature and will not be discussed in depth here.

The following tests were captured using video-based diagnostic techniques:

- Pump vibration (using VVA)
- Motor vibration (using VVA)
- Discharge piping vibration (using VMT)
- Vertical movement at discharge head base using (VMT)
- Discharge piping and flange vertical and horizontal movement (using VMT)

Experimental Modal Analysis

EMA is a critical diagnostic tool used to identify the natural frequencies, mode shapes, and damping characteristics of a mechanical system. In this method, an impact hammer is used to excite the structure, and the system's frequency response is measured, typically using accelerometers. Testing is conducted at various locations by either moving the impact hammer or "roving" the accelerometers across key points of interest. When the hammer impacts a fixed location, repeated strikes at that point allow for consistent excitation while the accelerometers are repositioned throughout the system. The resulting data is captured in terms of frequency response functions (FRFs), which include both amplitude and phase angle. These FRFs can be visualized on a 3D computer model, allowing for the animation of mode shapes and further insight into structural dynamics. Additionally, if an instrumented impact hammer is used (measuring the input force), modal damping characteristics can be extracted from the response data.

In this investigation, EMA was used to determine the structural natural frequencies of the pump and motor assembly, with particular attention to frequencies that could fall within, or near, the operating range or twice the operating speed. Since Operating Deflection Shape (ODS) testing was also planned, the EMA was not used to develop full mode shapes but rather to supplement the overall dynamic analysis with key frequency data. The two primary mode shapes relevant to this pump type, potentially within the critical speed ranges, are shown in Figure 2.

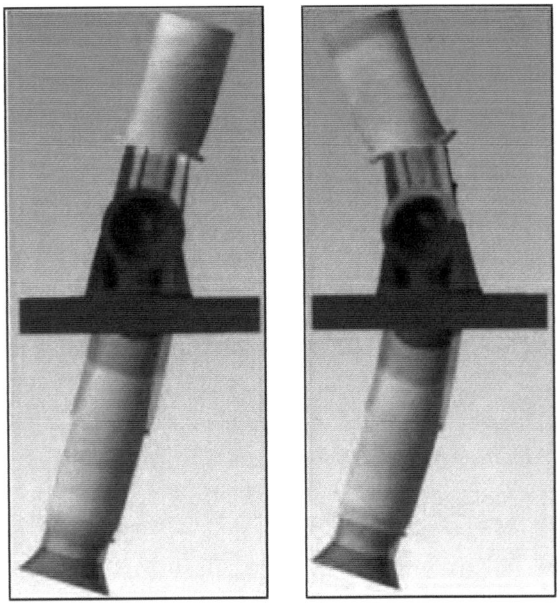

Figure 2. Typical mode shape for a large CWP-type pump. The left image shows the first global mode, referred to as a trunnion mode, and the right image shows the second global mode, referred to as the "C"-shaped mode.

The resulting data from the EMA were frequency response functions (FRF), which are a transfer function of the input to output signals. An example of an FRF from the EMA of the pump in the direction perpendicular to the discharge piping is shown in Figure 3. This FRF plot showed two natural frequencies that could get excited by the operating speed (one near 1x running speed and another near 2x running speed). The first natural frequency looked to be the trunnion mode, and fell very close to the operating speed (approximately 1% separation margin).

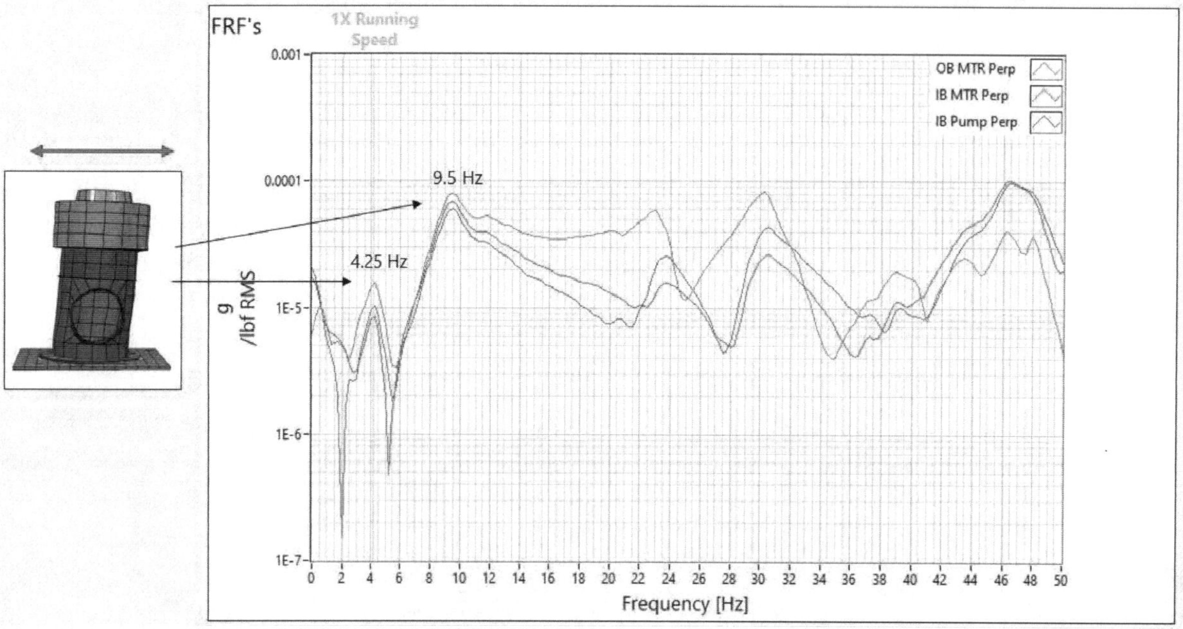

Figure 3: Pump FRF plot showing a natural frequency at 4.25 Hz (1% separation margin from 1x operating speed) in the direction perpendicular to the discharge piping.

Operating Deflection Shape Testing

ODS testing involves gathering extensive vibration data (amplitude and phase) over the entire system while it is operating. This data is then used to create a computer-animated 3D model that visually represents how the system is moving at various frequencies. The animation amplifies motion for clarity, helping to reveal dynamic behaviors that are otherwise difficult to observe. This visualization method provides insight into how the system is moving, what is not moving, excessive flexibility, separation between parts, and facilitates identifying the potential root causes. Onari and Boyadjis [3] discuss using ODS, along with other methods such as finite element analysis (FEA), to resolve vibration issues.

Figure 4 shows a frame from the ODS animation performed on the subject pump. Conducting a full ODS study typically requires several hours of on-site data acquisition, depending on system size and complexity. Post-processing then involves building the model, importing the vibration data, and assigning measured values to each node or substructure. Finally, interpolation equations are applied to generate the animated motion, allowing for comprehensive visualization of the equipment's dynamic behavior.

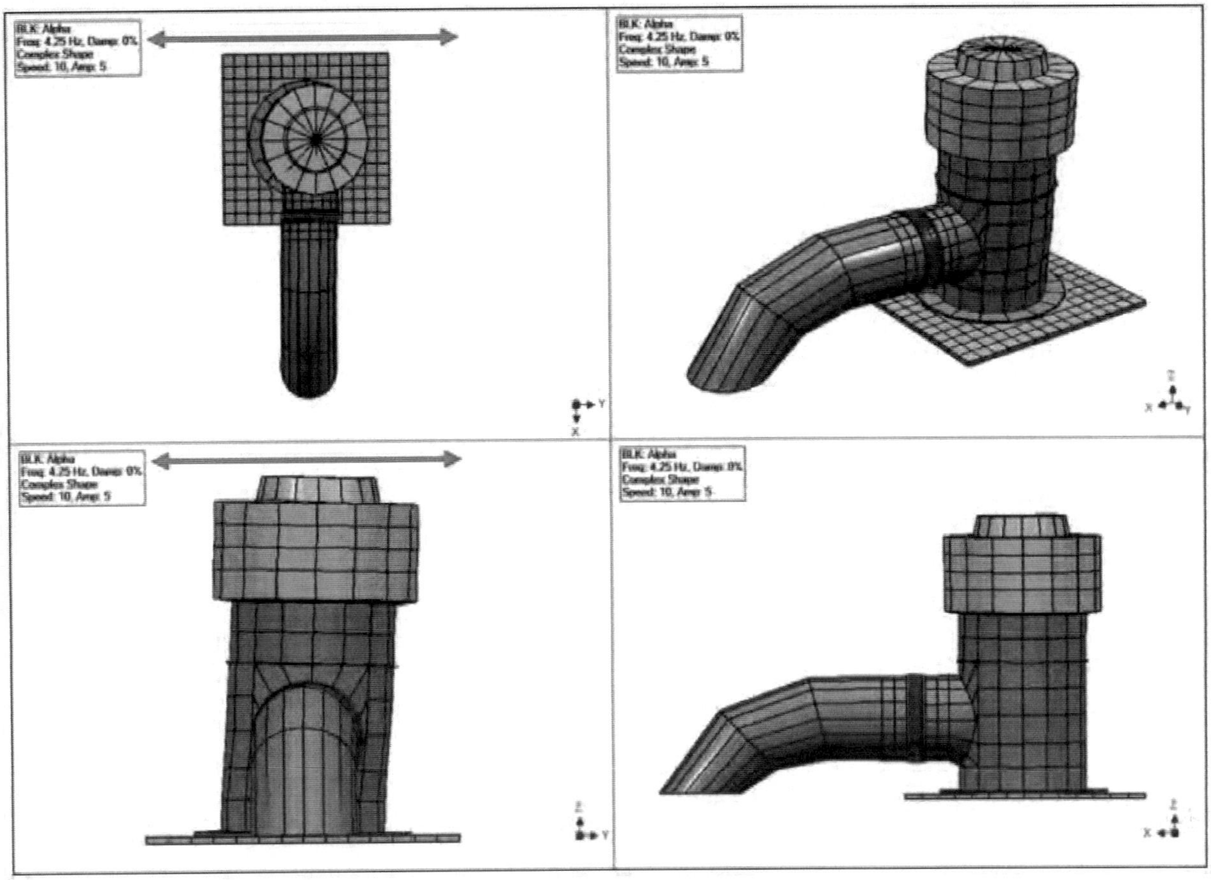

Figure 4: Still frame image of an ODS animation of the pump at 1x operating speed. The animated video showed excessive vibration perpendicular to the discharge piping.

Motion Magnified Video Testing

Motion Magnified Video (MMV), also referred to as Video Vibration Amplification (VVA), is a powerful visualization technique that, like ODS, displays exaggerated and slowed-down motion to reveal system vibration behavior. This method involves capturing a few seconds of high-speed video of the equipment during normal operation. The footage is then processed to produce a video representation of vibration motion at specific frequencies.

Unlike ODS, MMV does not incorporate rotordynamic data from proximity probes into its visualization. However, MMV offers the advantage of speed; videos can typically be generated within minutes of recording, making it a highly efficient and effective diagnostic tool. The method has been well-documented in recent literature by various researchers [4–11].

A still frame from the MMV analysis of the pump is shown in Figure 5. The animation corresponding to this frame highlights vibration motion occurring primarily perpendicular to the discharge piping, on the side opposite what is shown in the figure. The pump was confirmed to be vibrating at 1x running speed, with the dominant motion in the direction perpendicular to the discharge piping. However, identifying the root cause of this motion required further analysis.

Figure 5: Still frame image from motion magnified video of the pump at 1x operating speed.

It is known that the pump is vibrating at one times running speed, in a direction primarily perpendicular to the discharge piping; however, to get to the root of the movement and how to fix the issue, further investigation was needed.

To gain a more complete understanding, an additional VVA recording was made at the base of the pump. This revealed that the pump was securely mounted, and there were no indications of a soft-foot condition. Instead, it was observed that the pump base itself was flexing in the same direction as the dominant vibration, as illustrated in Figure 6.

Figure 6: Still frame image from motion magnified video of the pump baseplate flexing at 1x operating speed.

Video Motion Tracking

High-resolution videography was also used to track the overall movement of the pump, motor, and discharge piping during startup and typical operating conditions. This test tool method is newer than the other methods mentioned above, so it will be discussed in greater detail below. Marscher, Lerche, and Piotrowski [1] discussed this method along with other traditional methods for measuring offline to running alignment and movement. They summarized that the video-based tracking method is advantageous over traditional methods, such as laser-based, proximity probe-based, or survey transient-based methods, in that it is faster and simpler to implement. Additionally, video-based methods record all the pixels within the field of view of the cameras, allowing for measurement of all points simultaneously on the system. Other methods only take measurements where the tool is located at a certain point in time, and mounting arrangements can also be affected by thermal growth, which reduces the accuracy of the measurement. Additionally, a video of the movement can be magnified and played back, if needed, to see how the system is moving as a whole, and relative to other points on the system.

For this application, VMT was performed on both pumps, the one exhibiting high vibration and the one within acceptable limits. Video recordings were taken from views both perpendicular and parallel to the discharge piping, capturing motion of the pump, motor, and piping system.

The results revealed that the pump with elevated vibration exhibited significant downward movement of the motor, approximately 23 mils, after startup and under load, as shown in Figure 7. This behavior was not observed in the pump operating within normal vibration limits, supporting the conclusion that excessive vertical displacement may be contributing to the abnormal system response.

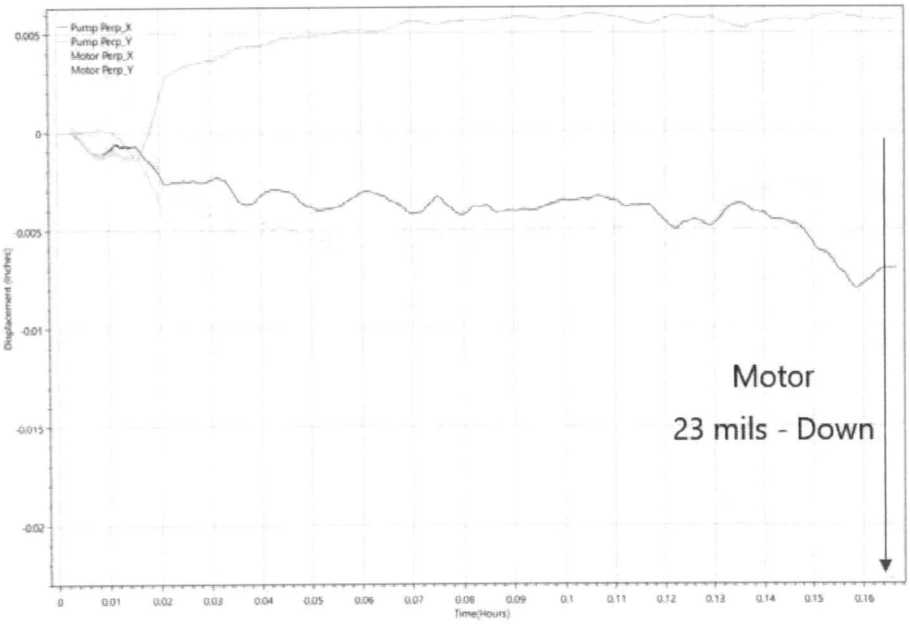

Figure 7. Video motion tracking (VMT) plot of pump and motor, showing motor movement 23 mils downward after start-up and operation.

The side view of the pump, showing the discharge piping, showed that the discharge piping moved 50 mils up, relative to the pump nozzle flange (Figure 8). This is relative movement at the discharge piping flange coupling.

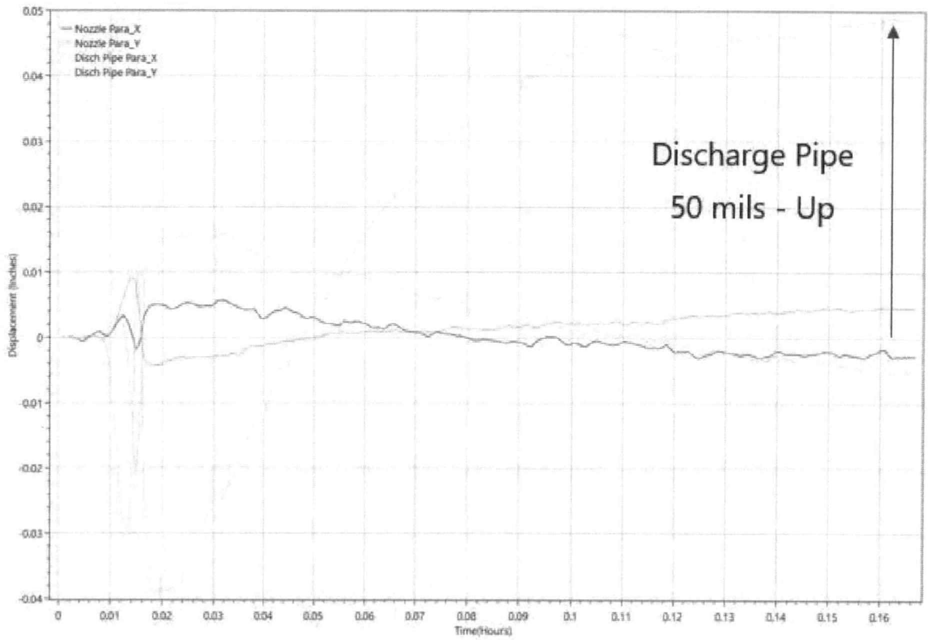

Figure 8. Video motion tracking (VMT) plot of discharge piping and pump discharge nozzle , showing discharge piping movement 50 mils upward, relative to pump discharge nozzle, after start up and operation.

Results:

This section captures the key results of the above testing that was performed. However, due to the amount of data that was gathered, only key data is shown.

The overall vibration points measured by accelerometers were mostly within reasonable limits for the Bravo pump; however, for the alpha pump, the vibration levels significantly exceeded ANSI/HI 9.6.4 acceptance levels. Figure 9 shows the overall vibration of the motor inboard bearing versus time, with the ANSI/HI 9.6.4 limit shown for reference. Figure 10 shows an FFT of the vibration at the motor inboard bearing. The dominant vibration was just above one times running speed at the trunnion mode natural frequency.

The proximity probe data showed an excessive 40 mils pk-pk radial vibration on the motor shaft of the alpha pump, while the bravo pump had only 4 mils pk-pk.

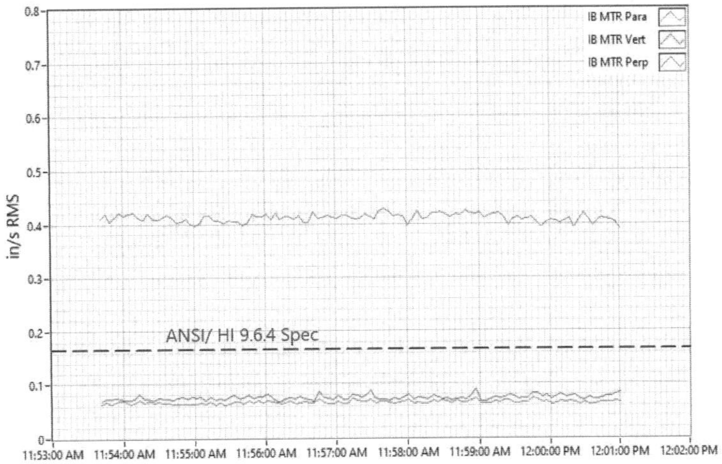

Figure 9. Overall vibration of the alpha pump at the motor inboard bearing versus time. Vibration perpendicular to the discharge head is above the HI 9.6.4 specification.

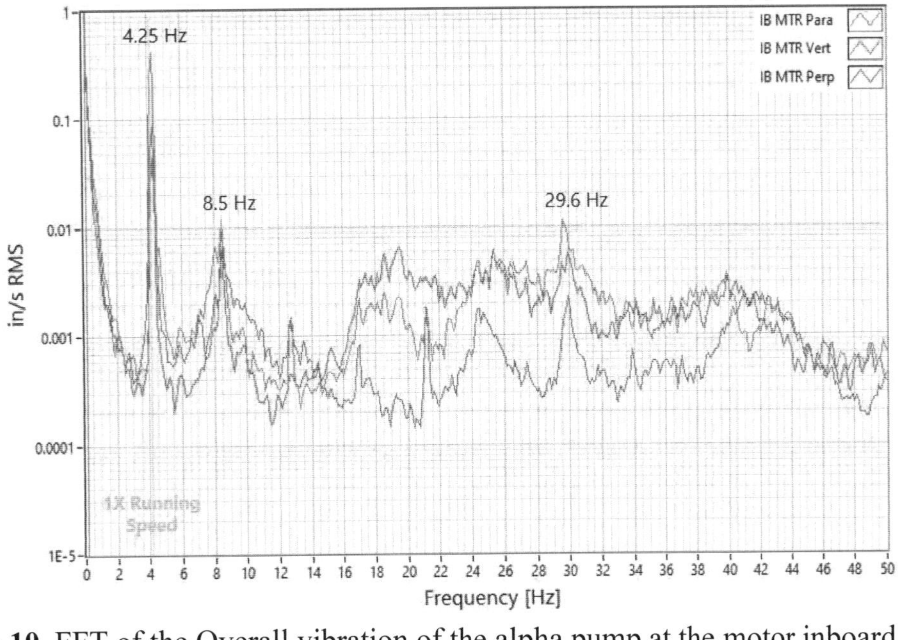

Figure 10. FFT of the Overall vibration of the alpha pump at the motor inboard bearing.

The EMA testing revealed a natural frequency (trunnion mode) on the alpha pump at one times running speed, with approximately 1% separation margin. The same mode shape on the Bravo pump had approximately a 7% separation margin from one times running speed. The "C"-shape mode had a 7% and 4% separation margin from two times running speed, for the alpha and bravo pumps, respectively.

Video-based result data is shown above in Figures 5 through 8. The VVA data provided similar frequencies and amplitudes as shown with the accelerometer data, but provided the additional benefit of being able to see where all the movement was originating from.

The VMT data in Figures 7 and 8 show the relative movement of the pump system from start-up to continuous operation, and the analysis of this data is discussed in the conclusion below.

Conclusions:

To conclude, the data has to be taken into account, along with other details such as previous vibration trends and service history of the pumps, and also other observations made onsite.

Looking at the overall vibration data, the alpha pump exceeded vibration levels, but the bravo pump had acceptable vibration levels. It was also shown that the alpha pump was operating closer to the trunnion mode natural frequency, which would also contribute to the higher vibration.

The alpha pump did show some movement after startup, as revealed by the video motion tracking system, but the movement was not excessive for the system.

The operating deflection shape testing and video vibration amplification testing both revealed that the alpha pump had excessive rocking motion side to side, in the direction perpendicular to the discharge piping. It is expected that the below-ground vibration of this trunnion mode shape is likely higher than the vibration measured at the top of the motor.

It was observed while onsite that the discharge piping had sunk down further into the ground over time, as observed in Figure 11.

Figure 11. Discharge piping on the alpha pump has sunk over time, causing the piping support to deform discharge piping.

The sinking of the discharge piping shifted the trunnion mode of the alpha pump to a lower frequency by increasing the effective mass of the system, without adding much, if any, stiffness. This explains why the vibration has gradually increased over time, as the trunnion mode natural frequency became closer to the one times operating speed of the pump. This was confirmed by adding temporary pipe supports downstream of the flexible discharge joint to help reduce the mass loading on the system, and shift the natural frequencies higher, and further away from running speed.

The above could potentially be used as a temporary solution to the vibration, but a long-term solution is needed to correct the sinking discharge piping and natural frequencies of the system. For this case, it would be possible to explore future solutions by performing a more in-depth finite element analysis (FEA) to better assess the below-ground response of the pump, and evaluate different stiffening and/or mass modifications to shift natural frequencies further away from the operating speed.

For this case, all the traditional methods of vibration troubleshooting, such as EMA, ODS, and vibration monitoring helped understand the issue, but the video-based methods of VVA and VMT helped to supplement the troubleshooting process to better understand the overall system and confirm whether or not other issues were contributing to the high vibration.

REFERENCES:

[1] Piotrowski, J., Marscher, W., Lerche, A., "*Characterizing Off-Line to Running Movement Using High-Resolution Imagery*", Proceedings of the Vibration Institute Annual Training Conference, Niagara Falls, NY, 2023.

[2] "*American National Standard for Rotodynamic Pumps for Vibration Measurements and Allowable Values*," ANSI/HI 9.6.4-2016.

[3] Onari, M., Boyadjis, P., "*Solving Structural Vibration Problems Using Operating Deflection Shape and Finite Element Analysis*", Proceedings of the 25th International Pump Users Symposium, Turbomachinery Laboratory, Texas A&M University, College Station, Texas, 2009.

[4] Chen, J.G., et al, "*Modal Identification of Simple Structures with High-Speed Video Using Motion Magnification*", J. of Sounds & Vibration, vol. 345, 9 June 2015

[5] Liu et al (2005), "*Motion magnification*", ACM Transactions of Graphics (TOG) – Proc. of ACM SIGGRAPH, TOG Homepage, v. 24, Issue 3, July, 2005.

[6] Marscher, W., and Lerche, A., "*Motion Magnified Video Applied to Vibration Modal Testing*", Proceedings of the Vibration Institute Annual Training Conference, Niagara Falls, NY, 2023.

[7] Marscher, W., and Lerche, A., "*Using Video to Magnify and Quantify Vibrating Motion*", Proceedings of the Vibration Institute Annual Training Conference, Lexington, KY, 2019.

[8] Marscher, W., Lerche, A., Frolov, S., "*What Drives Accuracy for Video Vibration?*", Proceedings of the Machinery Failure Prevention Technology (MFPT) Conference, Niagara Falls, NY, 2023.

[9] Timoner, SJ, and Freeman, DM, "*Multi-image gradient-based algorithms for motion estimation*", Optical Engineering 40.9, pp. 2003-2016, 2001.

[10] Wu, H-Y et al (2012), "*Eulerian video magnification for revealing subtle changes in the world*", ACM Trans. of Graphics (Proc. SIGGRAPH 2012), v. 31, August 2012.

[11] Marscher, W.D., and Onari, M.M. (2018), "*Video Motion Magnification vs. Operating Deflection Shapes for Machinery Diagnosis*", MFPT2018, Virginia Beach, May 2018.

UTILIZING MOTION MAGNIFIED VIDEO TO RAPIDLY TROUBLESHOOT AND CORRECT EXCESSIVE VIBRATION OF A ROD CONTROL COOLING FAN DURING AN EMERGENCY

Authors: Ryan Oleksy, Timothy Lebo, Andrew Lerche

VIATC TRAINING CONFERENCE & EXPO
NEWPORT NEWS, VIRGINIA | AUGUST 6-8, 2025

UTILIZING MOTION MAGNIFIED VIDEO TO RAPIDLY TROUBLESHOOT AND CORRECT EXCESSIVE VIBRATION OF A ROD CONTROL COOLING FAN DURING AN EMERGENCY

Ryan Oleksy, Project Engineer II
Timothy Lebo, Senior Staff Engineer & Business Development
Andrew Lerche, Senior Staff Engineer & Business Development

Mechanical Solutions, Inc.

11 Apollo Drive
Whippany, NJ 07981
(001) 973-326-9920
tim.lebo@mechsol.com

Mr. Oleksy is a Project Engineer that specializes in Finite Element Analysis of rotating machinery and mechanical structures. He has extensive training in various CAD software and excels at modeling and design. Mr. Oleksy has a background in vibration troubleshooting and field testing that benefits his understanding of applying practical solutions. He attained his B.S. and M.S. in Mechanical Engineering from Stevens Institute of Technology.

Mr. Lebo is an Engineer and Business Development Manager for Mechanical Solutions. He received his BS in Chemical Engineering from Rutgers University and has worked in a number of different roles in the Oil & Gas, Refinery, Chemical, Power, Marine, and Water and Wastewater industries including Operations, Design, Logistics, Applications Engineering, and Sales over the past 15 years. At MSI, his focus is on working with industry partners to eliminate vibration problems in rotating equipment preferably before they even happen.

Mr. Lerche is a mechanical engineer, and over the past 25 years has held roles in both technical and business development capacities. His technical experience has been focused in the areas of vibration and rotating machinery. He has a diverse background of mechanical engineering experience and skillsets including research and development, product design, finite element analysis, computational fluid dynamics, production and manufacturing engineering, laboratory and field testing, equipment troubleshooting,

engineering sales, and business development. Mr. Lerche has authored and presented more than a dozen technical papers in the area of vibration and turbomachinery, and holds two patents related to industrial compressors and valves. He holds a B.S. and M.S. in Mechanical Engineering from the University of Texas at Austin and the University of Texas at San Antonio, respectively. Mr. Lerche currently works at Mechanical Solutions, Inc., helping customers to resolve difficult problems involving vibration, as well as developing new business.

ABSTRACT:
Nuclear Power Plants rely on many rotating equipment components to properly and safely operate. Due to the risk of sudden, catastrophic failures caused by high vibration in these components, problems must be resolved in as little as 24 hours to maintain safe operation of the plant. This creates the need to quickly troubleshoot the vibration issue and determine an appropriate corrective action. For this case, Motion Magnified Video (MMV) was used to quickly identify the root cause, and assist in the corrective actions to resolve the excessive vibration. In recent years, MMV technology has played a key role of rapidly measuring, analyzing, and reporting comprehensive vibration information during emergent events allowing plants to take informed, corrective actions in short timeframes. This Case Study will detail a specific case of using MMV on a Rod Control Cooling Fan to help design a solution within 24 hours. Corroborating evidence from Operating Deflection Shape (ODS) testing and Finite Element Analysis (comparably slower methods) will be discussed and compared.

KEYWORDS:
Finite Element Analysis, Imbalance, Machinery Diagnostics, Measurement Accuracy; Motion Magnification, Nuclear Cooling Fan Operating Deflection Shape, Rotating Machinery

INTRODUCTION:
Nuclear plants generate power through nuclear fission, a sensitive process that requires constant supervision and depends heavily on successful operation of associated systems. Nuclear power plants maintain the highest standards for safety and will not function unless all systems are operational. Every component has an important role, so when specific equipment either fails or shows signs of failure, immediate action needs to be taken to provide corrective measures and ensure healthy plant operation. Elevated vibration is one of the first signs that certain factors have been introduced which negatively influence machinery and can potentially lead to failure, which turns into costly downtime. Factors that cause heightened vibration include, but are not limited to: rotor imbalance, misalignment, aerodynamic forces, installation issues such as soft foot, or structural and lateral resonances issues. Without knowing the direct cause for vibration, developing an adequate solution would be extremely difficult, if not impossible. Field testing is the first step in troubleshooting the issue. Traditional tools used to diagnose vibration range from accelerometers, proximity probes, laser alignment tools, operating deflection shape (ODS) testing, finite element analysis (FEA), and much more. These traditional methods, while surely provide accurate results, can take time to post-process and evaluate. Newer video-based methods (such as motion magnified video testing) have been developed that give real-time results to quickly pinpoint the issues at hand, allowing a "head-start" to solving the problem. These methods can be used in tandem to fully understand the system and ensure an adequate solution can be reached.

Background:

A rod control cooling fan (used to cool the control rod drive mechanism) was experiencing elevated vibration and it was suspected by the plant that some structural looseness existed at its pedestal and/or at the imbeds which the pedestal mounts to. This unit had not experienced heightened vibration in the past, so something within the system had changed since its original installation. Major components of this fan include: the pedestal, volute, pillow blocks, motor, and shaft, represented within the model shown in Figure 1.

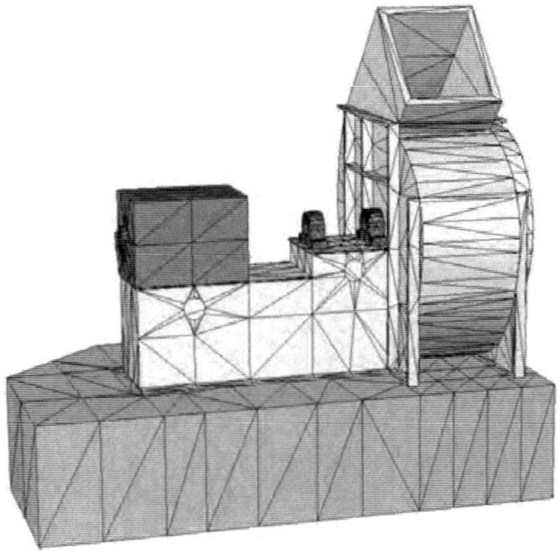

Figure 1: Rod Control Cooling Fan Model

Vibration Troubleshooting:

Given that the rod control cooling fan could not be shutdown, field testing methods were limited to motion magnified video testing (MMV) and operating deflection shape testing (ODS) as these are accomplished when the equipment is running. The main objectives of these tests were to either confirm or deny the existence of structural looseness at the pedestal, or locate the true source of vibration if that was not it. Regions of particular interest included the perimeter base of the pedestal and the pillow blocks.

Motion Magnified Video Testing:

Motion magnified video displays exaggerated, slowed-down motion of the captured area at various frequencies of interest. This method utilizes a high-speed camera to record a few seconds of the equipment during operation, which is then processed immediately to result in an animated vibration mode shape. More on this method has also been covered in recent literature by various authors [2-9].

In regards to the rod control cooling fan, video was recorded at multiple views around the pedestal, as well as a close-up of the pillow blocks. Initial inspection of MMV capture from the pedestal suggested the pedestal was rigidly attached to the surrounding concrete and no looseness at the base was present. However, when MMV was taken of the pillow blocks, it was noticed that the inboard pillow block had significant rocking and bouncing at running speed. The rocking would deform the plate it rested upon. A sample still image of this modeshape is shown in Figure 2, with arrows to call out the vibration direction of each respective component. This pillow block contained the fixed bearing which provides stiffness against both radial and axial motion of the shaft. Potential imbalance of the fan impeller could lead to this motion exhibited at the inboard pillow block, as this imbalance would create forces too large for the pedestal plate to keep the pillow block stationary. Immediately upon this discovery, the information was shared with plant personnel to determine the best corrective measure.

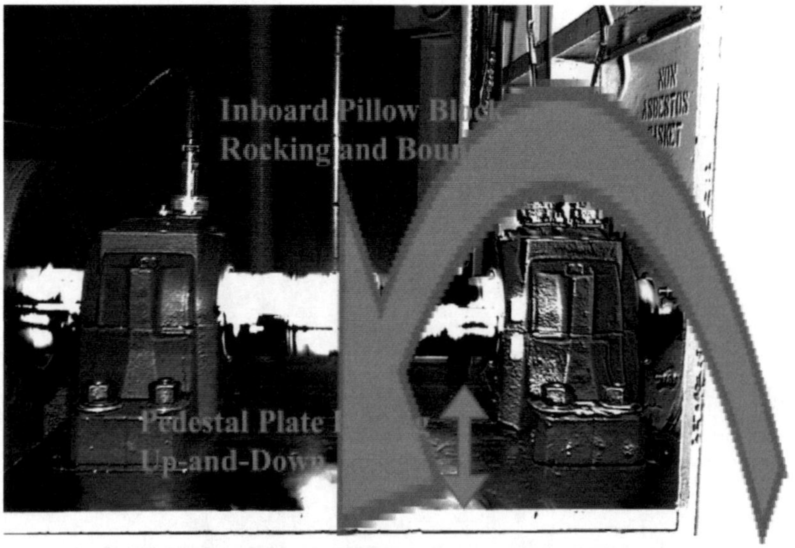

Figure 2: Still Image From MMV Capture at Running Speed

Operating Deflection Shape Testing:

Operating deflection shape testing involves gathering extensive vibration data (amplitude and phase) over the entire system, and then visualizing the data in a computer animated 3D model. Vibration at individual frequencies can be animated, with its motion amplified to better understand how the system is moving. Onari and Boyadjis discuss using ODS, along with other methods such as finite element analysis (FEA) to resolve vibration issues [1]. The visualization from ODS can confirm results from other testing methods, as well as uncover root causes that were previously hidden. With the MMV results, extensive focus was placed on the pillow blocks, but the entire fan still had data collected across it. This was to completely rule-out structural looseness at the base of the pedestal, and to ensure no vibration issues were present other than the pillow block rocking. The ODS model confirmed the rocking inboard pillow block and deformation of the plate it rests upon. Figure 3 shows a still image of an ODS animation, magnifying the vibration motion of the system at running speed.

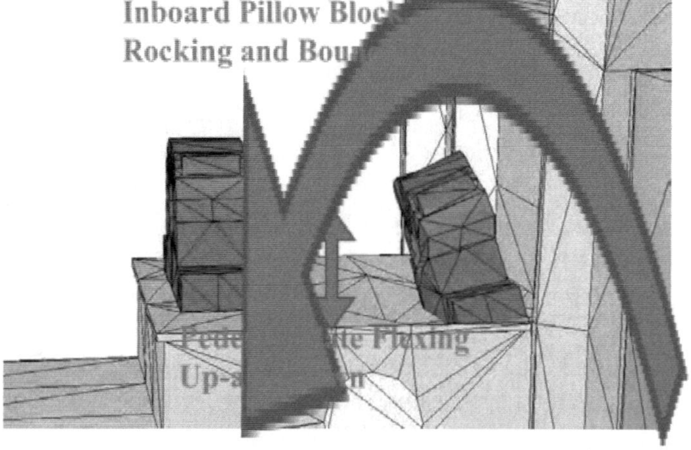

Figure 3: Still Image From ODS Animation at Running Speed

Field Testing Results:

Major findings from field testing were distinguishing the motion of the inboard pillow block, and the absence of structural looseness at the pedestal base. Vibration measurements (extracted from the ODS data) taken from the top of the inboard pillow block were read to be 0.55 in/s 0-pk in axial and vertical directions, which was inline with the plants vibration monitoring system. Vibration levels at other locations of the fan system were reasonable and did not draw concern. Since the fan impeller had not been given maintenance for a long time, it was likely residue had built up on it and caused an imbalance. The imbalance forces were strong enough to cause the inboard pillow block to push into and deform its supporting plate.

Typically, if an imbalance due to residue was recognized, the respective equipment would be shut down and respective maintenance would be performed. However, constant operation of this fan was necessary, so a solution that avoided shutdown was required. Since the plant was made aware of the imbalance immediately upon its discovery from MMV testing, an adequate solution was already being developed, ahead of a typical schedule where post processing with traditional methods is performed.

Solution:

To prevent deformation of the pillow blocks supporting plate, it needed to be stiffened in a non-invasive manner, as welding and drilling could not be accomplished next to a rotating shaft. Also, since the rod control cooling fan is inside reactor containment, heavy limitations were placed on the material selection. After some deliberation, a solution was reached that met all the requirements. It involved a clamp, that would go through the access holes in the pedestal and push down robust pieces of metal to prevent deformation of the plate. The pieces used were left over from another project at the plant and were suitable to go inside containment. The design is shown below in Figure 4.

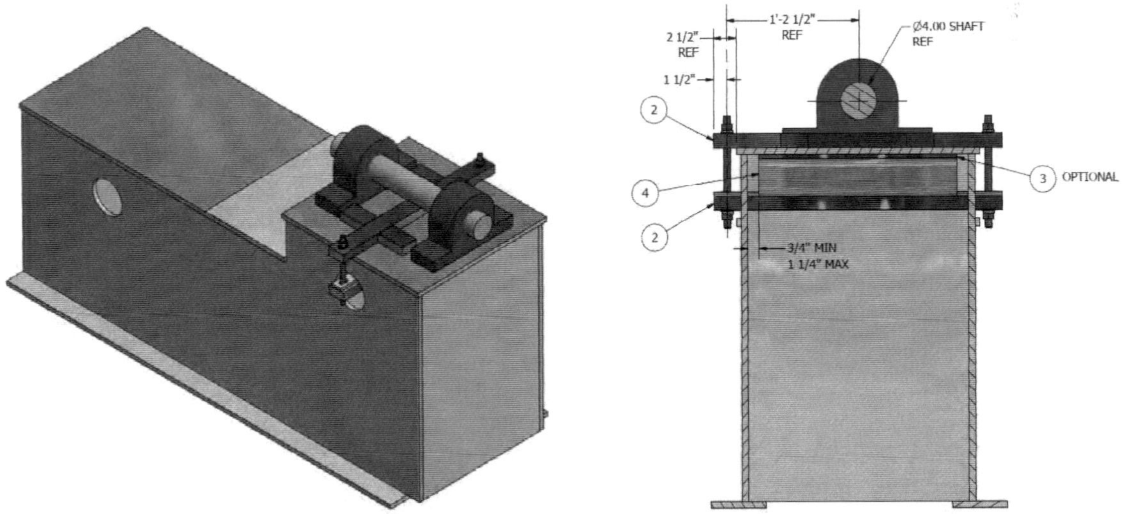

Figure 4. (Left) ISO View of Clamp Design. (Right) Cross-Section View of Clamp Design.

Finite Element Analysis Verification:

To ensure the design would be sufficient on the first implementation, a finite element analysis (FEA) was accomplished (Figure 5). As the deformation was localized to a few components, the FEA model consisted of only the pedestal, pillow blocks, and clamp. To properly sync the baseline model, the deflection of the inboard pillow block was extracted from the field testing data (4.8 mils axial deflection). This data point was taken from the top of the inboard pillow block, on the side that faces the outboard pillow block. An axial force was applied to the inboard pillow block and scaled accordingly to produce the same deflection. Then, a subsequent run was accomplished with the clamp in place to predict the reduction in deflection (Figure 6).

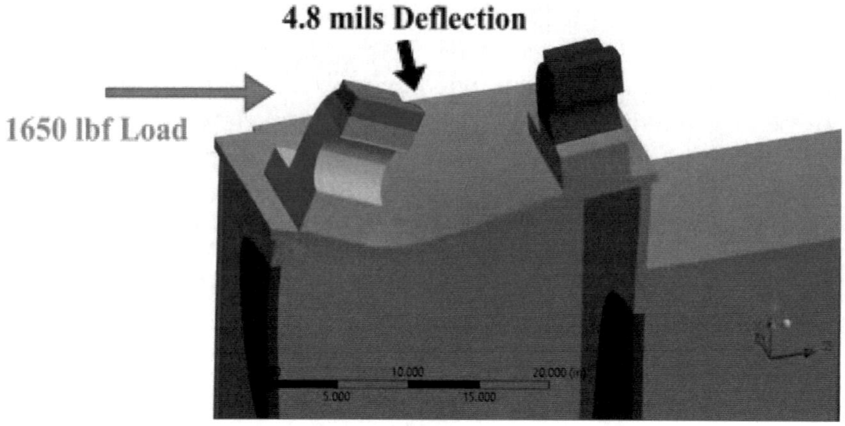

Figure 5: Baseline FEA Model

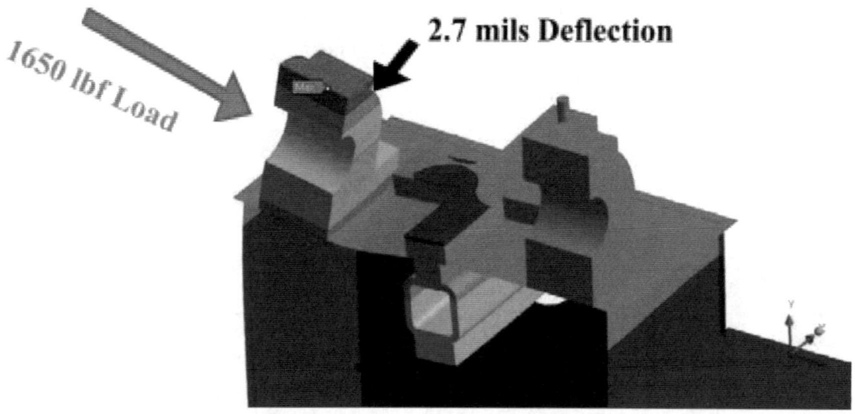

Figure 6: FEA Model with Clamp

This model gives a general undestanding of the behavior of the supporting plate when the pedestal is attached. Deflection was reduced by 44% and the clamp was predicted to be successful.

Conclusions:

After the design was validated through FEA, it was implemented in the field. When the clamp was attached, vibration reduced from 0.55 in/s 0-pk down to 0.28 in/s 0-pk, in both axial and vertical directions. The clamp proved successful and would act as a "bandaid" until the next outage when the fan could be shutdown and appropriately balanced to elimate the excessive forces acting on the inboard pillow block and supporting plate.

The diagnostic process was done ahead of typical schedules because of MMV testing. Having captured the inboard pillow block motion, the source of vibration became clear. Solutions could be produced before post-processing of traditional methods had even finished, which is what happened in this case. Plant personnel were immediately informed of the plate deformation discovery and provided the video to fully inspect the issue at hand. The clamp solution had started to be designed before the ODS was even finished. Upon its completion, the ODS model had further validated MMV results. When done correctly, both MMV and ODS testing methods can accurately distinguish vibration sources. However, MMV is processed immediately after its capture and does not require the same lengthy post-processing time as an ODS. MMV facilitates a "shortcut" to valuable information of the respective system. An ODS is still extremely vital to the full diagnostic procedure though, as it may uncover something hidden to MMV capture.

Furthermore, in this case FEA was used to support the solution. Given the fast development of the clamp design, time be could dedicated to ensure its effectiveness. Had only traditional methods been used, the vibration source would have been diagnosed after a much longer duration of time, at which point the plant could have been desperate to implement a solution without fully understanding the response it would create. Combining testing and analysis methods is the best approach to correcting a system.

References:

[1] Onari, M., Boyadjis, P., "*Solving Structural Vibration Problems Using Operating Deflection Shape and Finite Element Analysis*", Proceedings of the 25th International Pump Users Symposium, Turbomachinery Laboratory, Texas A&M University, College Station, Texas, 2009.

[2] Chen, J.G., et al, "*Modal Identification of Simple Structures with High-Speed Video Using Motion Magnification*", J. of Sounds & Vibration, vol. 345, 9 June 2015

[3] Liu et al (2005), "*Motion magnification*", ACM Transactions of Graphics (TOG) – Proc. of ACM SIGGRAPH, TOG Homepage, v. 24, Issue 3, July, 2005.

[4] Marscher, W., and Lerche, A., "*Motion Magnified Video Applied to Vibration Modal Testing*", Proceedings of the Vibration Institute Annual Training Conference, Niagara Falls, NY, 2023.

[5] Marscher, W., and Lerche, A., "*Using Video to Magnify and Quantify Vibrating Motion*", Proceedings of the Vibration Institute Annual Training Conference, Lexington, KY, 2019.

[6] Marscher, W., Lerche, A., Frolov, S., "*What Drives Accuracy for Video Vibration?*", Proceedings of the Machinery Failure Prevention Technology (MFPT) Conference, Niagara Falls, NY, 2023.

[7] Timoner, SJ, and Freeman, DM, "*Multi-image gradient-based algorithms for motion estimation*", Optical Engineering 40.9, pp. 2003-2016, 2001.

[8] Wu, H-Y et al (2012), "*Eulerian video magnification for revealing subtle changes in the world*", ACM Trans. of Graphics (Proc. SIGGRAPH 2012), v. 31, August 2012.

[9] Marscher, W.D., and Onari, M.M. (2018), "*Video Motion Magnification vs. Operating Deflection Shapes for Machinery Diagnosis*", MFPT2018, Virginia Beach, May 2018.

FLUID DRIVE UNBALANCE AND SYMPTOMS

Author: Alexander Vass, PE

VIATC TRAINING CONFERENCE & EXPO

NEWPORT NEWS, VIRGINIA | AUGUST 6-8, 2025

FLUID DRIVE UNBALANCE AND SYMPTOMS

Alexander Vass, PE
Rotating Equipment Engineer, Xcel Energy
Alexander.r.vass@xcelenergy.com

Alex is a Category III Vibration Analyst based in Denver, CO. He has 5 years of experience as a vibration analyst and reliability engineer performing route-based collection and analysis, ODS analysis, Modal analysis, and Dynamic Balancing for multiple clients. He began learning vibration analysis as a mechanical engineer for RedWolf Reliability in Fort Collins, CO. In this role, he served various customers in Refinery Operations, Upstream Oil and Gas, Food and Beverage, Mining, and Power Generation. Currently, he supports Xcel Energy's power generation fleet of gas, steam, and hydroelectric power plants across Colorado.

Abstract

This is a fluid coupling used to control speed of a boiler feed pump at the Hayden Power Plant in Hayden, CO. Since 1976, the machine had elevated vibrations across the fluid drive and motor. After a shim dislodged itself from underneath the fluid coupling, the vibration response had increased to unsafe operating levels, exceeding 1 in/sec pk. After unsuccessful soft foot adjustments, re-alignment attempts with multiple thermal offset configurations, and investigations into possible motor thrusting, dynamic balancing of the fluid drive oil drum was ultimately the solution to this legacy risk to reliable operation.

The balancing attempt was completed successfully on July 3rd. Initial balance quality per ISO 1940 was G50.1. ISO 1940 recommends operating this machine component with a balance quality of G6.3 or lower to prevent damage caused by centrifugal forces. The final balance quality achieved was G0.7

After balancing, the most significant vibration amplitudes were in the axial direction. These axial amplitudes were 0.08 in/sec rms overall and 0.07 in/sec rms overall on the Motor side and Pump side respectively. All axial and radial vibrations were below ISO 20816-3 alarm limits and remain acceptable for continued operation without restriction.

Key Words: Balancing, Fluid Drive, Misalignment.

Drivetrain Basics

The 2B Boiler Feed Pump machine train consists of a 5000HP motor, an American Standard fluid drive, and a Pacific boiler feedwater pump. Pictures of the machine train are displayed in Images 1, 2, and 3.

Image 1: View of 5000HP Allis-Chalmers Motor

Image 2: View of American Standard Fluid Drive

Image 3: View of Pacific Pump

Flow of feedwater to downstream heaters is controlled by the fluid drive. Image 4 shows the nameplate of the fluid drive. The fluid coupling is a hydrodynamic device using Mobil DTE 732 Turbine Oil to transmit rotating mechanical power between the drive shaft and driven shaft. The input and output impellers are contained within a doughnut casing. Pump speed is controlled by the scoop tube positioner, which controls the amount of oil imparting mechanical work within the doughnut onto the output impeller. The scoop tube is the mechanism which transfers fluid from the inside of the doughnut to the oil sump within the fluid drive casing. The more retracted the scoop tube is, the more fluid can impart work on the output shaft, and thus a greater output speed can be achieved.

Image 4: American Standard Gyrol Fluid Drive, size 212

Collection Setup

The entire machine train had up to 18 collection points, 6 for each machine. A table describing each velocity reading is below in Table 1.

Table 1: CommTest vb8 collection setup for velocity readings. Measurement locations collected during daily rounds are highlighted yellow.

Machine Name	Measurement	Fmin	Fmax	Lines	Resolution	Averages	%Overlap	Window	Collection Time (secs
MOBH	Velocity	1.25 Hz	2,000 Hz	1600	1.25	4	50	Hanning	2.0 s
MOBV	Velocity	1.25 Hz	2,000 Hz	1600	1.25	4	50	Hanning	2.0 s
MIBH	Velocity	1.25 Hz	2,000 Hz	1600	1.25	4	50	Hanning	2.0 s
MIBV	Velocity	1.25 Hz	2,000 Hz	1600	1.25	4	50	Hanning	2.0 s
MIBA	Velocity	1.25 Hz	2,000 Hz	1600	1.25	4	50	Hanning	2.0 s
FD MSH	Velocity	1.25 Hz	2,000 Hz	1600	1.25	4	50	Hanning	2.0 s
FD MSV	Velocity	1.25 Hz	2,000 Hz	1600	1.25	4	50	Hanning	2.0 s
FD MSA	Velocity	1.25 Hz	2,000 Hz	1600	1.25	4	50	Hanning	2.0 s
FD PSH	Velocity	1.25 Hz	2,000 Hz	1600	1.25	4	50	Hanning	2.0 s
FD PSV	Velocity	1.25 Hz	2,000 Hz	1600	1.25	4	50	Hanning	2.0 s
FD PSA	Velocity	1.25 Hz	2,000 Hz	1600	1.25	4	50	Hanning	2.0 s
PIBH	Velocity	1.25 Hz	2,000 Hz	1600	1.25	4	50	Hanning	2.0 s
PIBV	Velocity	1.25 Hz	2,000 Hz	1600	1.25	4	50	Hanning	2.0 s
PIBA	Velocity	1.25 Hz	2,000 Hz	1600	1.25	4	50	Hanning	2.0 s
POBH	Velocity	1.25 Hz	2,000 Hz	1600	1.25	4	50	Hanning	2.0 s
POBV	Velocity	1.25 Hz	2,000 Hz	1600	1.25	4	50	Hanning	2.0 s

These readings were collected using a Bently Nevada CommTest vb8 portable analyzer, with System1 as the analysis software. Acceleration time waveforms and Demodulated spectral readings were also collected at each point. Additionally, plant operations collect velocity readings on the machine train for their daily rounds. Typical collection points for plant operations are highlighted yellow in Table 1. For the purposes of troubleshooting and evaluating the machine train's condition, readings were recorded at all locations specified in Table 1. Locations of each collection point are labeled on Images 5, 6, and 7 below.

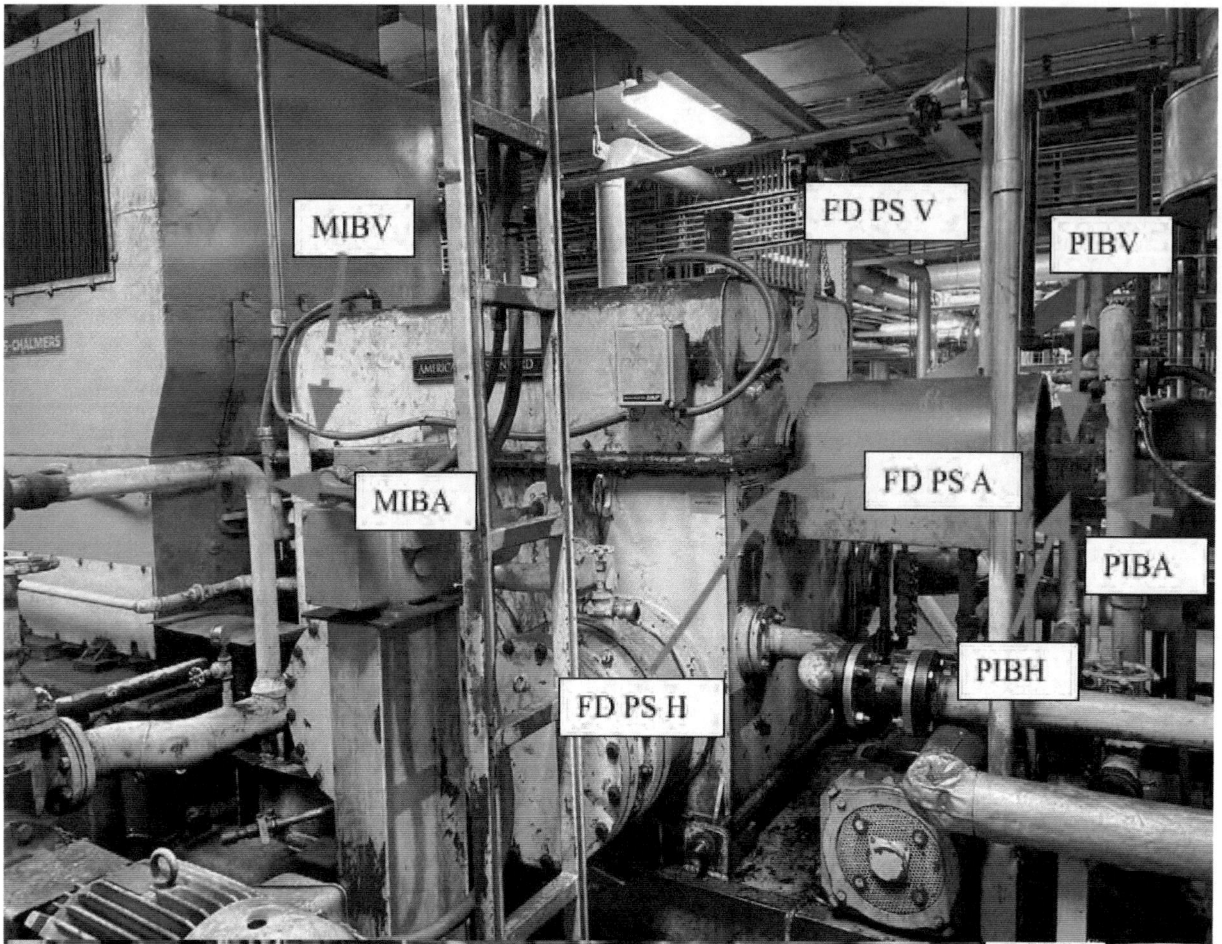
MIBV
FD PS V
PIBV
MIBA
FD PS A
PIBA
PIBH
FD PS H

POBV
POBH

The accelerometer used for these collection points was a model AC102 supplied by Connection Technology Center (CTC). It is a 100mV/g IEPE sensor with a dual-rail magnetic base. Specifications for the accelerometer are included in Figure 1, as provided by CTC.

Specifications	Standard	Metric
Part Number	AC102	M/ or M8/AC102
Sensitivity (±10%)	100 mV/g	
Frequency Response (±3dB)	30-810,000 CPM	0,5-13500 Hz
Dynamic Range	± 80 g, peak *Vsource ≤ 5V, 12V Bias	
Electrical		
Settling Time	<2.5 seconds	
Voltage Source	18-30 VDC	
Constant Current Excitation	2-10 mA	
Spectral Noise @ 10 Hz	14 µg/√Hz	
Spectral Noise @ 100 Hz	2.3 µg/√Hz	
Spectral Noise @ 1000 Hz	2 µg/√Hz	
Output Impedance	<100 ohm	
Bias Output Voltage	10-14 VDC	
Case Isolation	>10^8 ohm	

Specifications	Standard	Metric
Environmental		
Operating Temperature Range	-58 to 250 °F	-50 to 121 °C
Maximum Shock Protection	5,000 g, peak	
Electromagnetic Sensitivity	CE	
Sealing	Welded, Hermetic	
Submersible Depth	200 ft.	60 m
SIL Rating	SIL 2	
Physical		
Sensing Element	PZT Ceramic	
Sensing Structure	Shear Mode	
Weight	3.2 oz	90 grams
Case Material	316L Stainless Steel	
Mounting Thread	1/4-28 Blind Tapped Hole	
Connector (Non-Integral)	2 Pin MIL-C-5015	
Resonant Frequency	1,380,000 CPM	23000 Hz
Mounting Torque	2 to 5 ft. lbs.	2,7 to 6,8 Nm
Mounting Hardware Supplied	1/4-28 Stud	M6x1 or M8x1.25 Adapter Stud
Calibration Certificate	CA10	

Typical Frequency Response

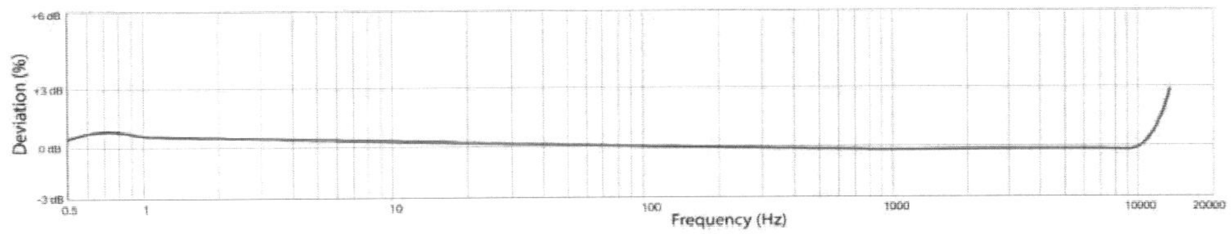

Figure 1: Accelerometer AC102 Specifications, as provided by CTC.

Overall velocity alarm setpoints were based on ISO 20816-3. A warning setpoint was set to 0.22 in/sec rms, while a danger setpoint was set to 0.35 in/sec rms. This standard was selected to help assess machine condition due to its applicability and lack of a known acceptable vibration for this fluid drive. Original acceptance testing was performed on the fluid drives in 1976, with readings provided in displacement. Although displacement readings would be best suited for assessing shaft movement within the fluid drive's plain bore bearings, Velocity alarm limits were used due to the equipment and historical readings available.

Background

Historically, the fluid drive had been operated with elevated vibrations. Previous attempts at decreasing vibration levels included multiple balancing attempts by multiple vibration technicians, re-alignments of the machine train, and a bearing lubrication system re-design. These attempts were unsuccessful. Because the machine train continued to operate with minimal damage observed over time, other emergent risks were prioritized where the consequences seemed to be less known.

During the Spring of 2024, the shimpack supporting the fluid drive became dislodged from underneath the machine. The location of the roughly 0.15" thick shimpack affected is displayed below. Despite installing the loose shim back into its original location, the fluid drive vibrations had increased significantly from baseline up to 1 inch/sec peak overall. The location of the shimpack is displayed in Image 8, with a closeup view in Image 9.

Image 8: Location of shimpack which became dislodged from underneath Fluid Drive.

Image 9: Close up view of location where shimpack was dislodged.

This event which created an unacceptable vibration response justified further investigation, including a detailed vibration analysis of the machine train.

Initial Analysis

In general, vibrations were dominated by 1x motor speed across the motor, fluid drive, and pump. The most significant vibrations were measured in the axial directions of both the motor side and pump side of the fluid drive. Initial velocity amplitudes were recorded across the machine train and are graphed against ISO 20816-3 alarm setpoints in Figure 2.

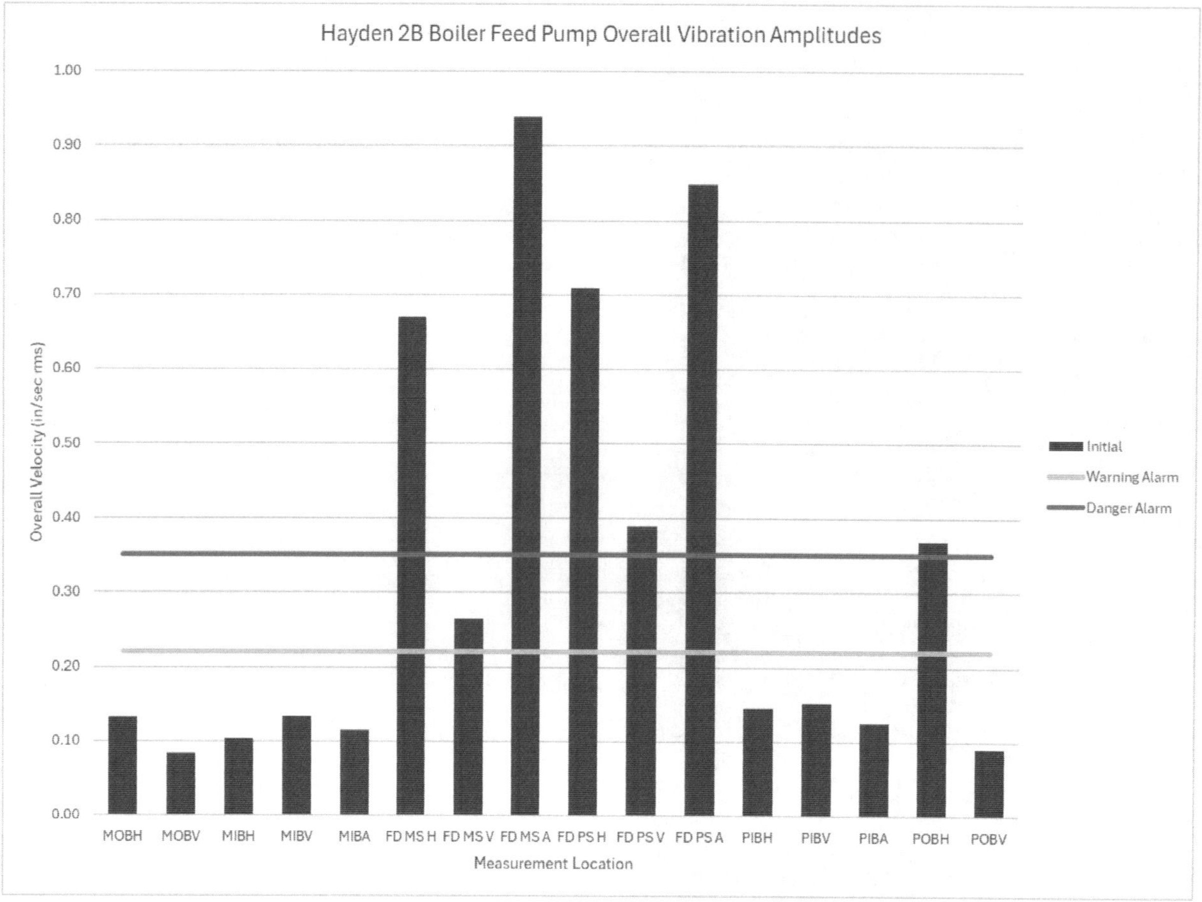

Figure 2: Initial velocity overall readings recorded before troubleshooting began.

The next two screenshots are motor spectra, with 2x AC Line frequency dominant in Figures 3 and 4. The sinusoidal time waveform shape and high frequency rotor bar pass frequency content present in the acceleration spectra is representative of an electrical vibration characteristic of the motor. True severity of this potential motor fault cannot be determined without additional trending or motor circuit evaluation.

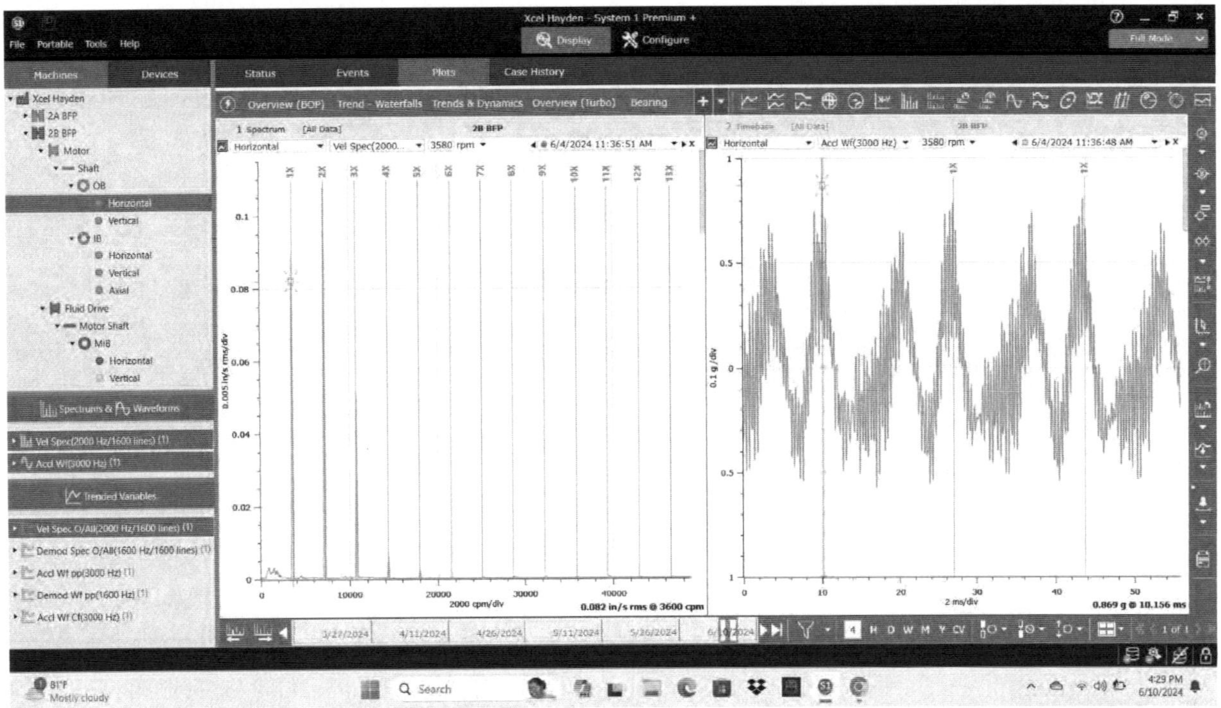

Figure 3: 2xAC line frequency is present radially on the motor at an amplitude slightly higher than 1x running speed.

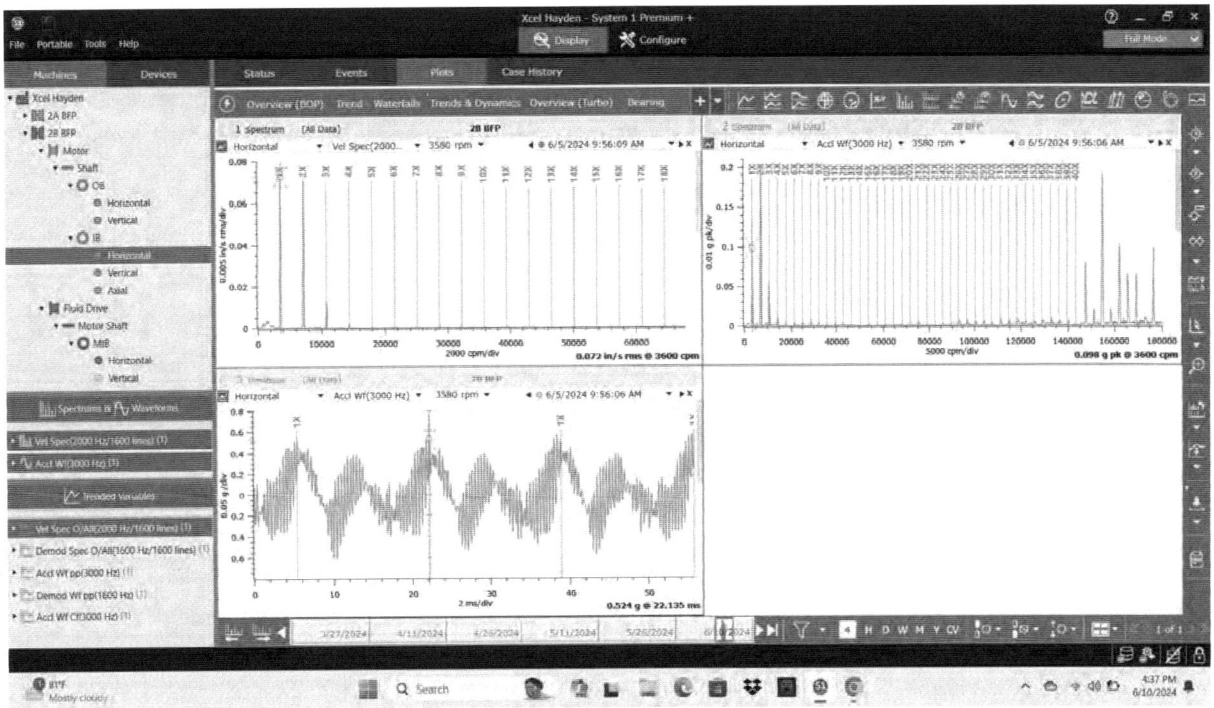

Figure 4: Velocity and Acceleration spectra with Rotor Bar Pass Frequency (RBPF) content and 2xAC line frequency.

Below in Figures 5 and 6 are the fluid drive spectra in the axial direction (inline with the shafts) on the motor side and pump side. 1x amplitude was dominant at both axial locations. Combined with the M and W shapes of the time waveforms, these amplitudes seemed to be caused by a misalignment of the fluid drive to the motor.

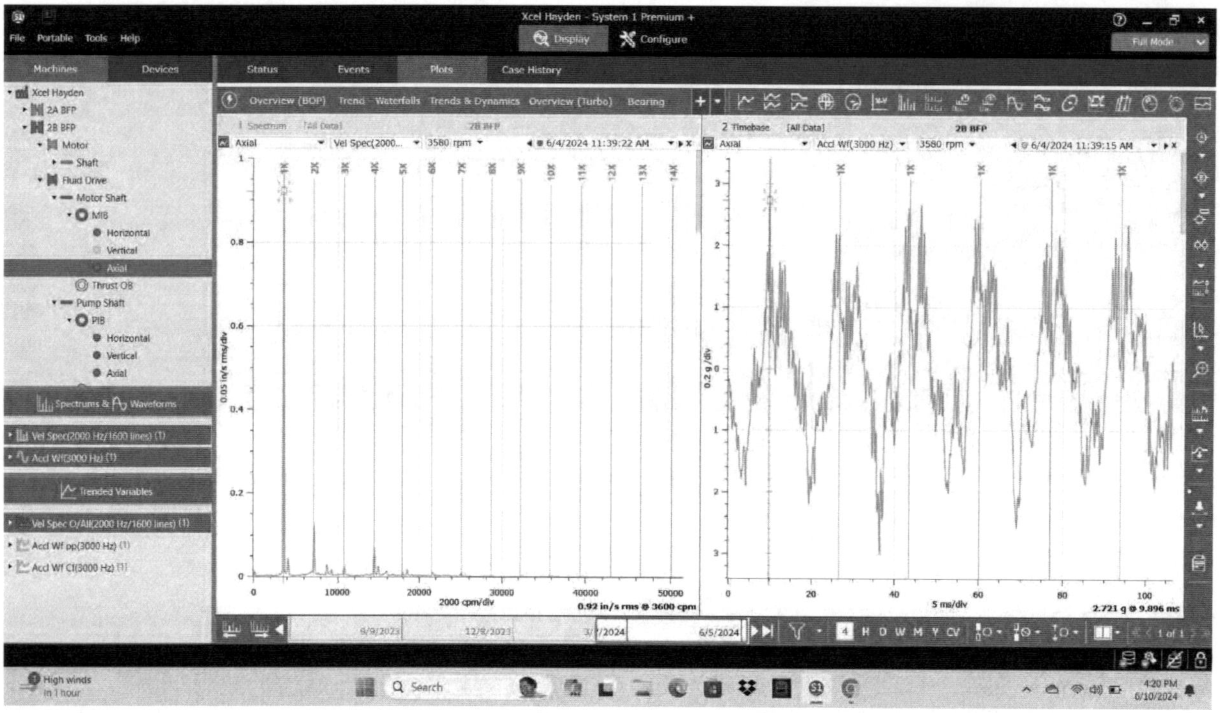

Figure 5: Fluid Drive Motor Side Axial vibration, with dominant 1x and M's and W's present in Time waveform (TWF).

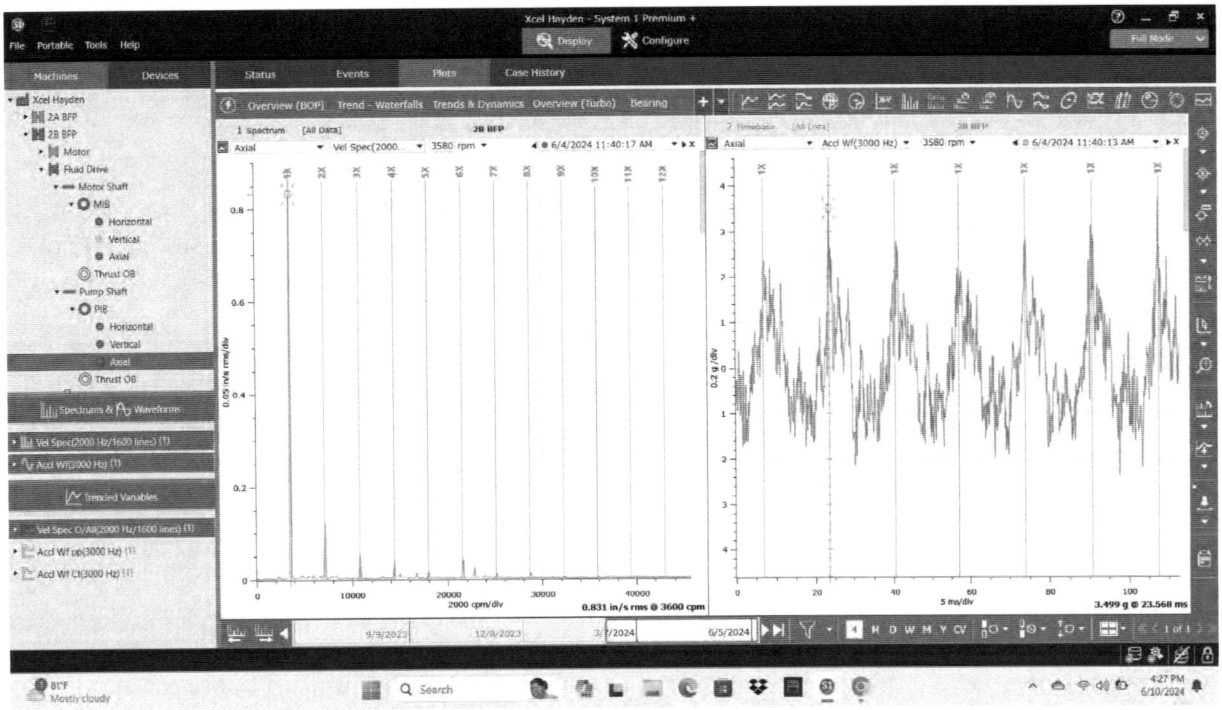

Figure 6: Dominant 1x vibration at driven side axial location.

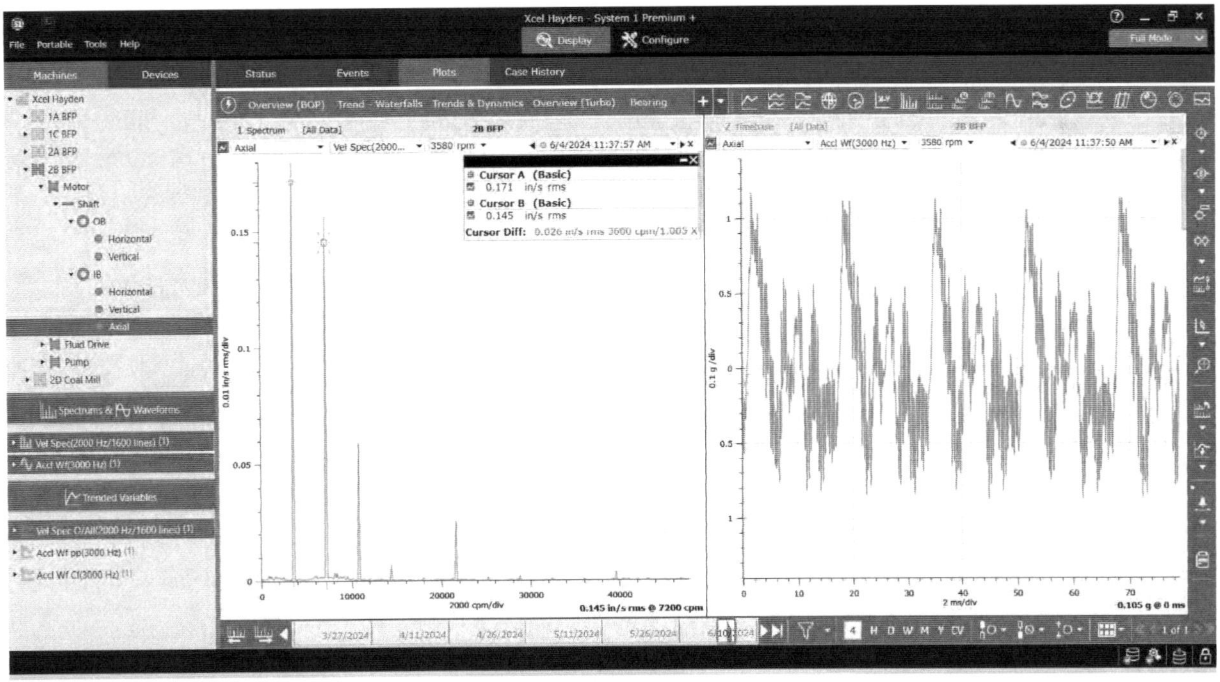

Figure 7: Motor IB Axial readings, with elevated 1x and possibly 2x or 2xAC line frequency.

Initial readings were collected June 4th with the pump recirculation line fully open, effectively adding minimal load to the machine. The motor had slightly elevated 1x running speed and 2x AC line frequency in the axial direction as well, as shown in Figure 7.

Phase analysis was also conducted across the motor and fluid drive. However, this proved inconclusive to determine if the forces acting upon the machine were rotational or stationary in nature. Because a rotational force wasn't initially clear and understanding the possibility of an elevated 1x being due to a misalignment, balancing of the fluid drive wasn't the first corrective action to be taken.

Due to the change in position of the fluid drive foot shims, elevated axial amplitudes on the fluid drive, and inconclusive phase analysis, the cause of the excessive vibration was attributed to a significant misalignment. The recommended action was to check the misalignment, and optimize the alignment as needed.

Misalignment

Mechanics were able to optimize the horizontal alignment at both couplings. However, the vertical offset needed was unclear due to a lack of previous alignment documentation. A handwritten document for a thermal offset calculation was found from 1982. These alignment numbers were used, but did not improve the vibration response. Therefore, an attempt was made to manually calculate new thermal offset values based on operating conditions of the 2A BFP machine train.

Because the unit was still online with a redundant, identical Boiler Feed Pump machine train, the 2A BFP was used to collect operating temperature readings. A digital infrared thermometer was used for collecting temperatures. The temperatures were recorded on the motor underneath the shaft, fluid drive underneath the shaft, and the pump support. Readings were not collected on the pump barrel because the barrel does not provide a normal force against the base which would impact the alignment of the pump shaft.

2A BFP Train (Hot):

$$T_{ma} = 115°F$$
$$T_{fda} = 165°F$$
$$T_{pa} = 110°F$$

2B BFP Train (Cold):

$$T_{mb} = 65°F$$
$$T_{fdb} = 65°F$$
$$T_{pb} = 90°F$$

Vertical Height between Pedestal base and shaft centerline:

$$L_{m} = 33''$$

$$L_{fd} = 40''$$
$$L_p = 43''$$

Growth Constant:

$$\alpha = \frac{6.5\times10^{-6}}{°F}$$

With the thermal growth constant, vertical height, and expected thermal difference between at rest and steady state operating conditions, one can calculate an expected vertical alignment offset due to thermal growth. The thermal growth equation (equation 1), is applied to the Motor, Fluid Drive, and Pump below:

$$\delta_x = \alpha L_x \Delta T_x \#(1)$$

Motor:

$$\delta_m = \alpha L_m \Delta T_m$$
$$\delta_m = 6.5\times10^{-6} * 33'' * (115°F - 65°F)$$
$$\delta_m = 10.7\ thou$$

Fluid Drive:

$$\delta_{fd} = \alpha L_{fd} \Delta T_{fd}$$
$$\delta_{fd} = 6.5\times10^{-6} * 40'' * (165°F - 65°F)$$
$$\delta_{fd} = 26\ thou$$

Pump:

$$\delta_p = \alpha L_p \Delta T_p$$
$$\delta_p = 6.5\times10^{-6} * 43'' * (110°F - 90°F)$$
$$\delta_p = 5.6\ thou$$

Fluid Drive Growth to Pump:

$$Offset_{fd\to p} = \delta_p - \delta_{fd}$$
$$Offset_{fd\to p} =- 20.4\ thou$$

Motor Growth to Fluid Drive:

$$Offset_{m\to fd} = \delta_{fd} - \delta_m$$
$$Offset_{m\to fd} =+ 15.3\ thou$$

The plant also decided to mobilize a contractor to help oversee the disassembly of the fluid drive and replace any damaged components as a result of the excessive vibration. Despite recording severe vibration amplitudes on the fluid drive, only minor damage to the pump side IB sleeve bearing was discovered. The most significant damage was on the motor IB bearing and thrust collar. Figures 8 and 9 below show the collar as new in 2018, and as found June 2024.

Figure 8: Motor Thrust Collar, New install in 2018.

Figure 9: Motor Thrust Collar Wear discovered in June 2024.

Both the motor IB bearing and slightly damaged pump side IB sleeve bearing were replaced, and the motor position was adjusted to prevent future thrusting onto the thrust collar. After the fluid drive inspection, alignment adjustment, and an addition of shims (0.015") underneath the fluid drive, vibrations remained excessive as of June 20th.

This alignment configuration did not improve the vibration response across the machine train. Although the pump side axial amplitudes decreased, motor side axial amplitudes remained significantly elevated. Radial amplitudes were elevated at some locations along the fluid drive as well, as seen in Figures 10 and 11. 2x AC Line frequency continued to be dominant on the motor, as shown in Figure 12.

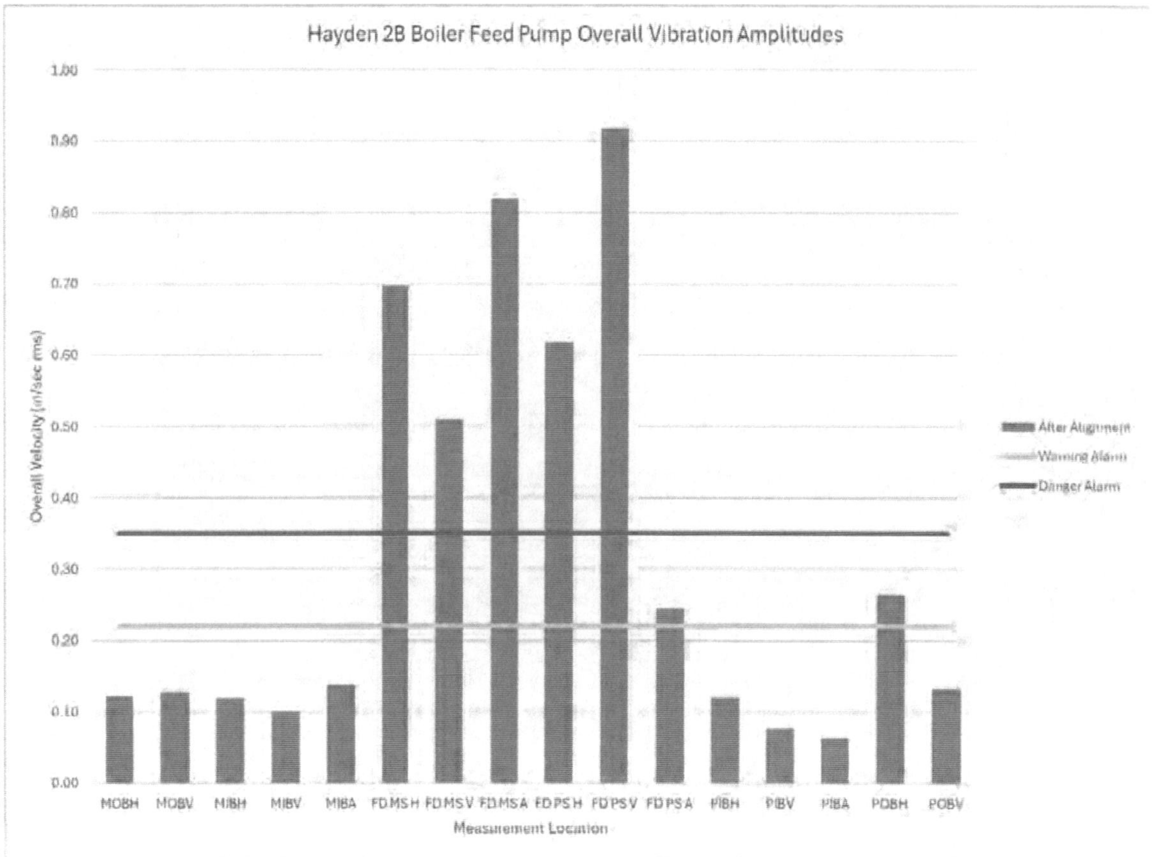

Figure 10: Overall velocity amplitudes recorded after alignment and bearing replacement.

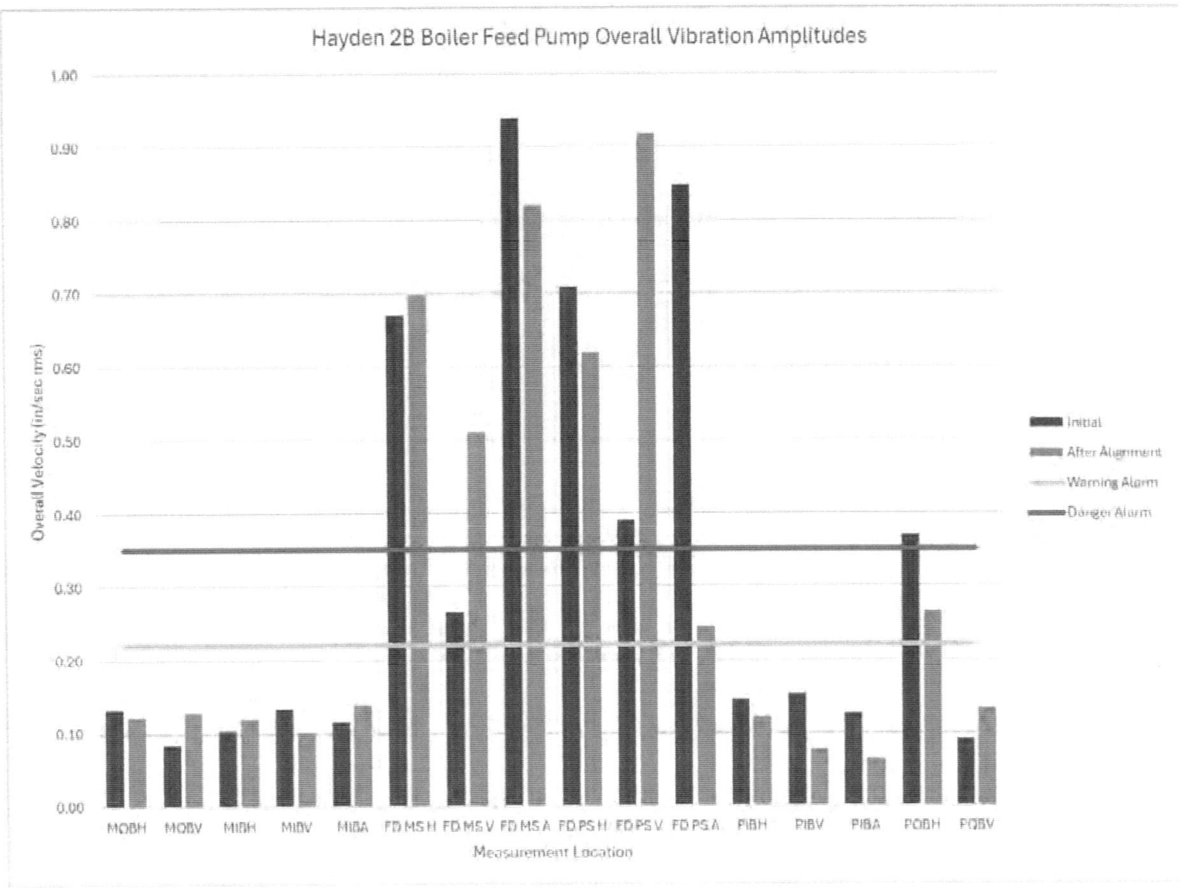

Figure 11: Amplitude comparison before and after the alignment attempt.

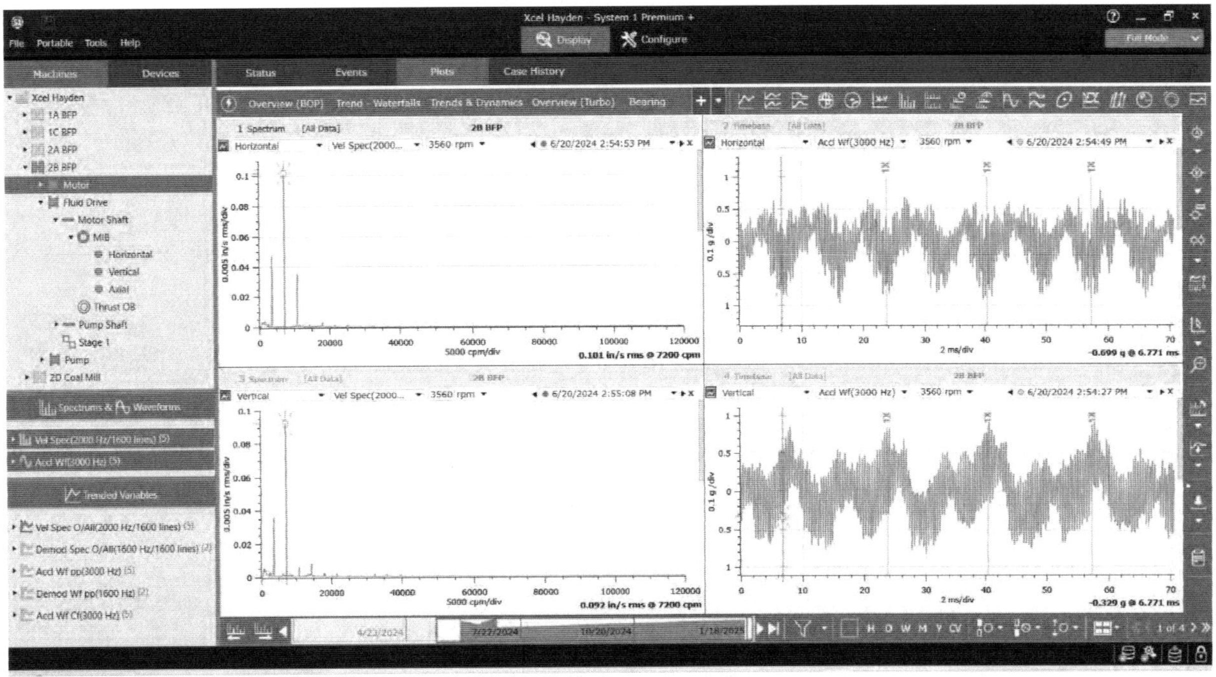

Figure 12: 2xAC Line frequency remains dominant after alignment on the motor.

The Fluid Drive had a dominant vibration response at 1x running speed (Figure 13). Considering the elevated axial response on the fluid drive, and elevated 2xAC line frequency content, the author concluded this was a symptom of motor thrusting due to the motor rotor oscillating about the axial and radial magnetic centers.

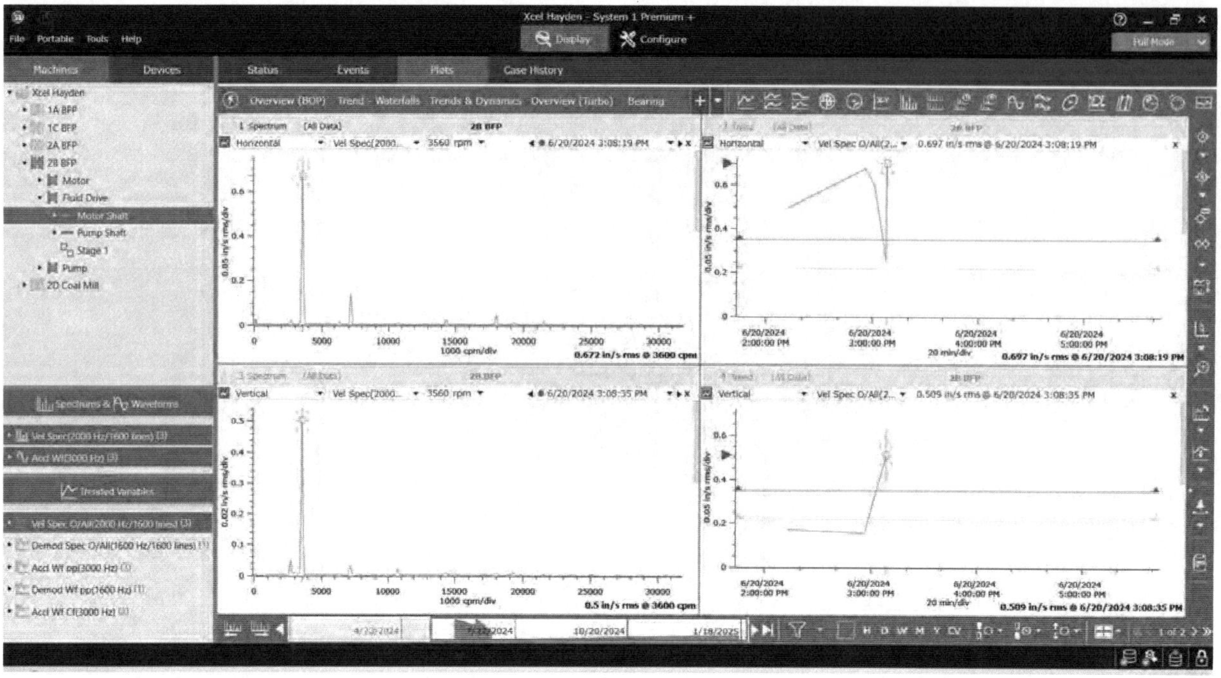

Figure 13: 1x running speed remains elevated on the Fluid Drive after alignment.

To rule out motor thrusting, the motor shaft maximum axial travel was measured by manually thrusting the shaft towards the OB and IB sides of the motor. The center of this travel was marked, and the motor was brought up to speed uncoupled to measure where the shaft settles axially. The rotor seemed to settle within ~5-10 mils of the geometric center. Therefore, this minimal difference between magnetic and geometric centers was determined to be inconsequential to the elevated vibrations seen on the fluid drive and electrical frequency content observed from the motor. Because 1x running speed was still the dominant vibration response on the fluid drive, the author recommended to attempt balancing the fluid drive's doughnut casing. Despite the previous attempts at balancing historically, the author still saw value in balancing because in a previous coast down plot (Figure 14) collected after alignment, there seemed to be no indicator of a natural frequency with the fluid drive which would cause elevated vibrations at 1x running speed.

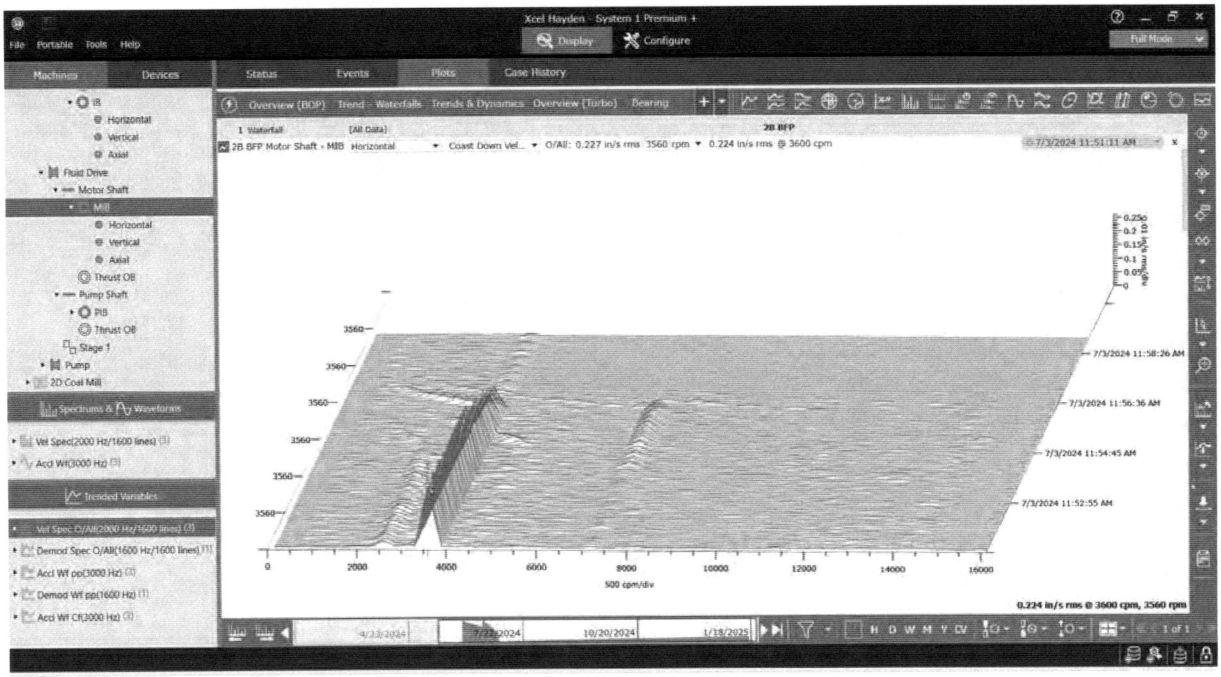

Figure 14: Coastdown plot from Fluid Drive Motor Side Horizontal location in effort to identify a possible resonant frequency near 1x running speed.

Balancing

The fluid drive balancing was conducted relative to the Motor shaft horizontal location because this was the location with the greatest radial amplitudes. Shown in Figure 15, 1x running speed amplitudes reached 3.5 mils on the fluid drive motor side horizontal location. This was excessive according to ISO 1940-1 / "Specification and verification of balance tolerances".

Permissible vibration due to unbalance, V_{per}, is calculated as follows in Equation 2:

$$Permissible\ Vibration = V_{per} = \frac{T}{W_t\ x\ R}\ x\ \frac{6.015GW}{N}$$

Where:

V_{per} = Permissible Vibration (mils, peak - peak)

G = ISO Grade

W = Weight of Rotor (lb.)

N = Speed (RPM)

T = Effect Vector (mils)

W_t = Trial Weight (oz)

R = Radius of Trial Weight (inches)

(2)

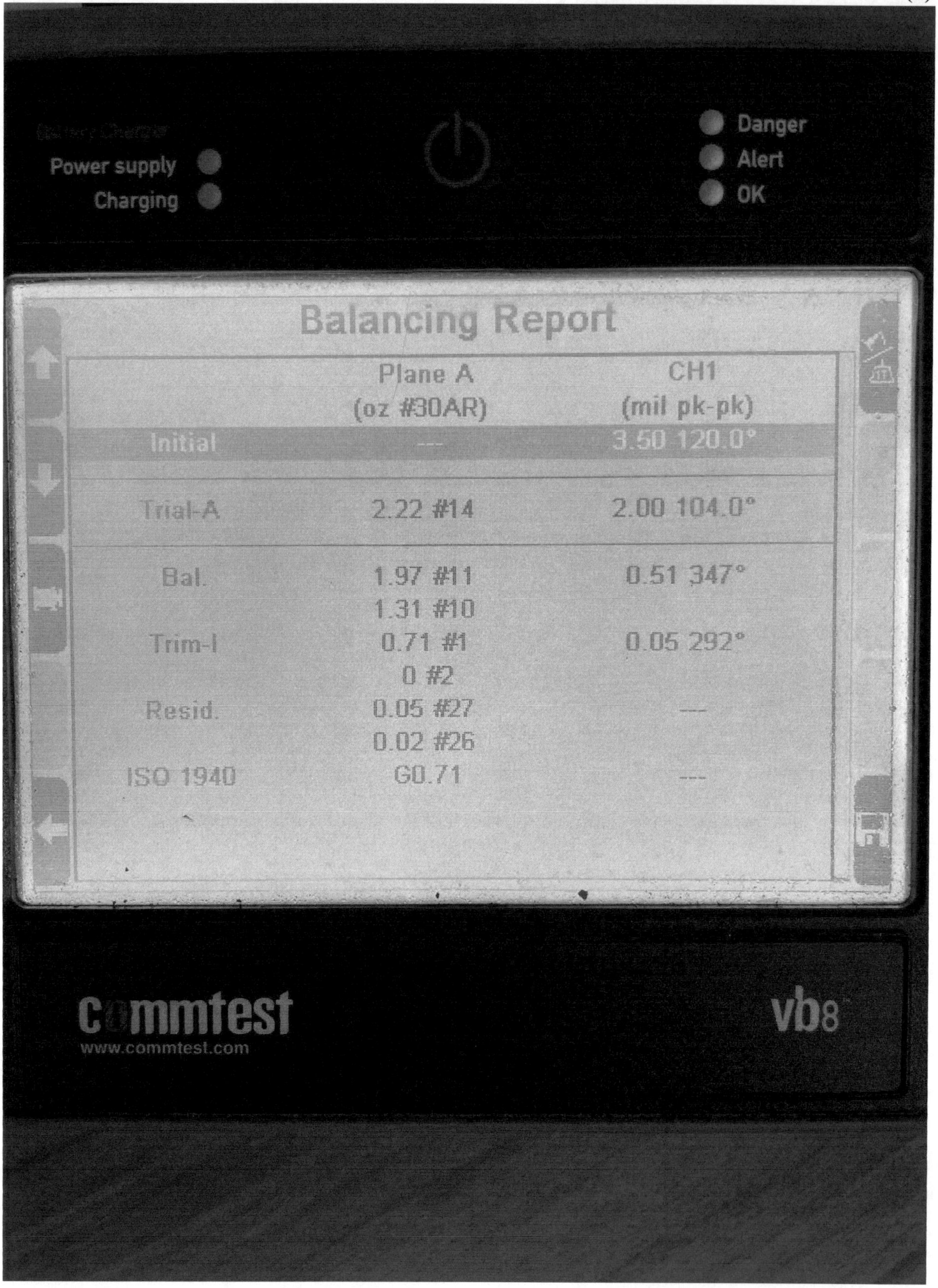
Power supply
Charging
Danger
Alert
OK
Balancing Report
Plane A
(oz #30AR)
CH1
(mil pk-pk)
Initial

3.50 120.0°
Trial-A
2.22 #14
2.00 104.0°
Bal.
1.97 #11
0.51 347°
1.31 #10
Trim-I
0.71 #1
0.05 292°
0 #2
Resid.
0.05 #27

0.02 #26
ISO 1940
G0.71
---
commtest
www.commtest.com
vb8

Figure 15: Commtest Balancing Report for 2B Boiler Feedwater Pump.

The calculated permissible vibration for the 2B BFP Fluid Drive is 0.44 mils. The calculated balance sensitivity is 1.33 oz/mil. Balancing was conducted by securing 5/8" brass washers on the motor side of the doughnut casing. There are 30 fixed locations on the doughnut casing where washers can be secured. These locations were marked with white paint pen*. The doughnut casing is shown in Figure 16. The balancing summary report is provided in Figure 17.

Figure 16: Doughnut Casing with bolts securing washers. Washer size and location were changed to optimize the in-situ balance.

Company: Xcel Energy - Hayden **Date:** 7/5/2024

Asset: 2B BFP - Fluid Drive **Analyst:** **Alex Vass**

Balancing Summary

Service Speed (rpm)	3597
Rotor Weight (lbs)	1000
Correction Weight Method	Single Plane
Previous Weights Removed	
Target Balance Quality G	6.3
Permissible Residual (U_{per}, oz-in)	10.5
Initial Balance Quality G	50.1
Final Balance Quality G	0.7
Residual (U_{per} oz-in)	1.2
Residual Force (lbf)	220
Permissible (V_{pe} mils p-p)	0.44
Current Vibration (mils p-p)	0.05
Balance Sensitivity (oz/mil)	1.33

		X	Y	Notes
Correction Radius (in.)		18	0	
Initial	Amplitude ($mils_{p-p}$)	3.5	0	
	Phase (deg.)	120	0	
Final	Amplitude ($mils_{p-p}$)	0.05		
	Phase (deg.)	292		
Total Correction Weight	oz	4.71		
	Location (deg.)	125°		

Additional Notes	Vibration measured at Fluid Drive Horizontal location on Motor (Driver) side. Before Balancing, loaded (6/20/24) - 0.897 inch/sec rms After Balancing, unloaded (7/3/24) - 0.027 inch/sec rms

Xcel Energy®

Figure 17: Balancing Summary.

An access port is available atop the fluid drive shell to install and remove weights as needed. Because the 1x vibration response was consistent with motor speed regardless of scoop tube positioning, the balancing washers were located on the drive side of the doughnut.
Vibration levels across the fluid drive before and after balancing are provided in Figures 18 and 19. Per ISO 20816-3, the warning and danger limits for overall vibration on the fluid drive are 0.22 ips rms and 0.35 ips rms respectively. All amplitudes across the fluid drive were acceptable after balancing. The sum of the masses installed equates to 4.71 oz located at 125° against rotation.

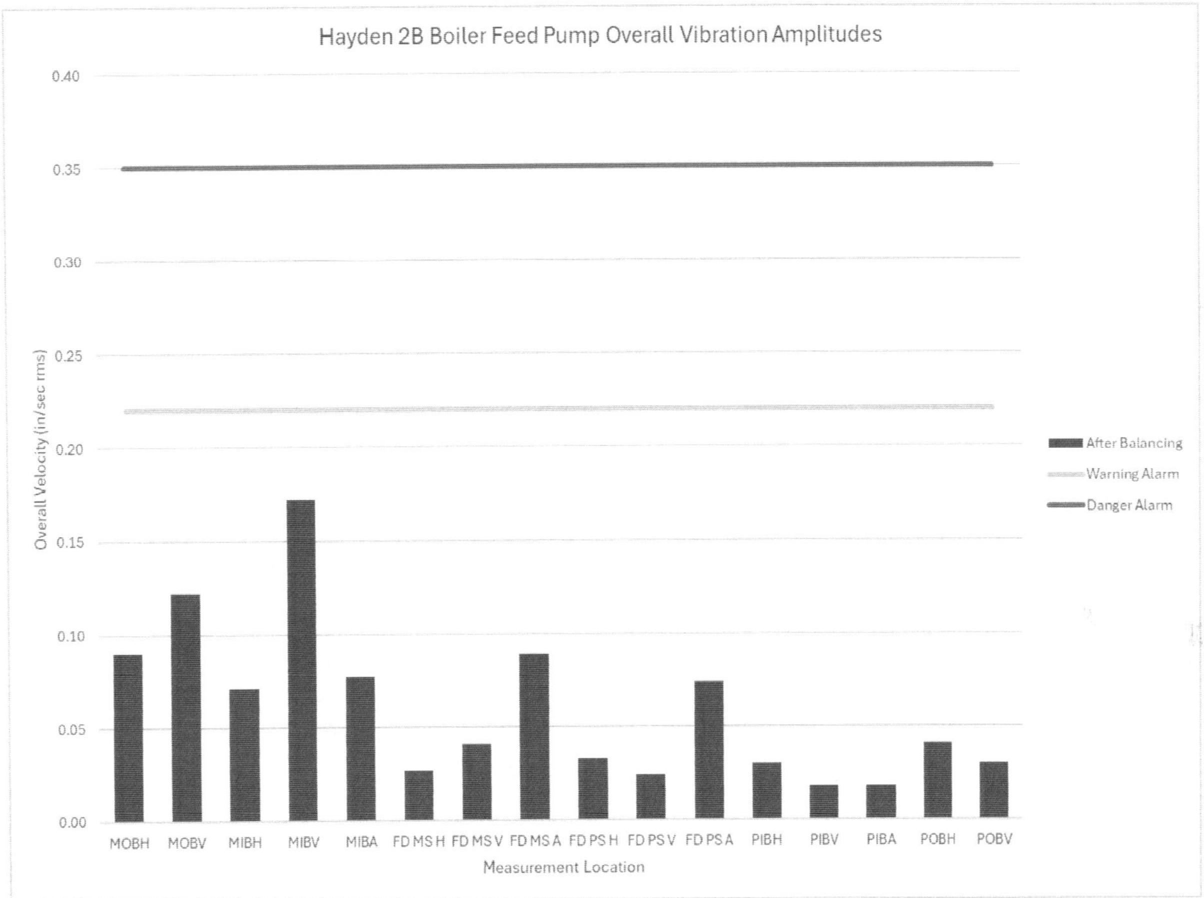

Figure 18: Vibration overall values across the machine train after balancing the fluid drive.

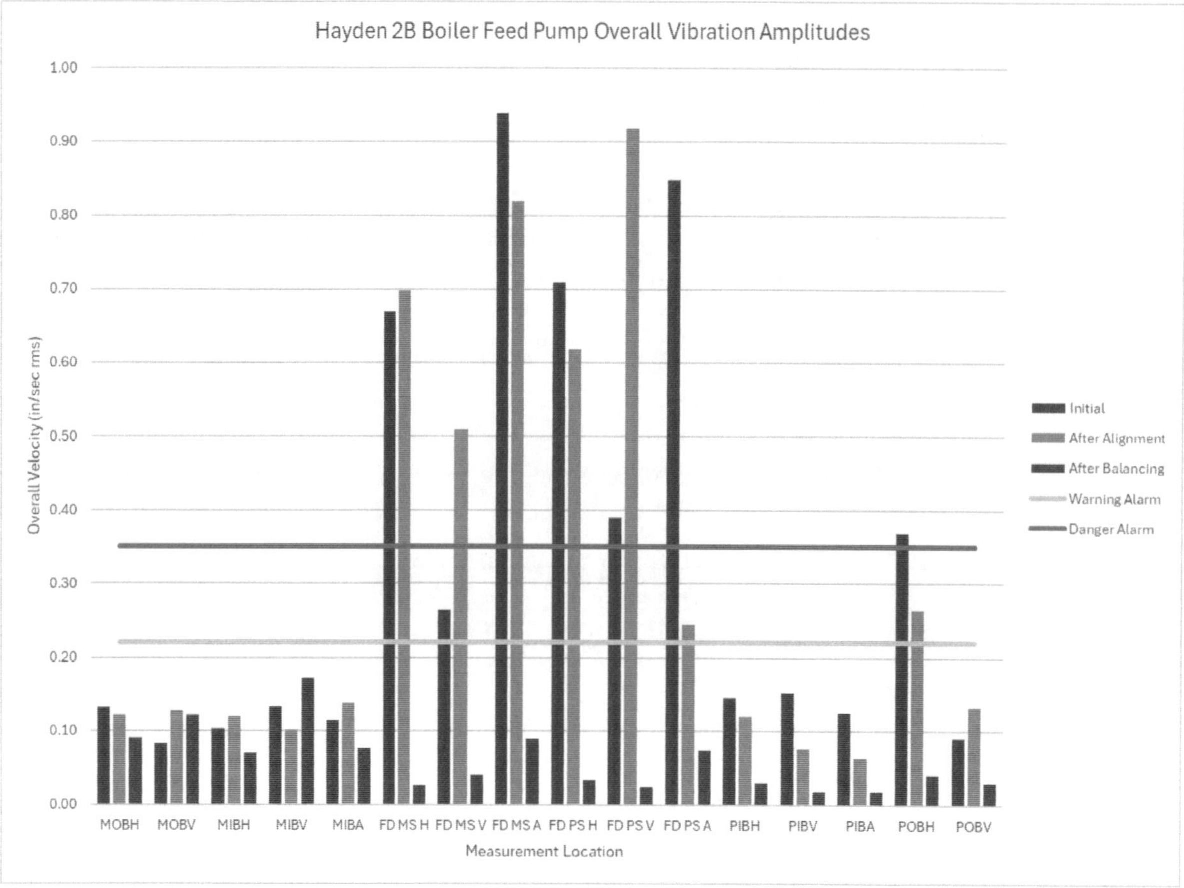

Figure 19: Progression of overall amplitudes after alignment attempts and fluid drive balancing.

During a visit to Hayden in January 2025, vibration amplitudes were recorded on the 2B BFP to evaluate the longevity of the balancing solution. Figure 20 is a waterfall of the daily vibration data collected by operations, which show 1x running speed vibration has remained low. Figure 21 plots the overall amplitudes at the same location since November 2023, and shows a similar low amplitude response since balancing activities were performed.

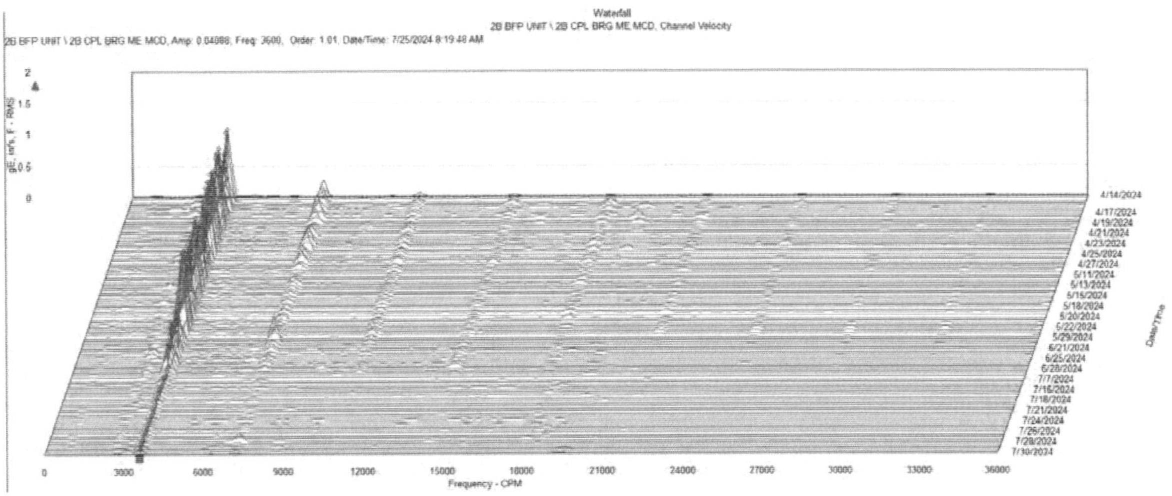

Figure 20: Waterfall plot of vibration data collected by plant operations from the Fluid Drive motor side.

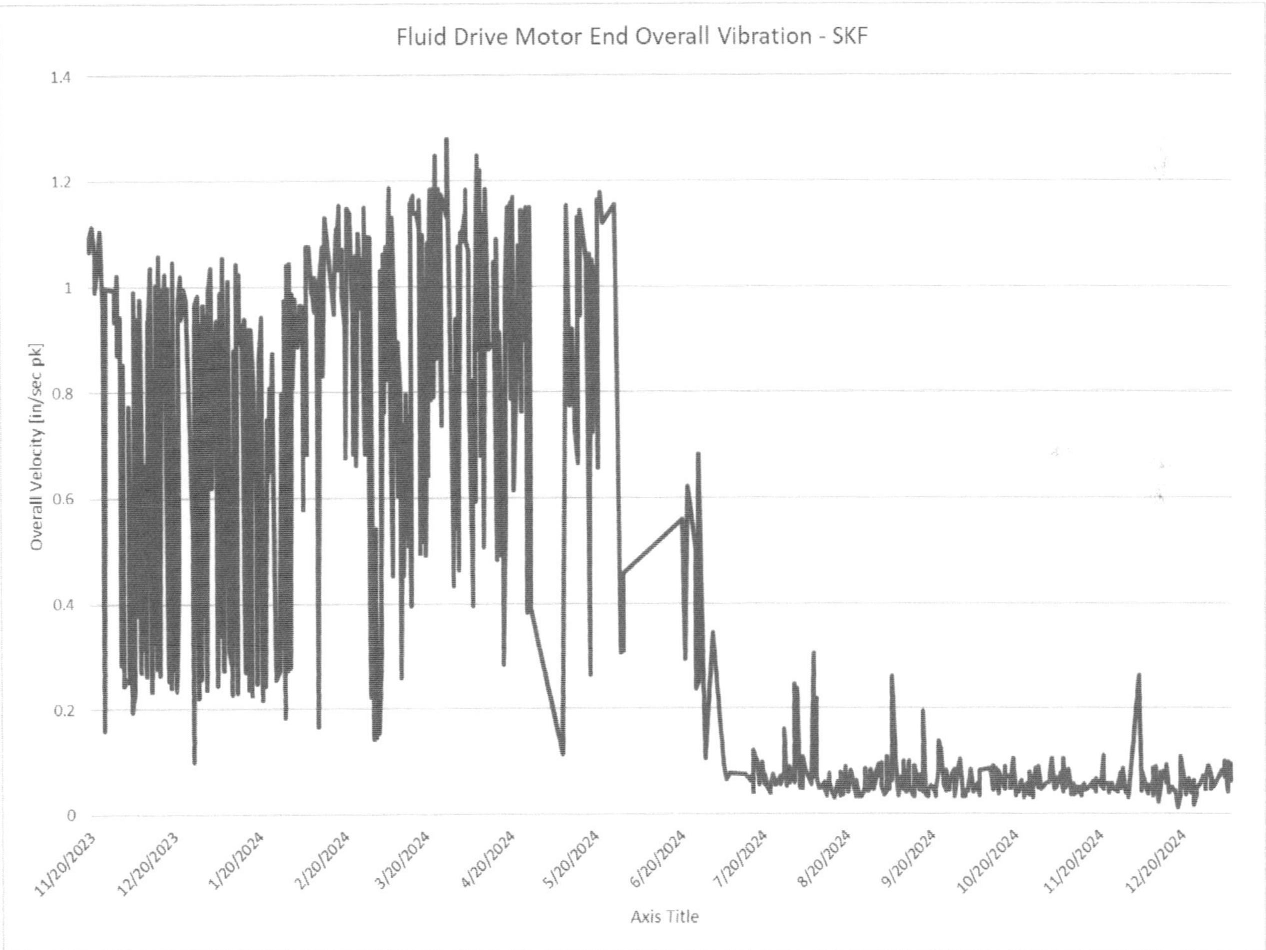

Figure 21: Overall vibration amplitudes collected by plant operations since November 2023. Vibrations have remained low since fluid drive balancing.

Conclusion

Since balancing, vibration levels have remained low across the machine train. Despite initial attempts at alignment, this exercise serves as evidence to tackle the greatest vibration responses generated across a complex machine train. An analyst can get caught up in specific frequencies and responses and may not correct the targeted response due to greater forces acting upon the machine. This fluid drive unbalance caused an increase in non-1x vibration content across the machine, which resulted in an extended troubleshooting process.

Acknowledgements

My appreciation goes out to the Hayden Plant maintenance and operations teams for allowing me to lead the troubleshooting exercise and work toward a solution.

VIBRIS: VIBRATION INTELLIGENCE BEARING RELIABILITY INTEGRATED SYSTEM CONCEPT

Authors: Christopher Harrison, Drumil Joshi, Andrew Riley

VIATC TRAINING CONFERENCE & EXPO

NEWPORT NEWS, VIRGINIA | AUGUST 6-8, 2025

VIBRIS: Vibration Intelligence Bearing Reliability Integrated System Concept

c
Senior Monitoring & Diagnostics Analyst at Southern Power Company
charriso@southernco.com

Drumil Joshi
Monitoring & Diagnostics Analyst 1- Renewables at Southern Power Company
djoshi@southernco.com

Andrew Riley
Monitoring & Diagnostics Performance Manager at Southern Power Company
andriley@southernco.com

***Christopher Harrison** serves as the Reliability Analyst for Southern Power's Monitoring and Diagnostic Center, bringing over 13 years of specialized experience with rotating assets within the power generation sector, encompassing both fossil fuels and renewables. As a certified CAT IV Vibration Analyst with additional certifications in thermography, Motor Circuit Signature Analysis (MCSA), oil analysis, and shaft alignments, Christopher's expertise is deeply rooted in enhancing long-term equipment reliability and performance. Throughout his career, he has demonstrated a robust commitment to optimizing operational efficiency and fostering innovative solutions. His collaborative approach and technical acumen make him a valuable asset to Southern Power, where he consistently contributes to the advancement of reliability practices and the cultivation of a forward-thinking work environment.*

Drumil Joshi *is a dynamic Data Scientist and AI Expert with a proven track record of blending innovation with real-world impact. Currently at Southern Power, he leads high-impact analytics projects that enhance renewable energy performance, including machine learning pipelines for wind turbine safety and proprietary tools for Battery Energy Storage Systems (BESS). With a Master's in Data Science from Indiana University and a portfolio of 8 international research papers, a published textbook, patents, and media features on NDTV and Authority Magazine, Drumil stands at the intersection of cutting-edge technology and strategic thinking. His expertise spans predictive analytics, AI-driven automation, and sustainable tech solutions. Passionate about democratizing data, he excels in translating complex insights into actionable strategies that resonate with both technical and non-technical audiences. Whether in boardrooms, classrooms, or control rooms, Drumil Joshi is recognized as a thought leader driving the future of intelligent systems.*

Andrew Riley *is the Performance, Monitoring, and Diagnostics Manager at Southern Power, where he oversees the reliability and performance of 55 power plants, including 15 wind sites. With over 15 years in the utility energy sector, Andrew specializes in predictive analytics, performance testing, and vibration analysis to enhance the reliability of over 13,000MW of capacity. An Auburn University Mechanical Engineering alumnus, Andrew is dedicated to fostering innovation and bridging academia with industry. He holds an MS from UAB and actively mentors young engineers, contributing significantly to the advancement of energy systems.*

Abstract

VIBRIS: Vibration Intelligence Bearing Reliability Integrated system is an advanced wind turbine condition monitoring platform that combines rigorous data security and contemporary machine learning. The system incorporates a token-based, secure authentication and multi-stage, hierarchical data acquisition process, which converts raw sensor output into a proprietary HEX-encoded data format, safeguarding confidentiality and optimizing data fidelity. The solution incorporates real-time SCADA input data into a PI Server, thereby normalizing and enriching the data stream for further analytical use. A collection of sophisticated anomaly detection algorithms like Isolation Forest, Local Outlier Factor, One-Class SVM, and K-Means Clustering operate in concert to produce composite anomaly scores that effectively filter out transient disturbances from persistent mechanical deviations. The experimental results confirm the model's superior capacity to forecast and pre-detect maintenance issues, thus setting a new benchmark for measuring reliability in wind turbine operation.

Key Words: Ensemble Anomaly Detection, Hierarchical Data Retrieval, Isolation Forest, K-Means Clustering, Local Outlier Factor, One-Class SVM, PI Server, Predictive Maintenance, Proprietary HEX Encoding, Reliability Assessment, SCADA Integration, Secure Data Extraction, Token-Based Authentication, VIBRIS, Wind Turbine Condition Monitoring

1.0 Introduction

Think of a well-tuned symphony where every instrument plays its part in harmony. In wind power, every component of a wind turbine, be it gearbox, blades, or support structure, must play its part flawlessly to produce clean power. However, as one misplayed note spoils a symphony, a small fault in one part of a turbine can lead to prolonged and costly downtime, repair, and even catastrophic failure. Not only are researchers today listening to these off-notes but are also using sophisticated machine learning to predict and diagnose failures before they become catastrophes. This introduction weaves together some of the forefront research, painting a picture of how machine learning and vibration analysis are revolutionizing wind turbine maintenance.

Wind turbines are subjected to severe and unpredictable operating conditions, with varying wind speeds and environmental loads generating complicated vibration modes. Amin et al. (2022) address this challenge by zooming in on the gearbox, a failure of which can spiral into exorbitant repair expenses. The authors created a framework that converts raw vibration data into visual maps through processes such as cyclostationary analysis and kurtogram imaging. Through the application of convolutional neural networks (CNNs) to the maps, the research attained early fault detection despite the presence of turbulent wind conditions. Their technique demonstrates that with the appropriate signal processing software, even slight indications of damage can be automatically detected, independent of expert human monitoring on a continuous basis.

But gearboxes are far from the only critical part that needs monitoring. Gas turbine engines, which share some of the same operating issues as wind turbines, have also benefited from sophisticated vibration monitoring. Matthaiou et al. (2017) considered the use of machine learning on gas turbine engine vibration data for anomaly detection. They addressed the issue of non-stationary data where the patterns of vibration change over time through techniques such as the wavelet transform and kernel principal component analysis. Their article, with its focus on novelty detection using one-class support vector machines, teaches us an important lesson: even when the only data available is that which describes a "normal" condition, machine learning can still identify anomalous behaviour that signals a fault.

Apart from rotating machinery such as gearboxes and engines, even wind turbine blades are vulnerable. Blades, being typically composite in character, are susceptible to defects such as delamination, cracks, and other material faults. Arcos Jiménez et al. (2018) tackle such susceptibility through the application of guided wave method coupled with wavelet denoising and autoregressive feature extraction. They demonstrate that even the onset of blade damage can be detected through the extraction of the most relevant features from vibration signals. Experimentally testing several classifiers from decision trees to neural networks their work demonstrates that feature selection is as vital as the machine learning algorithm itself. The study emphasizes that it is important to keep blades healthy not only for safety reasons but also because blades represent a significant portion of the overall cost of a turbine.

For offshore wind farms, even more risks are involved. Operating conditions are harsher, and maintenance costs are far higher than for onshore installations. Black, Richmond, and Kolios (2021) conduct a systematic review of condition monitoring systems for offshore wind turbines. They highlight that the combination of conventional vibration analysis with real-time supervisory control and data acquisition (SCADA) system data can give an enhanced view of the health of a turbine. By integrating sources of information, operators can create proactive maintenance programs that anticipate failure before it happens, minimizing the cost and risk of unexpected downtime.

Sometimes the problem is not simply finding one fault but identifying which of several possible faults. Ogaili et al. (2022) propose a new method to diagnose different kinds of blade cracks at the tip, mid-span, or root through vibration signals. They use feature ranking methods such as ReliefF and couple these with the classifiers k-nearest neighbours, support vector machines, and random forests. Their approach obtained classification accuracies as great as 97%, demonstrating that a competent machine learning model can successfully discriminate between several fault conditions. These sorts of accuracy are a welcome move towards systems not only alerting the operators to a fault but also identifying exactly where and what the fault is.

Laboratory tests also supplement these techniques in safely controlled laboratory environments. Granados et al. (2023) created a wind turbine vibration monitoring system with artificial vibration parameter modification and subsequent analysis of accelerometer data. The system was experimented with in safely controlled environments and utilized machine learning classifiers, including Bagged Trees, for predicting simulated failure modes. Among their methodological strengths is the practicability in the engineering context: simulations of several vibrations in experiments have validated that small-scale turbines are extremely suitable for testing methods of predictive maintenance. Their paper points out the contribution laboratory experiments can make in calibrating diagnostic models before they are put into practice.

The outcome of such studies is a paradigm shift in wind power equipment maintenance. Conventional techniques, i.e., manual inspection and time-based scheduled maintenance, are being replaced by data-driven methodologies. Advanced artificial intelligence techniques can detect intricate patterns within vibration signals that are imperceptible through direct observation. Scientists are merging advanced signal processing techniques with machine learning to make maintenance and predictive modelling for wind turbines more effective.

Machine learning allows the transition to be made from conventional, often unnecessary routine maintenance to ongoing monitoring and component condition-based maintenance strategies. This is a paradigm of reduced maintenance cost and enhanced sustainable energy generation, which is economic, environmental, and technological. In this background, the research tracked the innovation of high-tech condition monitoring systems with the goal of guiding tomorrow's wind energy infrastructure planning.

2.0 Methodology

2.1 OEM 1 Methodology

This research and its associated methodology have been developed as part of OEM 1, ensuring a tailored approach to secure and innovative data extraction and processing.

In order to protect critical operational insights while advancing academic discourse, this methodology has been deliberately obfuscated and reimagined. The following section outlines a multi-layered framework for the secure extraction, processing, and validation of sensor measurement data. Every step starting from encrypted authentication to hierarchical data retrieval and proprietary encoding has been designed with both security and innovation in mind.

2.1.1 Secure Access and Authentication

At the core of our approach lies a robust, token-based authentication mechanism. Rather than exposing sensitive credentials and endpoint details, the system initiates a secure POST request to a dedicated identity service. Using an encrypted payload formatted as x-www-form-URL encoded, the request includes anonymized tenant and client credentials. Successful validation yields a time-limited bearer token that authorizes subsequent interactions across the API suite.

2.1.2 Hierarchical Data Retrieval

The process of extraction adheres to a hierarchical sequence of endpoint interactions, with each layer building on the previous one:

- **Primary Indexing:** This procedure starts off with an abstracted indexing point which delivers a thoroughly picked group of device identifiers. These form the underlying basis of the mechanism for retrieval.

- **Sensor and Measurement Metadata Retrieval:** From the initial device reference, the system makes a secondary endpoint request to fetch sensor-specific metadata. This pulls anonymized sensor IDs, corresponding measurement codes, and significant temporal boundaries (such as the earliest measurement year and month available) to inform the next requests.

- **Dual Data Extraction consists of two concurrent POST requests:**

1. **Data Acquisition Request:** A request that is structured with identifiers related to the device, sensor, and measurement, along with a specified temporal extent (year and month), triggers the generation of a measurement file. This file, which is a proprietary compressed format is transferred to binary and includes raw sensor values as well as compressed metadata.
2. **Measurement Metadata Request:** Concurrently, the same request retrieves essential file metadata (e.g., creation dates and content length), thus validating data integrity and enabling update verification.

2.1.3 Proprietary Data Encoding and Schema Enforcement

To optimize storage and secrecy, the measurement data is packed in a specially encoded file:

- **Compressed Metadata**: XML-based metadata is compressed (using industry-standard algorithms) then encoded as a HEX string. The two-stage process provides compactness and obfuscation.

- **Sensor Measurements:** Raw measurement values which are expressed as single-precision floats are also HEX-encoded so the data remains in a secure, non-human-readable format.

A rigorously defined schema underpins all interactions. The request model imposes rigidly validated fields such as device identifier, sensor identifier, measurement identifier, year, and month, each with specific numerical ranges. The response schema also dictates the format of measurement metadata, device specification, and file content results, which improves data integrity at the expense of confidentiality.

2.1.4 Confidentiality and Operational Security

Here, all the functional complexities have been enhanced to minimize exposure risk for unauthorized data. Sensitive data is protected by the system through encryption of each transmission, obscuring endpoint identities, and employing stringent validation procedures. Not only does this end-to-end security model protect against potential intrusion, but it is a drastic shift from traditional data extraction systems.

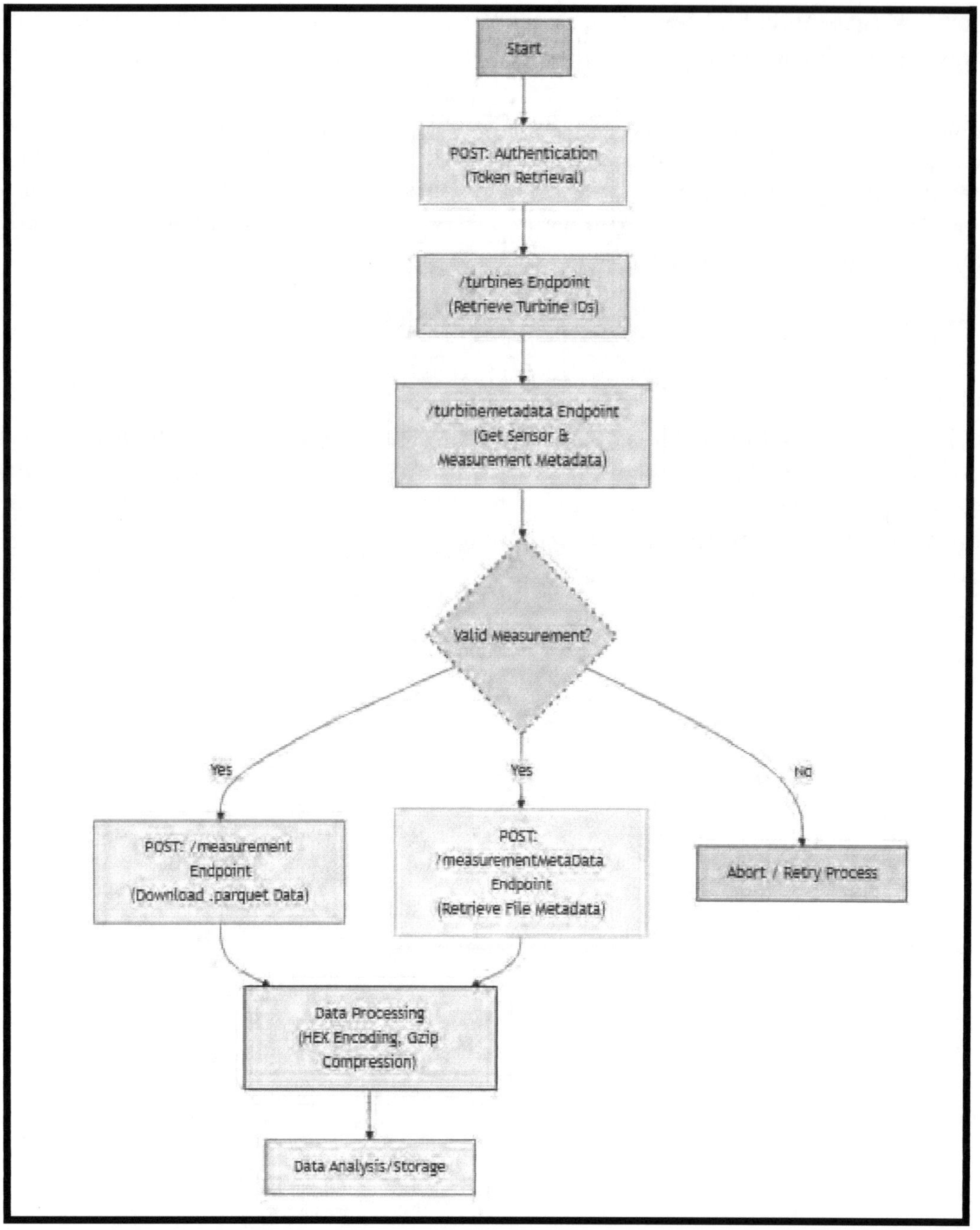

Figure 1: Encrypted Data Acquisition Flow Chart for OEM 1

2.2 OEM 2 Methodology

The research strategy, at the heart of the OEM 2 project, uses an innovative and nuanced paradigm in guaranteeing that sensitive operational information is safeguarded while challenging the limits of secure data handling.

In order to protect sensitive data and avoid unintentional revelation of key operating processes, the method outlined here employs abstract terminology and unconventional language that does not lend itself to facile copying.

2.2.1 Encrypted Access and Dynamic Authentication

Our technique relies on an authentication scheme with dynamics based on a token-based scheme. Rather than revealing traditional credentials or endpoint information, our technique follows a proprietary method involving the utilization of time-locked, encrypted tokens derived from secretive client parameters. Besides granting protection for access, the technique conceals the inner framework from third parties.

2.2.2 Multi-Tiered Data Retrieval Process

The process undertaken in retrieving information that is relative to turbines is carefully achieved through a systematic and layered retrieval process. It has been specially formulated to be concealed to hide its true nature and underlying processes:

- **Abstracted Enumeration:** The initial phase of the process consists of an opaque indexing service, which is responsible for generating a thoroughly designed and well-structured list containing turbine identifiers. The management of the pagination is done by unspecified parameters, i.e., only a small portion of the dataset is revealed, such that the information being provided is both protected and regulated.

- **Exposing the Identity of Hidden File Types:** For every turbine, there is an additional layer that is responsible for coming up with a list of file types that have been anonymized and cleared as acceptable. In particular, the process only considers such file types that comply with very stringent internal specifications that have been created for further processing. This helps to ensure that data retrieval that is ultimately produced is not merely highly targeted but is also protected from beginning to end.

- **Chronologically Ordered File List:** Following the previously outlined steps, the system initiates a time-limited request to get a comprehensive list of files. Through the utilization of an ISO-8601 timestamp in a non-transparent manner, the process ensures that only those files within a defined and given time range are processed. By this approach, the potential exposure of any pre-existing historical data is successfully limited and constrained.
- **Obfuscated File Extraction**: Finally, a single-file retrieval process is undertaken via an endpoint that has been specially secured to maintain its security. The file itself, as an encrypted binary stream, undergoes processing within a highly secure environment that seeks to safeguard sensitive data. In this controlled environment, the data in the file is concealed and not disclosed to any unauthorized individual who would seek to obtain access to the same.

2.2.3 Proprietary Data Encapsulation and Schema Enforcement

To add an additional layer to the measures taken for ensuring confidentiality, every segment of data extracted is encapsulated through the utilization of a proprietary encoding system that is designed specifically for this purpose. The metadata of the data along with the data itself are subjected to a transformation process that applies undisclosed compression algorithms along with HEX encoding processes. This multi-level process guarantees that the raw data is completely meaningless and inaccessible to any party not having proper access to our proprietary decryption processes. A schema strictly enforced underpins every interaction, guaranteeing input parameters (e.g., turbine names, file types, and temporal markers) are legitimized without revealing their exact form. Two-layer protection ensures data integrity and secures operational details.

2.2.4 Operational Secrecy and Data Integrity

Each aspect of this methodology is developed with confidentiality taken into consideration. Through the use of endpoint abstraction, encryption of all traffic, and obfuscation of terminology that is technical in nature, the system effectively deters intruder access and, simultaneously, makes reverse-engineering attempts untenable. Incorporating dynamic token authentication, nested levels of information retrieval, and proprietary encapsulation of data comprises an unconventional though powerful approach to handling sensitive turbine data.

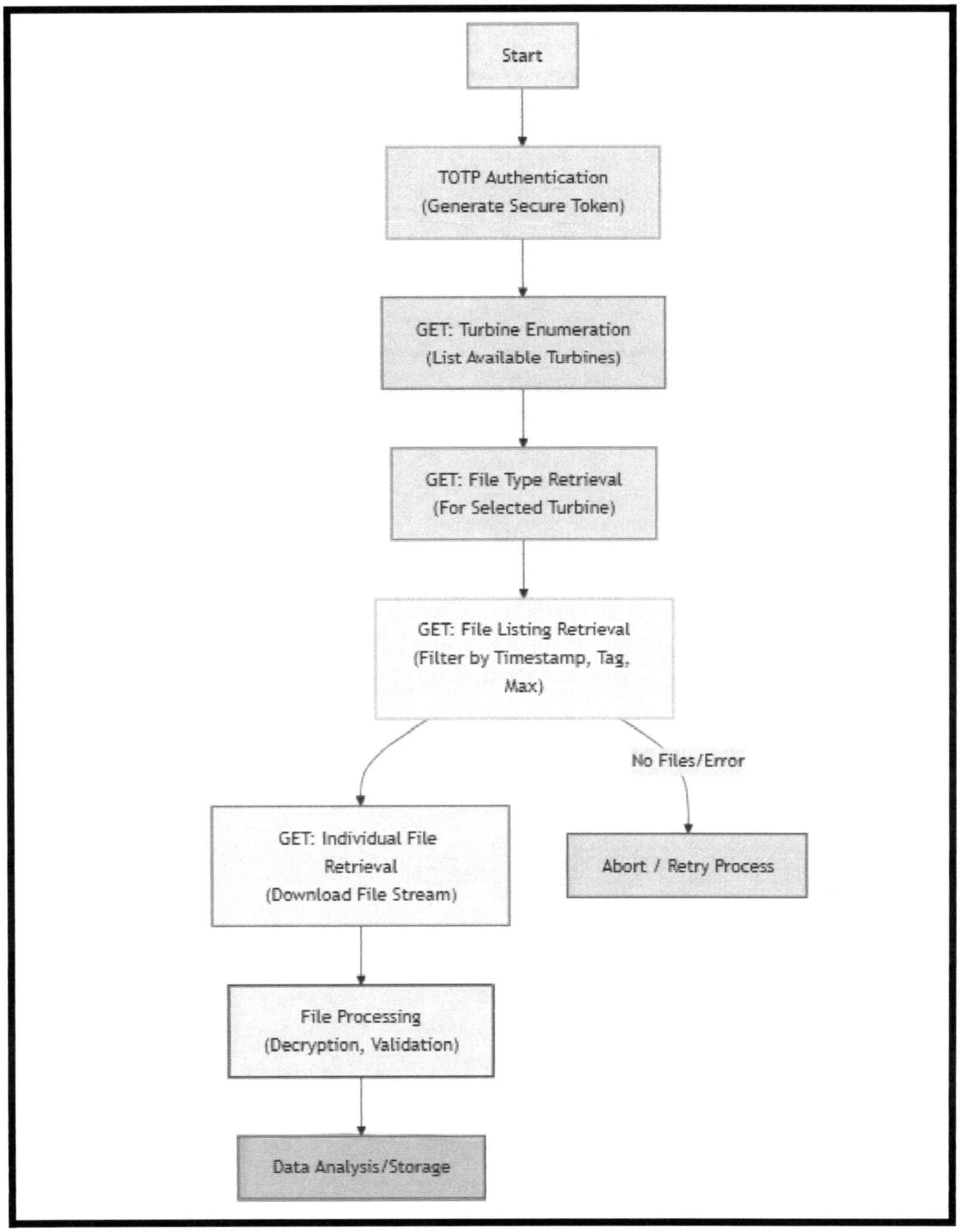

Figure 2: Data Acquisition Flow Chart for OEM 2

2.3 OEM 3 Methodology

OEM 3 focuses on the collection of real-time data from industrial sensors via a SCADA system, which in turn feeds this information to a PI Server a highly efficient time-series data historian. The SCADA system continuously monitors and controls industrial processes, gathering vital readings from sensors. These raw, real-time data streams are sent to the PI Server, where they are timestamped, aggregated, and stored long-term for later analysis. This integration guarantees the data is in a standard format, which is established through regular data types and consistent sensor readings, thus preparing it for additional processing, predictive maintenance, and more advanced analytical tasks.

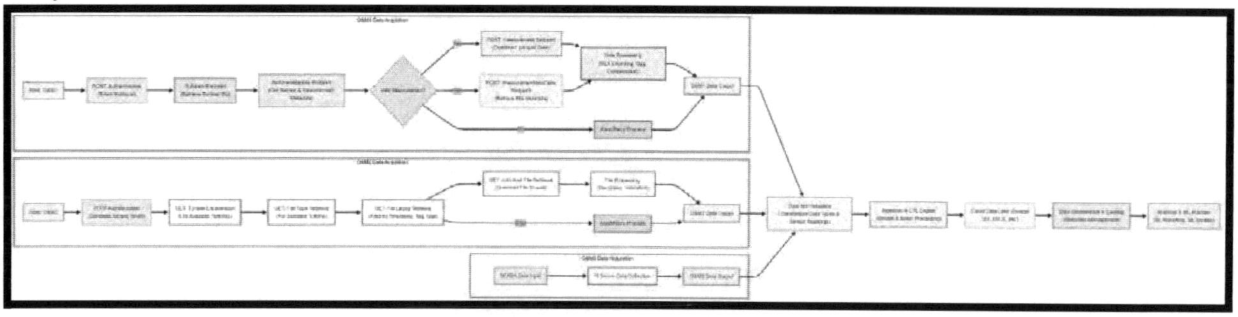

Figure 3: Combined Data Acquisition for all OEMs and cloud approach.

2.4 Machine Learning Framework

In this work, we introduce a proprietary framework for vibration-based condition monitoring of wind turbines. Owing to the strategic importance of this methodology, several key components have been deliberately abstracted to protect competitive insights. The framework is built upon three foundational pillars starting with Data Preprocessing & Feature Engineering, Tag Prioritization & Selection, and Multi-Model Anomaly Detection. Each operates synergistically to detect early signs of mechanical degradation while ensuring that critical sensor readings are standardized and secure.

2.4.1 Data Preprocessing and Feature Engineering

To ensure the integrity of our analysis, we correctly condition all raw vibration data. This method preserves original time stamps and translates sensor outputs into a stable normalized dataset. Key steps include:

- **Temporal Preservation & Numeric Standardization:** The original time index is safely retained while all other sensor values are converted to a normalized numeric representation. Non-conforming entries are sanitized automatically to deliver a continuous data stream.

- **Intelligent Imputation:** Unusual or missing values are corrected with an adaptive imputation technique based on moving averages. This technique maintains consistency of the data without exposing the underlying data patterns.

- **Feature Scaling with Confidential Norms:** Because the majority of sensor outputs fall in very narrow ranges, there is a proprietary scaling technique using standard normalization but with in-house adjustment. This assigns weights to feature contributions and keeps the actual scaling parameters proprietary.

These preprocessing steps are executed via a blend of open-source and custom-developed routines, ensuring that the final dataset is both consistent and secure for subsequent analysis.

2.4.2 Tag Prioritization and Selection

From a vast array of potential sensor tags, our framework selectively focuses on those most critical to assessing turbine mechanical health. The selection process is guided by domain expertise and proprietary criteria:

- **Core Vibration Indicators:** A select group of RMS-based vibration tags takes precedence because they give clear information about how major components are performing. High RMS values are early warnings of problems like imbalances, misalignments, or abnormal wear. The names of the actual tags and thresholds are not revealed.

- **Critical Gear Engagement Metrics:** Harmonic frequencies, especially those related to 1X and 2X mesh frequencies are tracked by tags that pick up specific harmonic frequencies. These signals give early warning of problems like misalignment, backlash, or tooth wear, based on private research that has demonstrated their capability to predict problems.

- **Extra Mechanical Health Measures:** Other tags track finer details of how things work (like bearing rotation and cage movement). While many tags are present, unnecessary or less useful ones are left out to prioritize the most useful information.

The rationale behind this focused selection is validated by rigorous testing and industry-standard recommendations, while specific details remain confidential to safeguard intellectual property.

2.4.3 Anomaly Detection Using a Multi-Model Approach

At the core of our framework is a robust, ensemble-based anomaly detection strategy that leverages complementary machine learning algorithms. Each model is tuned using proprietary parameters and integrated into a consensus-based system to capture both transient and persistent deviations in vibration patterns.

2.4.3.1 Isolation Forest

- **Mechanism:** Isolation Forest isolates single observations through the random selection of features and split values. The anomalies, being less in number and less similar, get isolated with fewer splits than the normal instances.

- **Reasoning:** Its concentration on detecting unexpected patterns rather than examining all the data renders it extremely good at working with complicated fields. It is particularly effective at catching sudden, unanticipated issues within turbine vibration data so that they can be easily identified as potential mechanical malfunctions.

2.4.3.2 Local Outlier Factor (LOF)

- **Mechanism:** LOF measures how many points are close to one another by looking at how dense each point is compared to its k-nearest neighbors. Points that are much less dense than their neighbors are considered outliers.

- **Rationale:** LOF is particularly good at spotting slight variations in sets of similar data. In turbine monitoring, this is helpful in picking up slight but significant variations from background noise, so that even slight changes in vibration patterns are detected.

2.4.3.3 One-Class SVM

- **Mechanism:** One-Class SVM learns a decision boundary around normal running conditions in an n-dimensional space with the help of a kernel function. Any observations that fall outside this boundary are flagged as outliers.

- **Reasoning:** This method is especially helpful when the system's normal state is complicated and not straightforward. It can adjust to different operating conditions, making it perfect for wind turbines, where changes in load and the environment can affect vibration signals. The One-Class SVM can reliably find differences that might show early signs of component wear.

2.4.3.4 K-Means Clustering (Distance-Based Detection)

- **Mechanism:** K-Means groups data into clusters and calculates how far each point is from the center of its cluster. Points that are far enough away (e.g., the points in the 90th percentile) are identified as unusual.

- **Rationale:** K-Means is usually used for grouping things, but it also helps us see how much something is different from normal behavior. This method is good at finding slow changes that could show long-term problems, like wearing out gear or slowly getting out of alignment.

2.4.4 Ensemble Consolidation

The results from various anomaly detection models are blended together to deliver a composite anomaly score for each data point. The use of multiple models not only improves the accuracy and reliability of fault detection, but it also obscures the individual contribution of each algorithm thereby further safeguarding proprietary information.

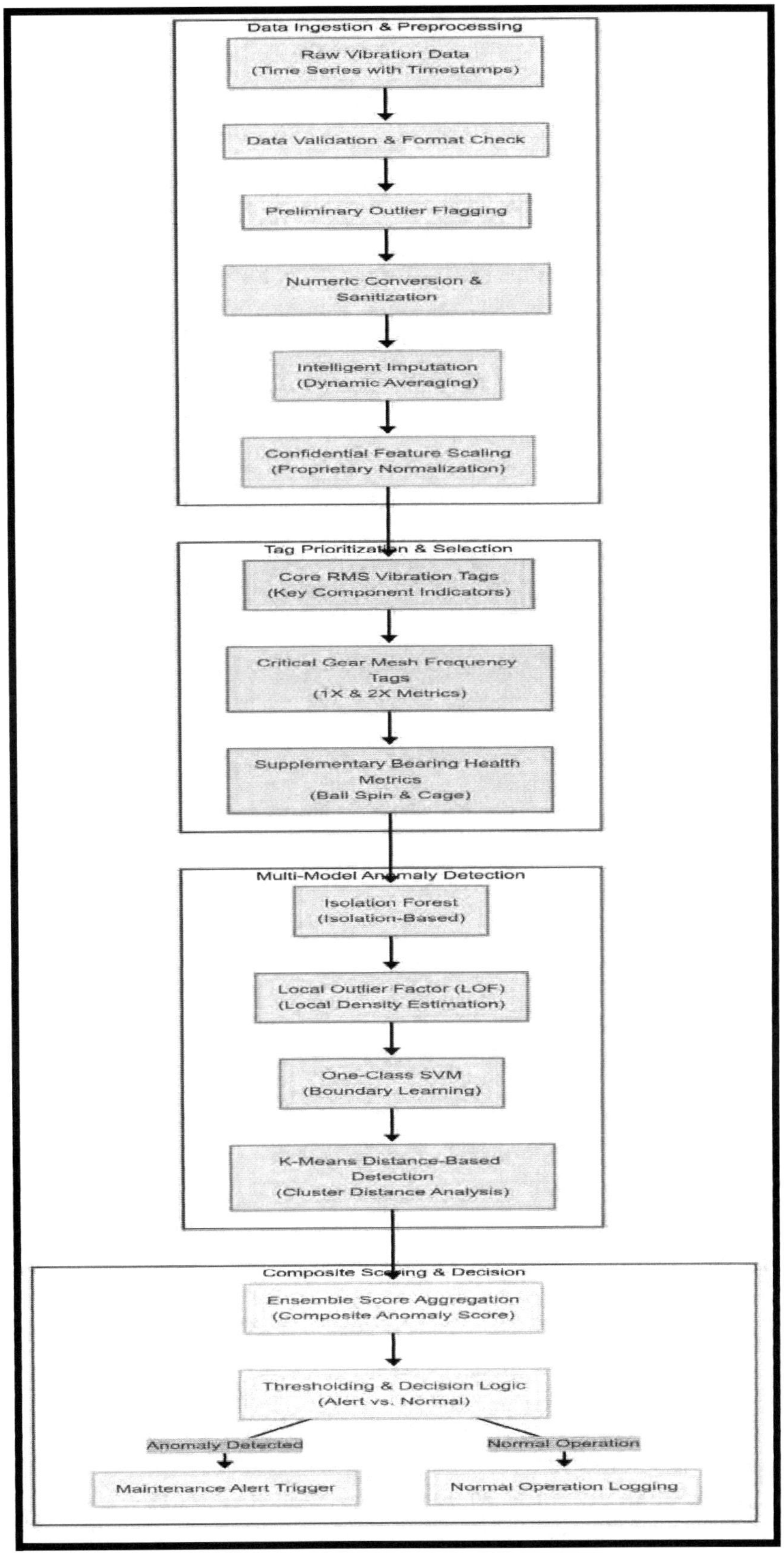

Figure 4: Machine Learning Holistic Framework

3.0 Results & Discussion

3.1 Isolation Forest

The Isolation Forest plot employs features such as DE_H (Drive End Horizontal vibration), GE_H (Generator End Horizontal vibration), and HS_H (High-Speed Shaft vibration) to recognize vibration anomalies across the dataset. Anomaly scores are separated in the plot, with certain data points—especially at the higher ranges of GE_H and HS_H—identified as outliers. This graphical setup highlights the Isolation Forest's capacity to identify abrupt, extreme deviations that can be associated with phenomena such as shaft imbalance, looseness, or abrupt resonance. The clear cluster of outlier points indicates periodic mechanical stress, probably not within normal operation. This renders it very useful at highlighting transient faults, allowing maintenance crews to intervene before brief moments of abnormality escalate into more systemic damage.

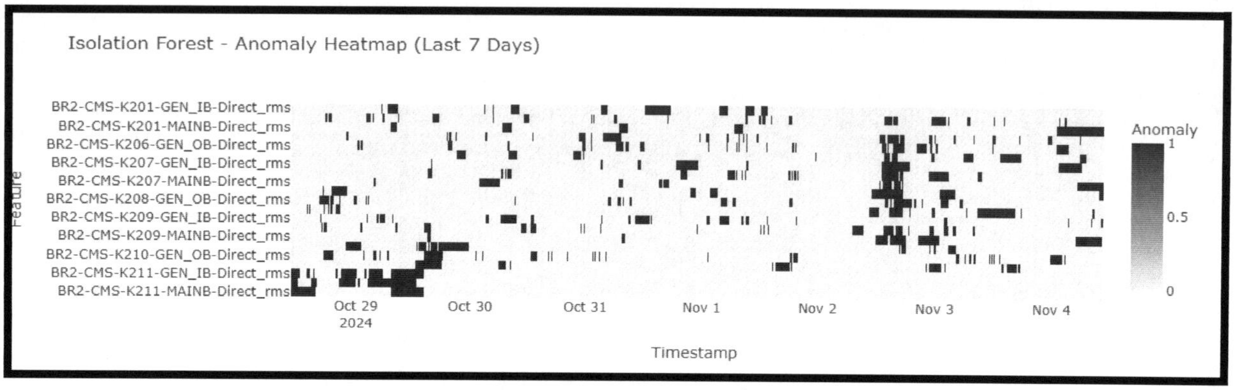

Figure 5: Isolation Forest Result Heatmap (Last 7 Days)

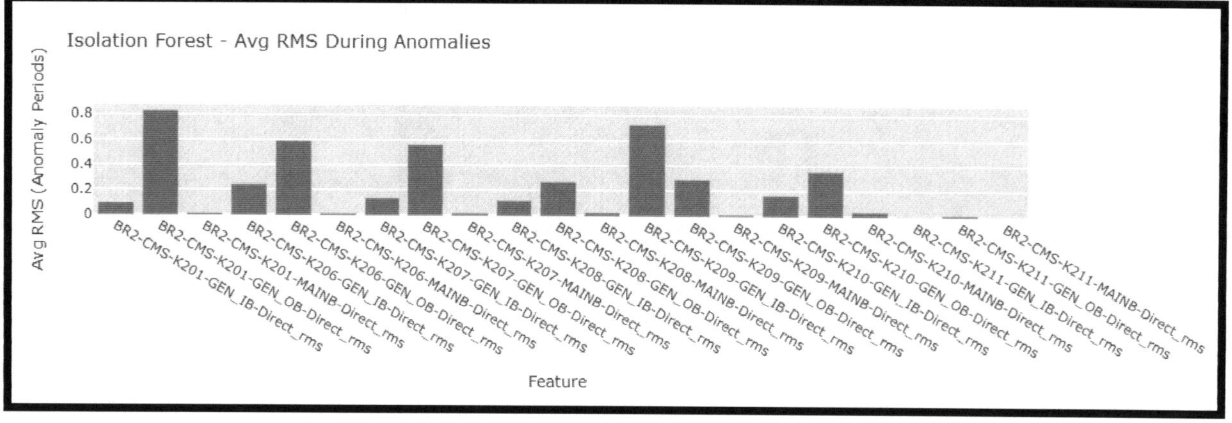

Figure 6: Isolation Forest Bar Chart for Average RMS During Anomalies

3.2 Local Outlier Forest

The LOF visualization plots the same underlying vibration characteristics—DE_H (Drive End Horizontal), GE_H (Generator End Horizontal), and HS_H (High-Speed Shaft)—but in a density-sensitive fashion that picks up on nuanced operation drifts. In this graph, data points are loosely distributed compared to their locality, especially where HS_H values begin to diverge from the cluster formed by DE_H and GE_H. These low-density points, as determined by their anomaly scores, are the local anomalies that may signal bearing misalignment, unbalance in its early stages, or early wear on drivetrain components. Unlike Isolation Forest, LOF detects context-dependent anomalies, finding turbines that do not necessarily act quite abnormally on a global level but are somewhat out of step with peer performance. The visualization effectively conveys how subtle data separation indicates real-world degradation that has not yet reached catastrophic levels but is worth monitoring.

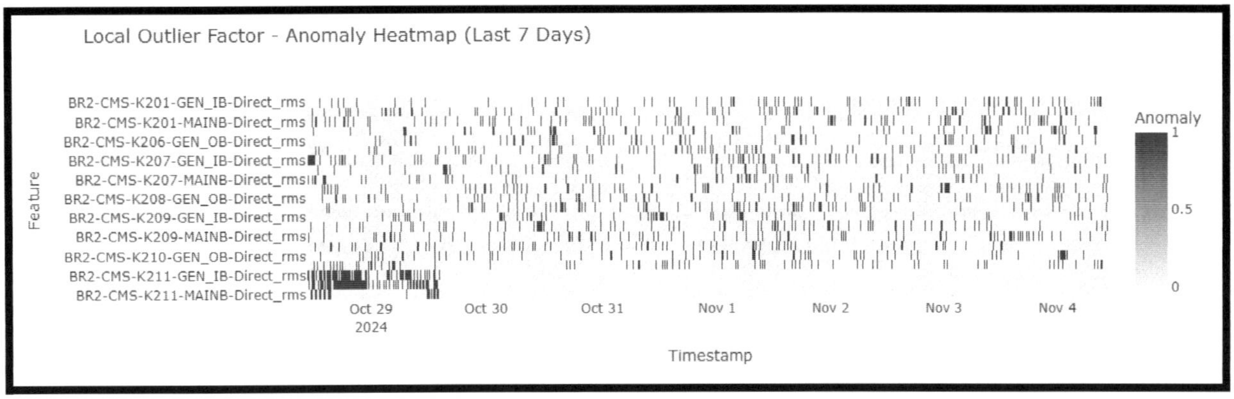

Figure 7: Local Outlier Forest Result Heatmap (Last 7 Days)

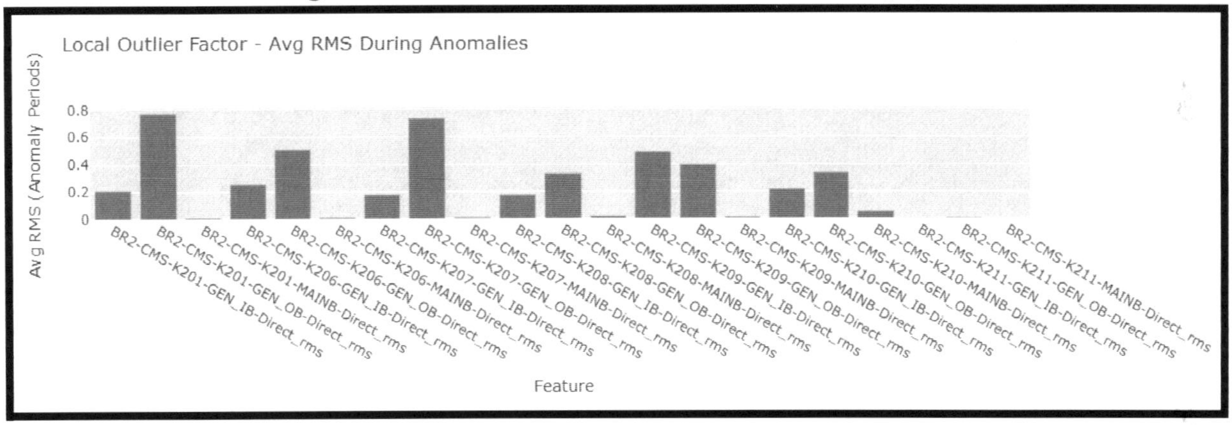

Figure 8: Local Outlier Forest Average RMS During Anomalies

3.3 One Class SVM

The One-Class SVM plot takes RotorSpeed (Turbine Rotor Speed in RPM) and a combined anomaly score of DE_H (Drive End Horizontal), GE_H (Generator End Horizontal), and HS_H (High-Speed Shaft) to establish the boundary of "normal" turbine behavior. In the plot, most points are within the decision boundary, representing stable operating behavior. However, some of the data, particularly at high levels of vibration for moderately high RotorSpeed, are outside of this area, showing that the system picked up anomalous mechanical activity during otherwise normal load conditions. This has particular application to wind power usage, where turbines experience dynamic wind speeds and weather loading. The plot is valuable in that it does not simply flag points—it defines a safe operating envelope, helping to reveal previously unnoticed deviations that can signal the incipient beginnings of gear mesh issues, torsional looseness, or developing misalignment.

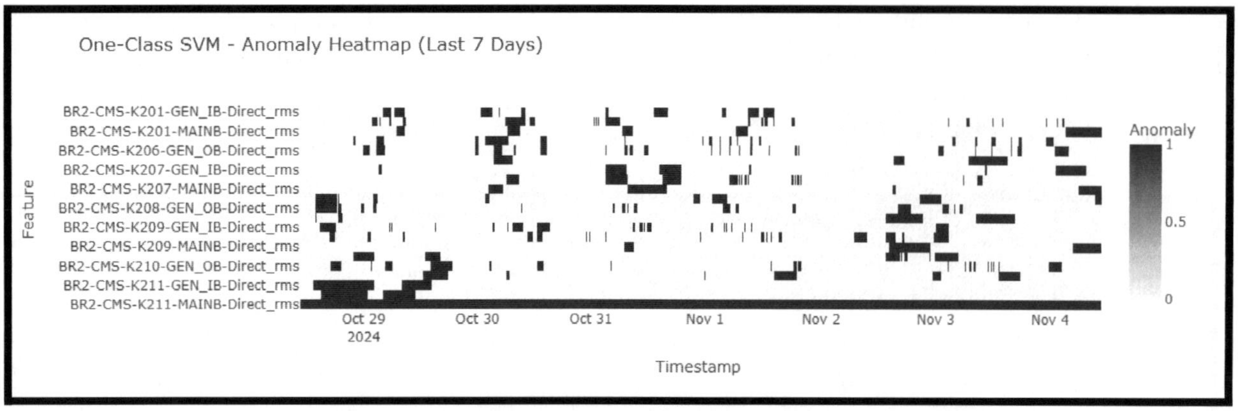

Figure 9: One Class SVM Result Heatmap (Last 7 Days)

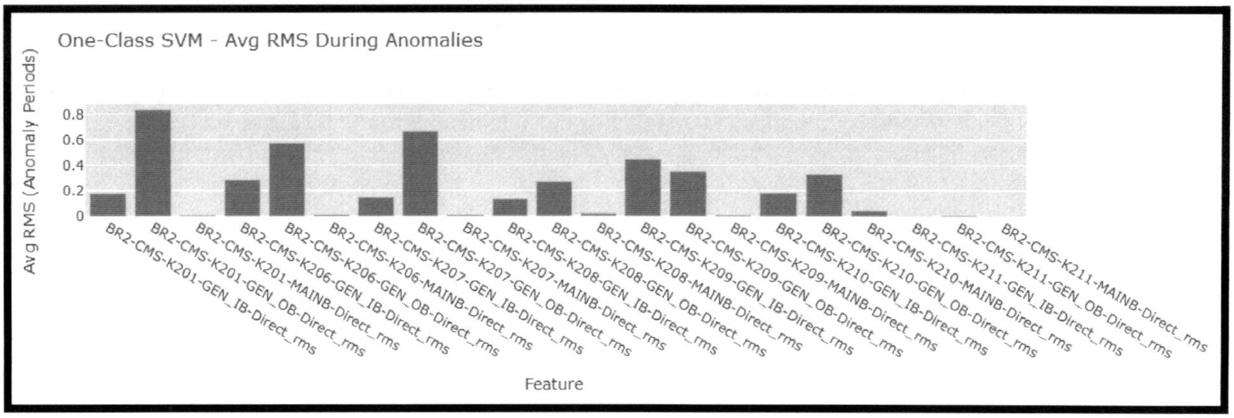

Figure 10: One Class SVM Bar Chart Average RMS During Anomalies

3.4 K-MEANS

In the K-Means clustering plot, attributes like DE_H (Drive End Horizontal vibration), GE_H (Generator End Horizontal vibration), and RotorSpeed (Rotor Speed in RPM) are utilized to separate the data into clear operating clusters. Most of the turbines cluster neatly into well-defined clusters, representing stable behavior, but outliers are present on the fringes, especially where the vibration increases without an increase in RotorSpeed. This drift from the center of the cluster uncovers long-term degradation trends, like shaft misalignment, gear wear, or resonance accumulation, that happen gradually over time. The graphical layout, with color-coded clusters and dispersed outliers, shows how turbines that were once in a stable behavioral zone are now deviating. These outliers are not just different, but chronically inconsistent, indicating parts that are quietly drifting out of tolerance and in need of urgent preventive maintenance remedies.

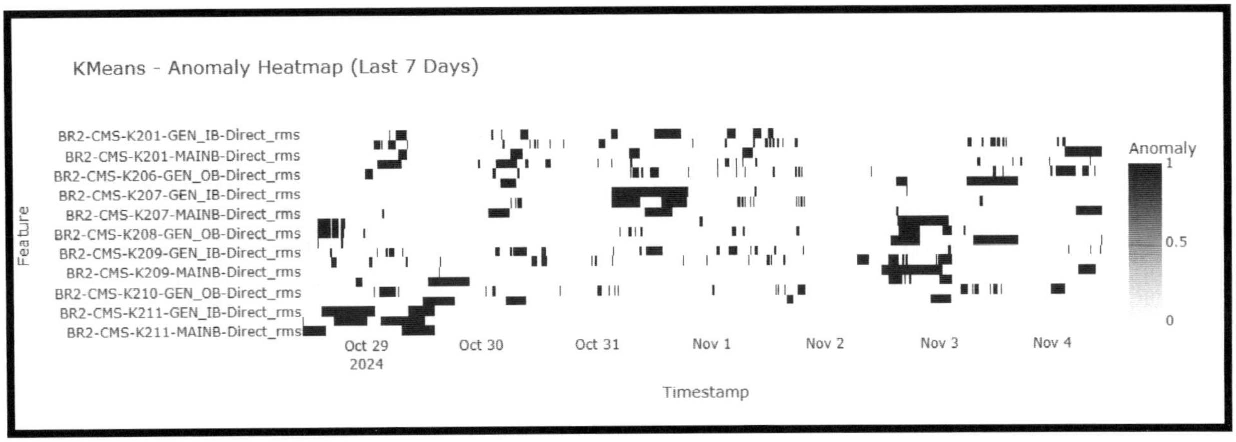

Figure 11: K Means Result Heatmap (Last 7 Days)

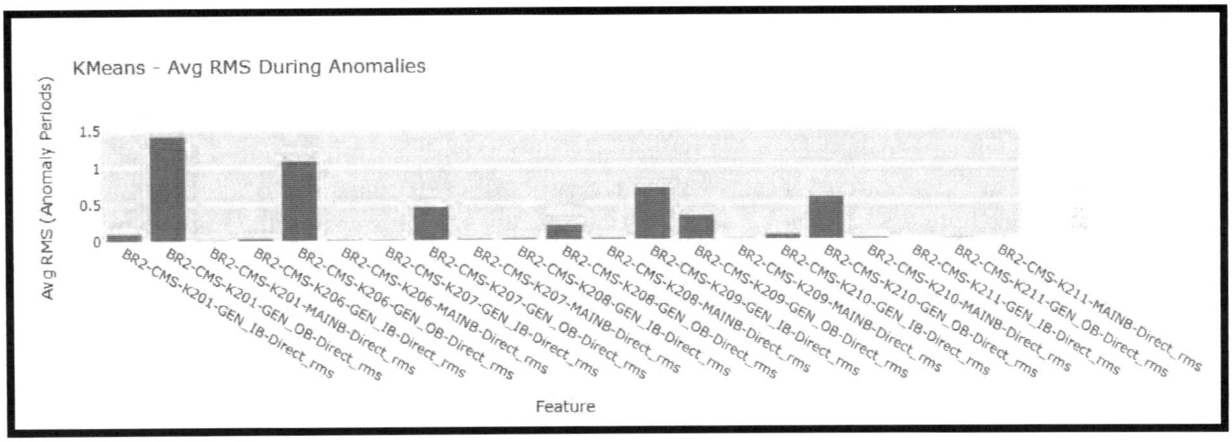

Figure 12: Kmeans Average RMS During Anomalies

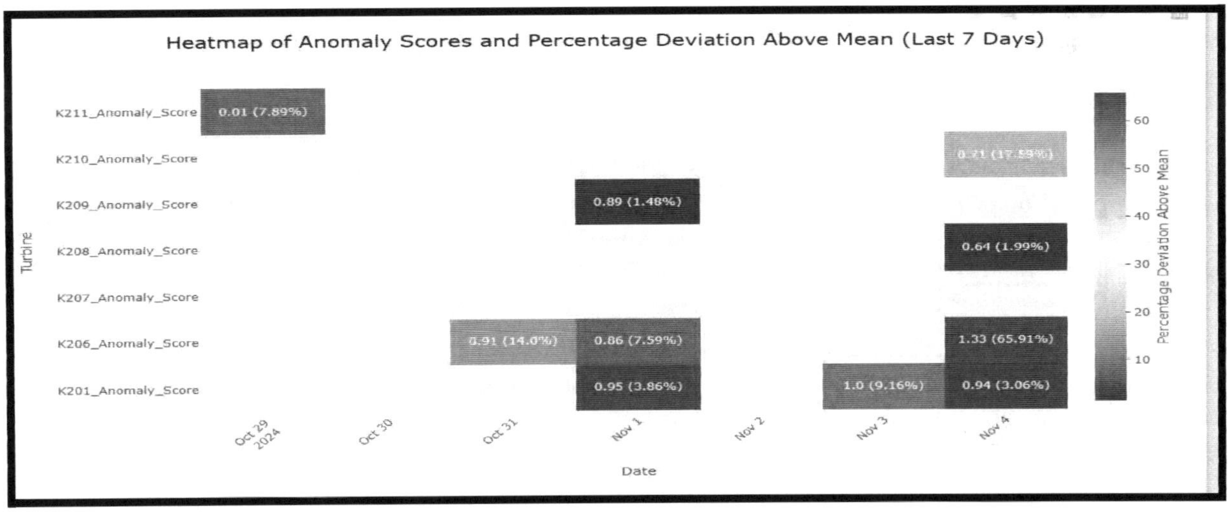

Figure 13: Combined Heatmap of Anomaly Scores and Percentage Deviation Above Mean (Last 7 Days)

The given heatmap presents the daily anomaly scores of wind turbines K211 to K206 and their percentage deviation from the weekly mean over a seven-day range (October 29 to November 4,

2024). They are computed by an ensemble-based anomaly detection system, which merges a multitude of models each capturing various facets of sensor behavior to compute composite scores capturing sudden and gradual deviations. The increased score detected for K210 (2.84 with 17.59% deviation on November 2) is probably due to an abrupt and temporary change in sensor readings, which can indicate a transient mechanical stress or a change in operating conditions. In contrast, the persistent score of 4.0 for K206 (with a 3.53% deviation on November 3) suggests the presence of a long-lasting anomaly, indicating the likelihood of a persistent issue. These findings are heavily driven by a rigorous methodology in which raw vibration data is pre-processed meticulously through normalization, intelligent imputation, and proprietary feature scaling so that only high-quality, normalized data is fed into the anomaly detection. It is the processes of secure, hierarchical data retrieval and encoding that render the input data consistent and reliable. It is this reliability that enables the detection algorithms to effectively distinguish between normal operation variability and the notable deviations that should be additionally analyzed.

4.0 Conclusion

VIBRIS platform transforms condition monitoring by coupling an ultra-secure, multi-level data retrieval process with a cutting-edge ensemble-based anomaly detection system. Proprietary HEX encoding and the hierarchical retrieval scheme not only safeguard sensor data's integrity and confidentiality but also build a solid foundation for high-accuracy analytical procedures. The compound anomaly score, derived from an ensemble of synergistic machine learning models, has been shown to have the capability to delineate short-term variability from long-term mechanical degradation. This breakthrough in predictive maintenance highlights the potential for VIBRIS to greatly improve operational efficiency and reliability in wind turbine systems. Future work will place emphasis on enhancing real-time adaptive monitoring functionalities and enlarging the scope of the framework to address a variety of industrial applications to drive the frontier of secure and intelligent condition monitoring.

References

[1] Amin, A., Bibo, A., Panyam, M., & Tallapragada, P. (2022). *Vibration based fault diagnostics in a wind turbine planetary gearbox using machine learning*. Wind Engineering, 47(1), 175–189.
[2] Matthaiou, I., Khandelwal, B., & Antoniadou, I. (2017). *Vibration Monitoring of Gas Turbine Engines: Machine-Learning Approaches and Their Challenges*. Frontiers in Built Environment, 3:54.
[3] Arcos Jiménez, A., Quiterio Gómez Muñoz, C., & García Márquez, F. P. (2018). *Machine Learning for Wind Turbine Blades Maintenance Management*. Energies, 11, 13.
[4] Black, I. M., Richmond, M., & Kolios, A. (2021). *Condition monitoring systems: a systematic literature review on machine-learning methods improving offshore-wind turbine operational management*. International Journal of Sustainable Energy, 40(10), 923–946.
[5] Ogaili, A. A. F., Jaber, A. A., & Hamzah, M. N. (2022). *A methodological approach for detecting multiple faults in wind turbine blades based on vibration signals and machine learning*. Curved and Layered Structures.
[6] Granados, D. P., Ortega Ruiz, M. A., Acosta, J. M., Gama Lara, S. A., González Domínguez, R. A., & Kañetas, P. J. P. (2023). *A Wind Turbine Vibration Monitoring System for Predictive*

Maintenance Based on Machine Learning Methods Developed under Safely Controlled Laboratory Conditions. Energies, 16, 2290.

DISK AND IMPELLER MODES OF VIBRATION: TEST VS PREDICTION

Authors: Brian Pettinato, Mark Pechulis, Manish Thorat

VIATC TRAINING CONFERENCE & EXPO
NEWPORT NEWS, VIRGINIA | AUGUST 6-8, 2025

VIATC TRAINING CONFERENCE & EXPO
NEWPORT NEWS, VIRGINIA | AUGUST 6-8, 2025

DISK AND IMPELLER MODES OF VIBRATION: TEST VS PREDICTION

Brian Pettinato
Manager, Aero and Structural Dynamics
Ebara Elliott Energy

Mark Pechulis
Engineer, Solid Mechanics
Ebara Elliott Energy

Manish Thorat
Senior Engineer, Machinery Dynamics & Acoustics
Ebara Elliott Energy

Brian Pettinato is a Senior Manager of Aero and Structural Dynamics at Ebara Elliott Energy in Jeannette, Pennsylvania. He has been with Elliott since 1995. His primary area of expertise is machinery dynamics. Mr. Pettinato received his B.S. (1989) and M.S. (1992) Mechanical Engineering degrees from the University of Virginia. Mr. Pettinato is a fellow member of ASME and a member of STLE. He is a registered Professional Engineer in the state of Pennsylvania.

Mark Pechulis is a Development Engineer in the Aero and Structural Dynamics Department, Solid Mechanics Group at Ebara Elliott Energy in Jeannette, Pennsylvania. He has been with Elliott since 2013. His specialties include modal testing and acoustic measurements. Mr. Pechulis received his M.S. (1994) Mechanical Engineering degree from Lehigh University.

Manish Thorat is a Senior Engineer and Machinery Dynamics & Acoustics Group Lead, R&D at Ebara Elliott Energy. He received his MS degree in Mechanical Engineering from Texas A&M University (2010) and BE Mechanical Engineering from Mumbai University (2007). He has been with Elliott since 2010. His areas of expertise include rotordynamics, vibration analysis, testing and evaluation of seal designs, squeeze film dampers and fluid film journal bearing design.

Abstract

Multistage centrifugal compressor manufacturers must ensure reliable operation of their equipment by avoiding impeller resonance. This is done by designing the equipment with a separation margin between the predicted excitable modes and their upstream and downstream sources of excitation. The excitable modes are those modes that are not phase cancelled or otherwise attenuated. Impeller modes include the hub and cover tip modes, the disk nodal diameter modes, and the blade leading edge modes.

In this paper, blade leading edge modal test data were collected for a wide range of centrifugal impeller sizes and compared to the finite element analysis predicted modes. The results led to model refinement and a reduction in the required separation margin after replacement of the assumed statistical uncertainty with the tested uncertainty. This resulted in a greater application envelope. This paper discusses the approach to develop the separation margin and also to investigate the testing repeatability by collecting data from identical impellers. Example test and finite element analysis modal results are presented along with a detailed examination of one impeller.

Key Words: Impeller, Turbomachinery, Vibration

Introduction

Turbomachinery plays a critical role in several industries. Centrifugal compressors are used in oil and gas recovery, gas / oil separation processes, gas gathering and boosting, gas transportation, gas treatment, and gas processing. Rotating blade and disk failures cause a fairly small percentage of machine shutdowns but can require extensive repairs and downtime. Plant downtime due to equipment failures is expensive and compressor suppliers design their impellers to avoid resonance during operation. Turbomachinery rotating speeds and upstream sources of excitation are often a limiting factor in the aerodynamic design and therefore a constraint to improving aerodynamic performance. High cycle fatigue in the centrifugal compressor impellers is always a concern.

Kushner (Reference [1], [2], and [3]) discussed interaction resonance, phase cancellation, and the principles of failure avoidance in bladed disks. Pettinato, et al. in Reference [4] demonstrated phase cancellation principles through testing while also characterizing mistuning levels in an industrial impeller. Failures can be minimized by carefully designing the equipment to either avoid resonant conditions or avoid excitable modes through selection of stationary and rotating blade counts. An impeller typically has potential for excitation of either blade or disk modes. Nodal diameter mode shapes, commonly referred to as nodal diameters (NDs), contain lines of zero out-of-plane displacements, which cross the entire disk. As discussed by Kushner, the most excitable modes are those that follow Equation (1) and Equation (2).

When not at a disk critical speed,

$$\begin{aligned} m &= |z \cdot b \pm h| \\ h &= y \cdot s \\ f_r &= h \cdot \omega \end{aligned} \qquad (1)$$

Whereas when at a disk critical speed and for the number of rotating blades, $b > 1$,

$$h = m \qquad (2)$$
$$f_r = m \cdot \omega$$

where

b	= Number of rotating blades	h	= Harmonic of speed
s	= Number of cyclo-symmetric stationary vanes	m	= Number of diameter nodal lines
		y & z	= Integers > 0
f_r	= Natural frequency at speed, Hz	ω	= Rotating speed, Hz

Modes that meet equations (1) and (2) are excitable. By contrast, modes that fail to meet these equations experience considerable phase cancelation and are relatively difficult to excite. Modes of concern range from the 0-diameter mode up through the number of blades divided by two-diameter modes. A common practice for the centrifugal compressor stage is to select the number of stationary upstream return channel and downstream diffuser vanes such that the 0-diameter and 2-diameter modes do not satisfy either equation and are thus phase cancelled. The one-diameter disk mode (ND1) is usually coupled to the shaft and is not easily excited and therefore typically not a concern even when Equations (1) and (2) are satisfied. Equations (1) and (2) are often represented graphically by the zig-zag diagram [5], which is also known as a SAFE diagram [6]

Impeller mode shapes and their corresponding frequencies are discussed in the following sections. The impeller blade leading edge modes are excited by upstream return channel vanes, whereas disk modes are excited by either the upstream return channel vanes or downstream diffuser vanes. For resonance to occur, not only does the excitation frequency have to match the resonant frequency, but also the mode shape must match the excitation pattern as described.

To reduce the chances for blade and disk high cycle impeller fatigue in multistage compressors, they should be designed and operated to avoid resonance of the most excitable modes. Original equipment manufacturers (OEMs) require reliable Finite Element Analysis (FEA) models to estimate nodal diameter and blade leading edge modal frequencies and an understanding of the variance to prediction that actual parts will demonstrate.

Data Collection

Two types of impeller modes that must be evaluated include the disk nodal diameter (ND) modes and blade leading edge (LE) modes. The number of modes that must be identified within each of these two types depends on the number of impeller blades in the wheel. The number of ND and LE modes of interest is up to the number of blades divided by 2 and rounding down to the nearest integer. For example, an impeller with 15 blades would have modes of interest spanning from 0 to 7 nodal diameters and similar for the LE modes.

Finite Element Modeling

The Cyclic Symmetry method is used in the FEA model to determine the mode shapes and frequencies. This method requires only a sector of the impeller to reduce the model size and its simulation time. The sector size depends on the number of blades in the impeller. For example, a 13 bladed impeller will have a sector equal to 360 divided by 13 which is a 27.69-degree section of the circular impeller. An additional advantage of this cycle symmetry is that it classifies the modes based on nodal patterns. Welds are modeled explicitly where the blade and hub and cover join. Bonded contacts are used to simulate the welds and frictionless contacts are used between the blade and hub/cover unwelded interface. The FEA is modelled with the impeller installed on the rotor. In general, the sector model had roughly 125,000 elements and 200,000 nodes. The mesh size was chosen based on a mesh sensitivity study and variation in frequency was less than 0.2%. Figure 1 shows an example of the sector FE model using a cyclic sector.

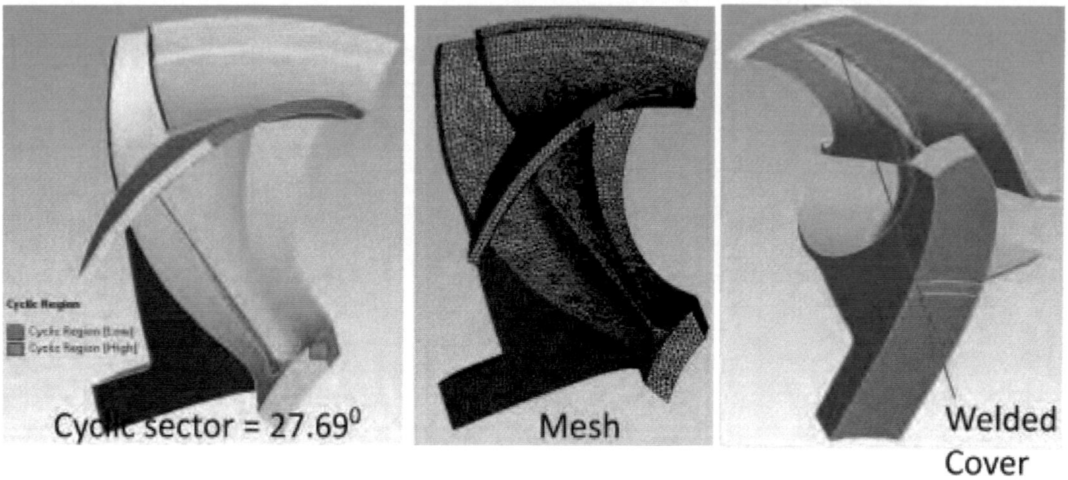

Figure 1: FEA Impeller Sector Modelling

The impellers selected for this study ranged from under two feet to over six feet in diameter and their availability was based on the timing of their fabrication. The manufacturing methods used to construct the impeller were single-piece milled, two-piece welded, and brazed. Data from thirty impellers was used in the analysis. Figure 2 shows an example of some of the impellers tested.

Figure 2: Some of the Multistage Compressor Impellers Tested

Modal Testing

Modal testing requires at least one exciting stimulus with force transducer and at least one sensor measuring the motion. The exciting stimulus can be any directional force including impact hammer, shaker, acoustic, or other source. The motion sensor can be any transducer that measures motion including accelerometer, strain gauge, laser, eddy current, optical, or other transducer. To measure the entire structure, one of the following must occur: (i) excitation is provided at multiple locations while measuring at a single location, (ii) motion is detected at multiple locations while exciting at a single location, or (iii) excitation is provided at multiple locations while measuring at multiple locations. One method of modal testing that is often used is a single fixed accelerometer with force applied successively at different locations with an instrumented hammer. The testing conducted herein utilized a roving instrumented hammer and either one or two stationary miniature accelerometers for data collection on the impellers (Figure 3).

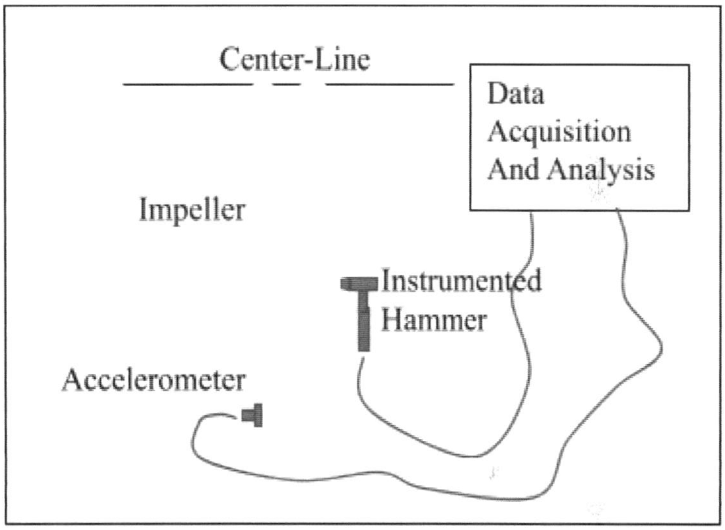

Figure 3: Modal Testing by Roving Hammer Method

Initial testing was performed to help identify the blade leading edge mode approximate frequencies. The measurement range was then setup for the remainder of the data collection. Next, hammer impact locations were marked out on the impeller cover and blades while the accelerometer was secured to the impeller tip with epoxy. The impeller was impacted axially in the direction of the shaft at up to 200 locations. Figure 4 shows a typical impeller tested and the data collection model with modal hammer impact locations. Figure 5 show a typical forced response function (FRF) where each peak represents a mode of vibration.

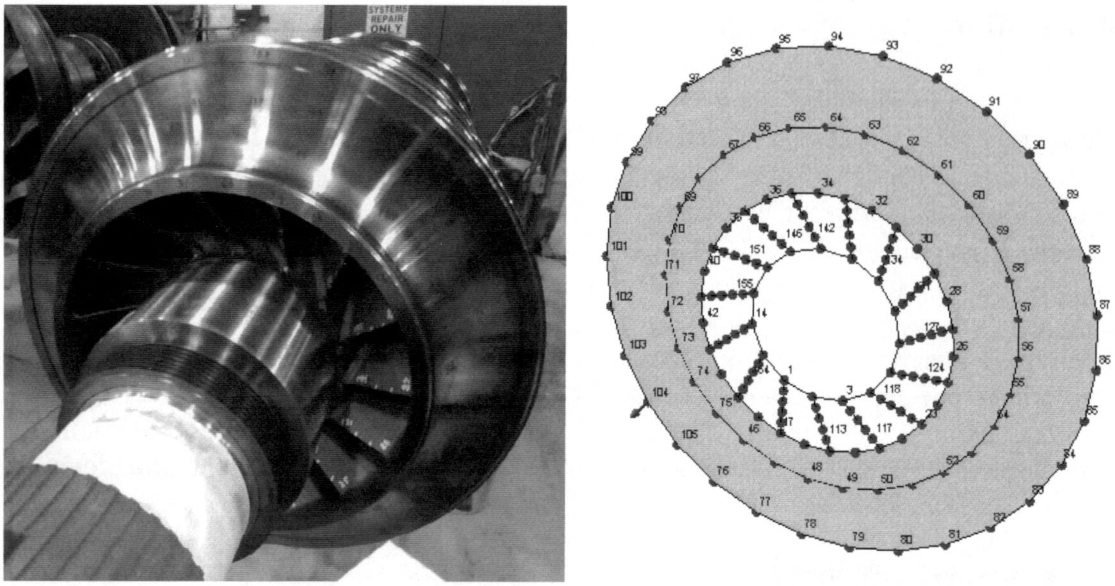

IMPELLER AXIAL VIEW IMPELLER TEST MODEL

Figure 4: Impeller Modal Testing Model

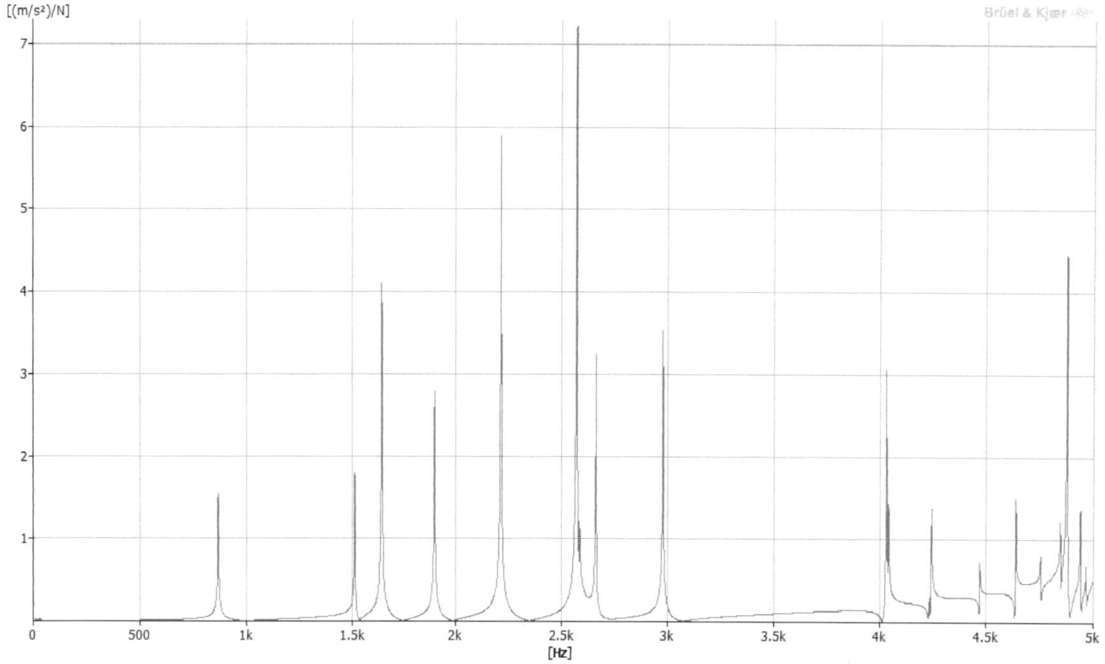

Figure 5: Measured Forced Response Function

Impeller Modes

The mode shapes are classified based on the number of nodal diameters (ND) and nodal circles (NC). Disk ND modes are defined by nodal lines (red lines in the following figures) across the diameter of the impeller whereas NC modes are defined by circular patterns on the impeller.

Displacements are zero on the nodal diameter/radial lines and on the nodal circle lines, and modal displacement occurs everywhere else for a given mode. Circle modes and combined circle plus diameter modes usually do not pose any problems provided that stationary vanes and rotating blades are not equal to or multiples of each other. Diameter modes up to the number of blades divided by 2 rounded to nearest integer are typically considered when designing the equipment. Examples are provided for circle modes ND0 and LE0, 1-diameter modes ND1 and LE1, and 2-diameter modes ND2 and LE2.

Zero-Diameter Disk (ND0) and Blade (LE0) Modes
The zero-diameter disk mode is depicted in Figure 6. The mode consists of zero nodal lines and one nodal circle. This mode does not pose any risk to turbomachinery operation when the stage is designed with an appropriate number of upstream return channel vanes in accordance with Equation (1) when the number of stationary vanes and rotating blades are not equal to or multiples of each other. The entire impeller moves in an up and down motion like the opening of an umbrella for the ND0 animation. The predicted and measured motion are very similar.

FEA ND0 MODE ANIMATION

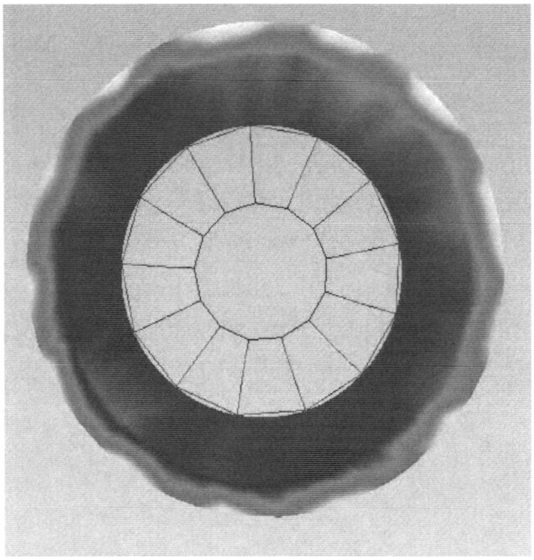
MEASURED ND0 MODE ANIMATION

Figure 6: Impeller ND0 and LE0 Mode Shapes

The zero-diameter blade mode is depicted in Figure 7. The leading edges of the impeller blades move up and down in unison with little participation from the cover. Note that the blades are shown as lines for the measured case to reduce the complexity of the model.

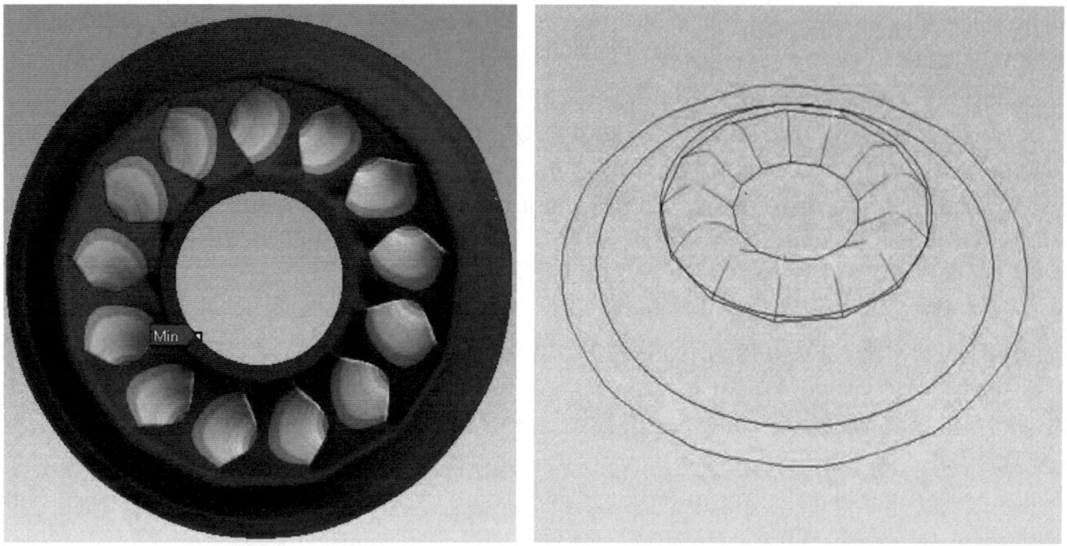

FEA LE0 MODE ANIMATION   MEASURED LE0 MODE ANIMATION

Figure 7: Impeller ND0 and LE0 Mode Shapes

One-Diameter Disk (ND1) and Blade (LE1) modes

One-diameter modes have a single nodal line passing through the center of the impeller. The displacement is zero at this nodal line. The response of the one-diameter disk mode from upstream or downstream sources of excitation tends to be low as it is coupled to the shaft as a twisting motion around the nodal diameter line with the entire impeller moving in a rocking motion along the nodal diameter line for the ND mode animation. Figure 8 compares the one-diameter disk mode shape to the measured results.

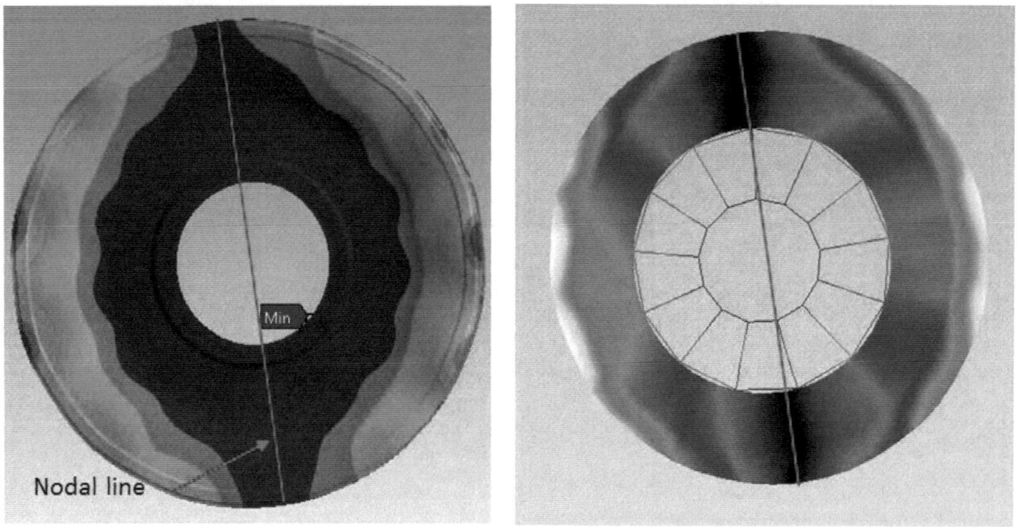

FEA ND1 MODE ANIMATION   MEASURED ND1 MODE ANIMATION

Figure 8: Impeller ND1 Mode Shapes

In contrast to the one-diameter disk mode, the one-diameter blade mode does not couple to the shaft and the response of the one-diameter blade mode is quite dependent on the number of upstream or downstream vanes in the stage design and whether or not there is phase cancelation in accordance with Equation (1). Figure 9 compares the one-diameter blade mode to the measured results. For the LE1 mode the blades move in a similar rocking motion, but there is relatively little motion on the cover or hub.

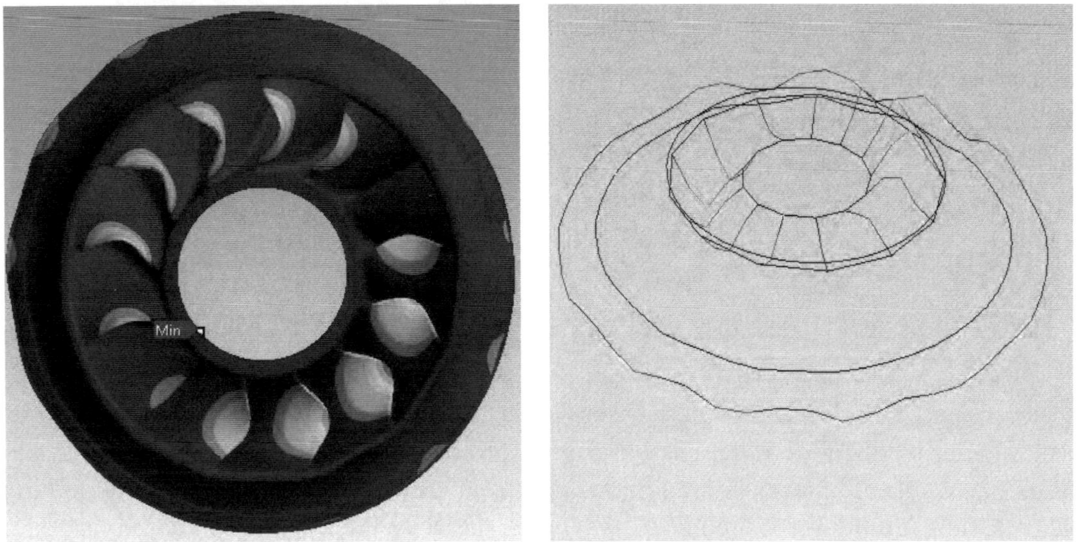

FEA LE1 MODE ANIMATION MEASURED LE1 MODE ANIMATION

Figure 9: Blade LE1 Mode Shapes

Two-Diameter Disk (ND2) and Blade (LE2) modes

Two-diameter modes have two nodal lines passing through the center of the impeller. The displacement is zero at these two nodal lines. The two-diameter mode is typically very easy to excite and stages must be carefully designed to avoid exciting this specific mode. The impeller moves in a sort of saddle shape where opposite ends move in unison up or down in phase while the other two sides move in the opposite direction for the ND animation. Figure 10 shows the predicted and measured 2-diameter disk modes. To better illustrate the mode, the quarter section marked with the + symbols can be considered as moving out of phase with the – symbols.

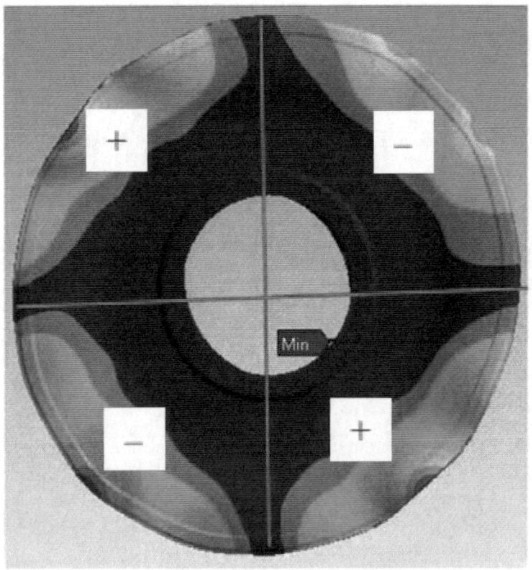

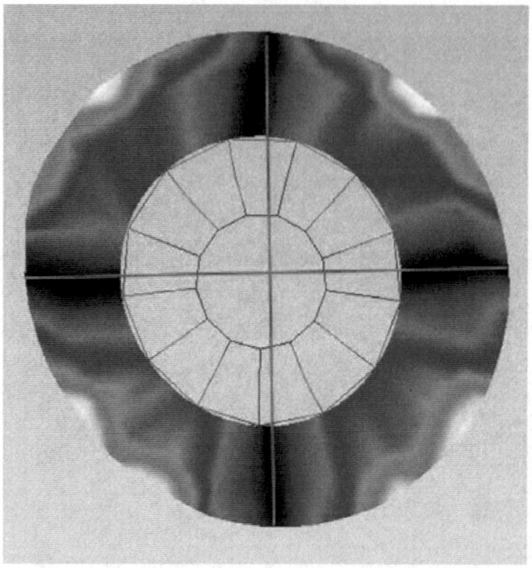

FEA ND2 MODE ANIMATION MEASURED ND2 MODE ANIMATION

Figure 10: Impeller ND2 Mode Shapes

The two-diameter blade mode is depicted in Figure 11 comparing the FEA and measured modal animations. Two sets of blades respond in one direction while the other two sets respond in the opposite direction. Unlike the 2-diameter disk mode, the cover and hub are relatively stationary.

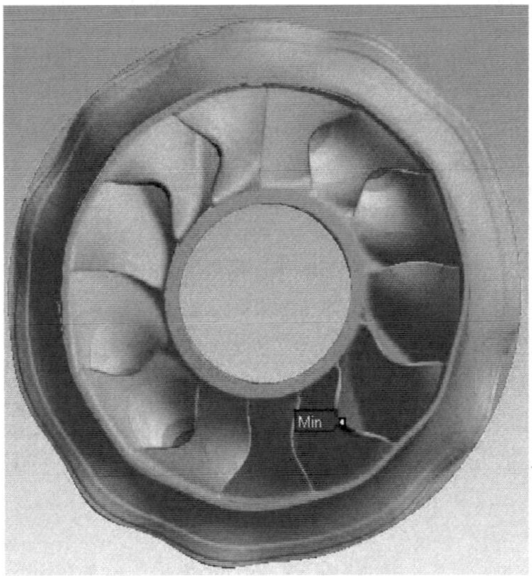

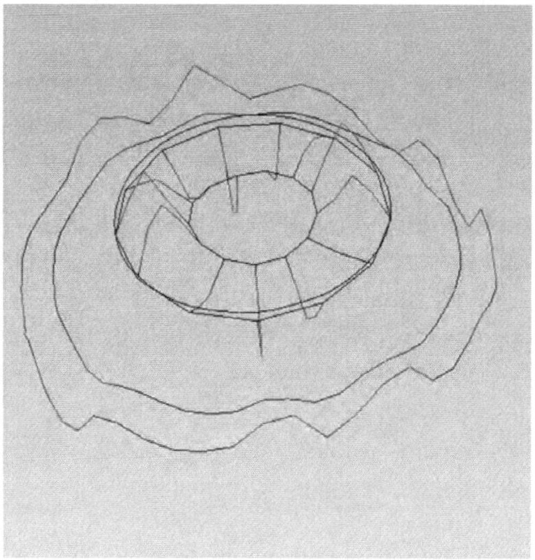

FEA LE2 MODE ANIMATION MEASURED LE2 MODE ANIMATION

Figure 11: Blade LE2 Mode Shapes

Similar patterns exist for ND3-ND6 (and LE3-LE6) for this 13-blade impeller example.

Test Results

A proper and safe design criteria for ND and LE resonance avoidance considers the sum of a fixed margin from the wake passing frequency at running speed plus an uncertainty in the predicted values based on a comparison of measured data to FEA models. The goal of impeller manufacturers should be to better define this uncertainty and enhance the mode predictive capabilities. Modal test data was collected and compared to FEA models for various impeller frame sizes.

Measured vs FEA Data Results Comparison

Figure 12 shows a typical comparison plot of the FEA model predicted and measured ND modal frequencies for one of the impellers that had eleven blades. The % difference between the test and FEA is provided in Figure 13. In this case, the FEA is taken as the datum because it is done in a standard way and is essentially without variance whereas the manufactured impellers did have variance. Similar data comparisons were made for all 30 of the impellers for both the LE and ND modes. This data was tabulated and a statistical analysis was performed in order to predict uncertainty bands where the frequency data would lie in accordance with a confidence value.

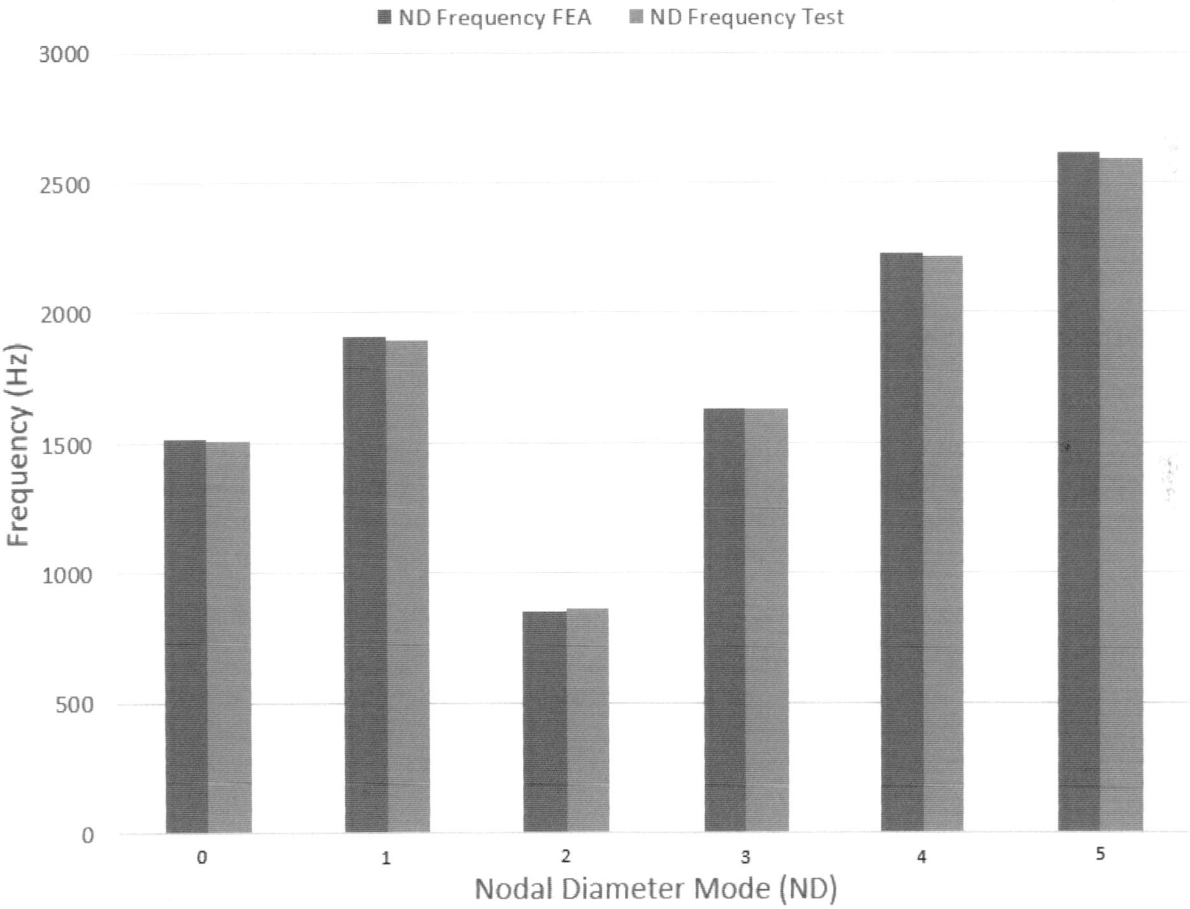

Figure 12: Measured and FEA Disk Modes for an 11-Blade Impeller

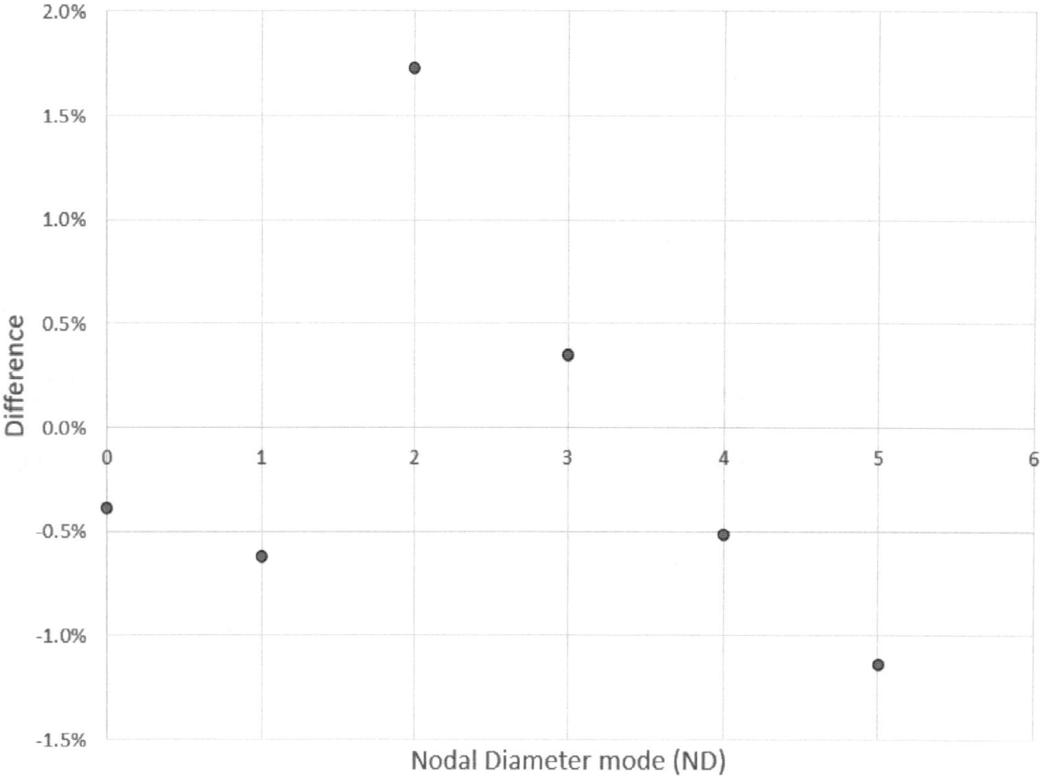

Figure 13: ND Mode Frequency %Difference for an 11-Blade Impeller

Statistical Analysis

Blade leading edge frequency data was analyzed using statistical methods to predict the future uncertainty in the observations and to establish the uncertainty margins. A Prediction Interval is an interval constructed using a set of sample data to contain the future observation(s) as described in References [7] - [11]. This Prediction Interval can be defined as where one can expect a future value to fall (e.g. 97.5% certain that the next measurement will fall within the calculated range)

$$X_n \pm t_{\alpha,n-1} \cdot \sigma \cdot \sqrt{1 + 1/n} \tag{3}$$

where

X_n	= sample mean
$t_{\alpha,n-1}$	= t-statistic value (one-tailed)
σ	= sample standard deviation
n	= number of samples
1-α	= confidence level

The sample size is taken into consideration by the t-statistic and the prediction uncertainty relates directly to the t-statistic coefficient such that additional samples will result in a lower uncertainty provided that the standard deviation does not increase. It should be noted that this paper is a snapshot in time and additional data continues to be gathered.

Using the sample mean and standard deviation for the blade leading edge frequency modal data from the thirty impellers along with the t-statistic corresponding to the number of independent samples and desired confidence interval (1-α), a prediction interval for the FEA model was defined. This process was completed for each of the ND and LE impeller modes and an overall prediction interval was calculated. Figure 14 shows the results for one of the blade leading edge modes. The 0% datum is the FEA prediction. 0σ represents the mean of the data. The data for this specific mode is skewed higher than prediction. The analysis of the data resulted in a reduction and shift of the original uncertainty and also more confidence in the application of the FEA predicted models. It also resulted in more flexibility in the impeller design selection process.

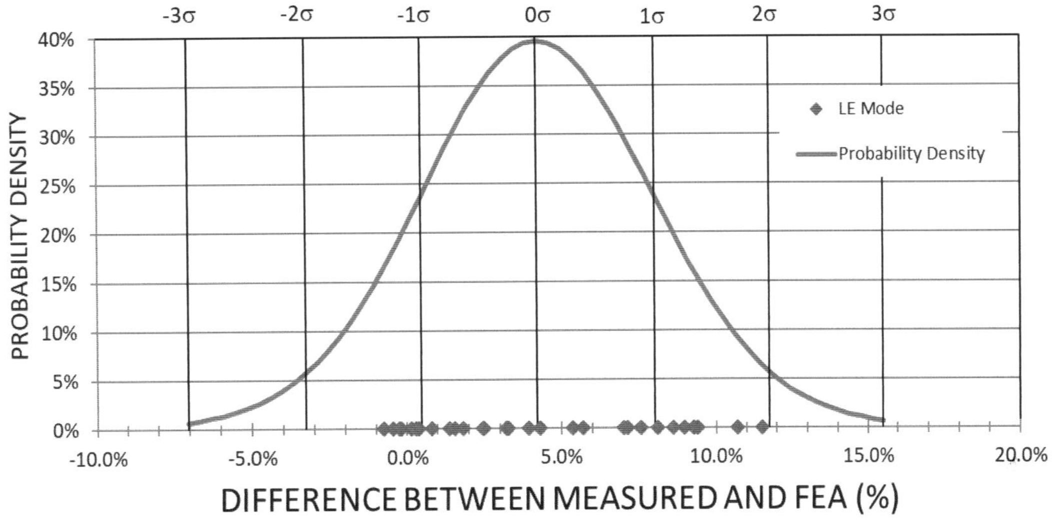

Figure 14: Blade Leading Edge Results for Thirty Impellers and One Mode

Measurement Repeatability

In addition to comparing the measured data to the FEA model modes, data was collected on three pairs of identical impellers to estimate the repeatability including manufacturing variation.

Two rotors with several identical impeller models were fabricated and available during the execution of this study. Impellers from each of the rotors were selected and ND and blade LE data was collected and compared to determine the repeatability as an estimate of the uncertainty that would contain the manufacturing and testing error. Figure 15 shows that the percent difference from the average value for the mode was +/- 1% which was actually better than expected.

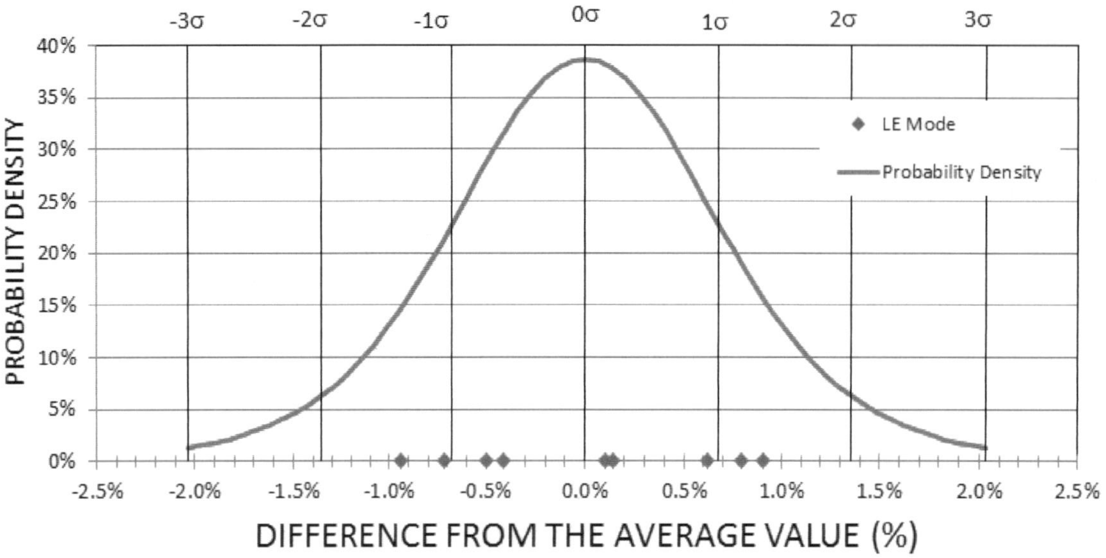

Figure 15: Repeatability Plot for Three Identical Impellers

Comparison of Impeller to Annular Disk

An impeller is intuitively expected to have more complex behavior and less agreement to FEA than a simple structure such as a continuous annular disk. An annular disk was constructed with a similar 3-diameter disk mode as the 11 bladed impeller previously discussed for comparative purposes. Figure 16 shows the 11 bladed impeller and annular disk, both of which measured 18 inches in diameter.

Figure 16: 11-Bladed Impeller and Annular Disk

Figure 17 shows the % differences of the different modes for the two disks. The annular disk was more biased toward higher than FEA frequencies as compared to the impeller for this specific case. The annular disk also had less scatter with all measured modes being 2.5 to 3.1% high. One notable difference is that the annular disk was not mounted to a shaft, which might have affected results. The lack of shaft mounting resulted in no 1-diameter mode.

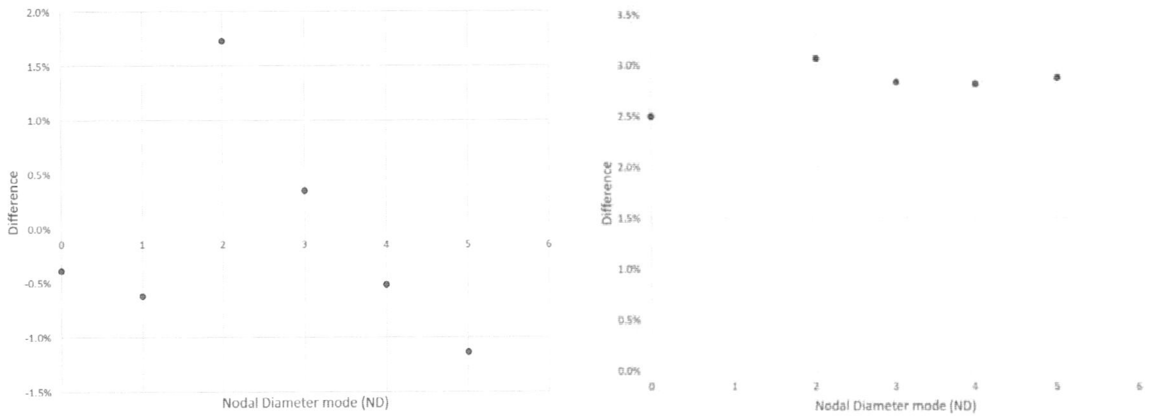

(a) Impeller (b) Annular Disk

Figure 17: % Difference to FEA

Table 1 provides the values of modal assurance criterion for the different modes as compared to FEA indicating how well the mode shapes matched the prediction. It is of some interest that the impeller had better cross MAC than the annular disk for the 2-diameter mode. The 3-diameter modes are shown in Figure 18. As expected, the impeller had poorer cross MAC than the disk in the 3-diameter mode. If the impeller was a 12-bladed impeller, there could be reasonable expectation for similar MAC results given that a 3-diameter mode is evenly divisible into 12 sectors. MAC results became increasingly lower for each nodal diameter higher than two.

Table 1: Cross MAC

	Impeller			Annular Disk		
ND	f_n, Test (Hz)	f_n, FEA (Hz)	MAC	f_n, Test (Hz)	f_n, FEA (Hz)	MAC
0	1510.4	1516.2	0.971	1213.3	1183.7	0.998
1	1897.8	1909.7	0.809	Not mounted to a shaft		
2	864.5	851.2	0.959	672.1	652.1	0.799
3	1634.4	1628.7	0.737	1690.9	1644.3	0.969
4	2208.6	2224.0	0.612	2954.8	2873.8	0.915
5	2573.0	2615.9	0.244	4418.0	4294.2	0.853

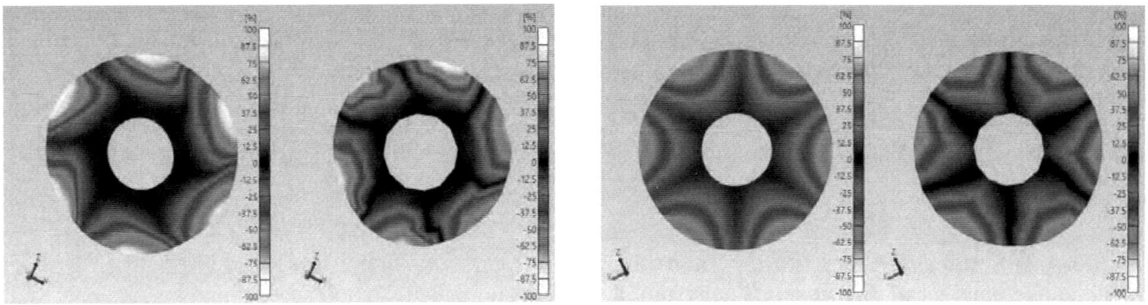

(a) Impeller FEA (b) Impeller Tested (c) Disk FEA (d) Disk Tested

Figure 18: 3-Diameter Modes

Conclusions

Impeller manufacturers need reliable finite element prediction models to estimate nodal diameter and blade leading edge modal frequencies. Increased confidence in these models can lead to a smaller separation margin requirement between the modal frequencies and multistage compressor operating speeds and therefore result in greater flexibility during the impeller selection process. As part of this effort, modal data was collected on thirty impellers of various designs and sizes to measure the nodal diameter and blade leading edge modal frequencies to compare to finite element model predicted frequencies. In this study, examples of nodal diameter and blade leading edge modal animations were presented and it was discussed how they were used to determine prediction intervals to define the overall uncertainty when selecting impellers in the design process. In addition, the uncertainty from manufacturing and tests data collection was validated by performing measurements on identical impellers. From the prediction intervals calculated, the impeller blade leading edge frequency separation margins from minimum and maximum continuous operating speed can be defined.

Acknowledgements

The authors would like to thank Ebara Elliott Energy for permission to publish this paper.

References

[1] Kushner, F., 1980, "Disc Vibration · Rotating Blade and Stationary Vane Interaction", ASME Journal of Mechanical Design, *102*, pp. 579-584. https://doi.org/10.1115/1.3254788

[2] Kushner, F., 2004, "Rotating Component Modal Analysis and Resonance Avoidance Recommendations", Tutorial, Proceedings of the 33rd Turbomachinery Symposium, Turbomachinery Laboratory, Texas A&M University, College Station, TX, pp. 143-161. https://doi.org/10.21423/R1WW7Z

[3] Pettinato, B., Griffin, J., Wang, Y., Feiner, D., Echols, B., and Cushman, M., 2013, "Experimentally Based Statistical Forced Response Analysis for Purpose of Impeller Mistuning Identification," Proceedings of the 2nd Middle East Turbomachinery Symposium, Turbomachinery Laboratory, Texas A&M University, College Station, Texas. Presented in Doha, Qatar, March, 2013. https://turbolab.tamu.edu/wp-content/uploads/2018/08/METS2Lecture10.pdf

[4] Kushner, F., Pettinato, B., Pechulis, M., and Thorat, M., 2024, "Modal Analysis and Resonance of Bladed Disks," Proceedings of the Turbomachinery and Pump Symposium. https://hdl.handle.net/1969.1/1582470

[5] Wildheim, S. J., 1979. "Excitation of Rotationally Periodic Structures," *ASME J Appl Mech*, V 46(4), pp 878-882. https://doi.org/10.1115/1.3424671

[6] Singh, M., P., Thakur, B. K., Sullivan, W. E., and Donald, G., 2003, "Resonance Identification for Impellers," *Proceedings of the Thirty-Second Turbomachinery Symposium,* Turbomachinery Laboratory, Texas A&M University, College Station, TX. pp. 59-70. https://doi.org/10.21423/R1XM17

[7] Pettinato, B., Pechulis, M., Thorat, M., and Chundru, R., 2024, "Modes of Vibration for Closed Impellers: Test vs Prediction," Proceedings of the Asia Turbomachinery and Pump Symposium.

[8] Luko, S., Neubauer, D. V., "Statistical Intervals – Part 2", ASTM Standardization news September/October 2011

[9] EPA statistical analysis of groundwater monitoring data at RCRA facilities, March 2009

[10] Wikipedia.org. "Prediction Interval"

[11] https://www.statisticshowto.com/tables/t-distribution-table/, Statistics How To, t-distribution table (one-tailed), copyright 2020

USING CELL PHONE VIDEOS FOR OPERATIONAL DEFLECTION SHAPES

Author: Maryon (Skip) Williams

VIATC | TRAINING CONFERENCE & EXPO

NEWPORT NEWS, VIRGINIA | AUGUST 6-8, 2025

USING CELL PHONE VIDEOS FOR OPERATIONAL DEFLECTION SHAPES

Maryon (Skip) Williams
Senior Engineer, Condition Analyzing Corporation
swilliams@cacvibe.com

Skip Williams has conducted condition monitoring including vibration and infrared thermography of ships for more than 34 years. He is a Certified Vibration Analysis Level IV and Professional Engineer (NJ). He has M.S and Ph.D. degrees from Rutgers University and B.E.E. with Highest Honor from Georgia Tech. He patented an artificial piezoelectric heart assist device operating in mechanical resonance with the beating heart.

Abstract

This article describes how cell phones can be used to collect Operating Deflection Shapes (ODS's). The effects of various frame speeds are demonstrated. Aliasing is discussed as to how it relates to a cell phone video. The existence of aliasing greater than 2*Fmax is shown by calculation and example. Case studies from ships are presented.

Key Words: aliasing, cell phone, ODS, video

Background

Video Operational Deflection Shapes (ODS's) started with early examples of measuring heartbeat motion of babies in cribs [1]. Since then, video ODS analysis has become a recognized tool for measuring vibration on machines [2]. A picture is worth a thousand words, and a movie is worth a thousand pictures. The method is useful to show the vibration of machines when the use of normal sensors may be too time consuming, may mass load a machine, or may be unsafe or may require access to inaccessible locations.

An ODS has two purposes: 1) show the relative motion of different parts of a machine, i.e. phase differences and 2) show the magnitude of the motion. Relative magnitudes are all that is required to show where a machine is moving the most. A high-speed camera, as well as the cell phone in your pocket, can be used to acquire ODS data. In practice this method is invaluable for showing clients how machines are vibrating when raw vibration data alone may be unconvincing.

Data collection

In almost half the world's countries, over 90% of the population owns at least one cell phone [3]. The most common cell phones are iPhone (by Apple) and Galaxy (by Samsung). There are hundreds of different models of cell phones. Smart phones have built-in cameras with features such as high resolution, high frames rates, slow motion modes, gigabytes of memory, and stabilization.

Unlike a high-speed camera, a cell phone usually requires no artificial lighting, A cell phone camera acquires the image simultaneously contrary to popular opinion. The playback shows negligible phase differences within a frame.

Cell phone video files are compressed to decrease size while preserving the quality for viewing. For the iPhone 14, the camera format setting can be either Most Compatible or High Efficiency, codec H.264[1] or codex H.265,[2] respectively. H.265 is the more modern method and gives better quality than H.264. A video file has a metadata section, which can be read from many editing programs[3]and which gives information such as frames per second and compression method.

The default frame rate for an iPhone is 30 fps. An iPhone in "slo-mo" mode can also record a video at 1080p HD at 240 fps (frames per second) or 120 fps. The Galaxy A14 in slow motion mode records at 120 fps. The middle 84% of the frames are displayed at 30 fps, which can be adjusted by sliders on the iPhone frame viewer (Figure 1).

The bin size of the vibration spectrum is $\Delta f = fps/n$ where n is the number of frames. Since $n = fps*T$, $\Delta f = 1/T$. For instance, a bin size of 3 cpm requires a video recording time of 20 seconds. A bin size of 6 cpm requires a data recording time of 10 seconds.

---

[1] H.264, Advanced Video Coding (AVC) or MPEG-4 Part 10

[2] H.265, High Efficiency Video Coding (HEVC) or MPEG-H Part 2

[3] VLC

[4] 1080p HD means the image is 1,920 x 1,080 pixels. p means progressive scanning in which the lines are one after the other as opposed to interlaced. HD means High Definition as opposed to SD or Standard Definition with lower resolution.

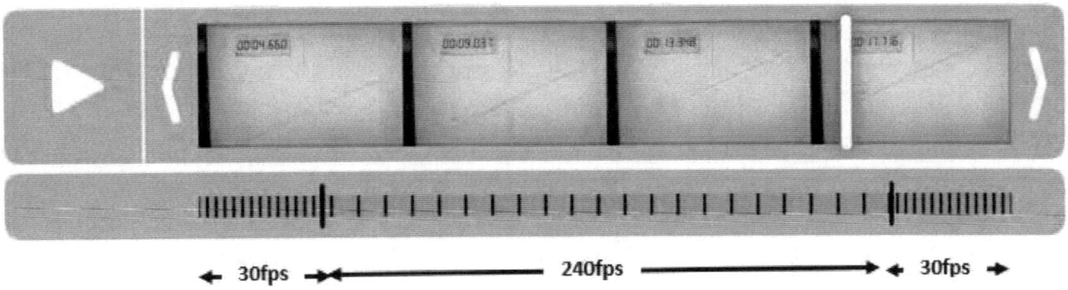

Figure 1: Frame viewer from iPhone 14

Image stabilization

An ODS can be affected by camera movement or by movement in the background of a machine. ODS software programs can reduce background vibration, but it might be difficult to remove all of it. On a ship, the steel deck often vibrates at the same frequency as the machine itself, making it difficult to isolate the machine vibration from the deck vibration (Figure 2). Solutions include positioning the camera on a support frame isolated from the machine, using a cell phone stabilization feature, and using a camera tripod incorporating a Bluetooth switch[5] to start/stop the recording without touching the camera.

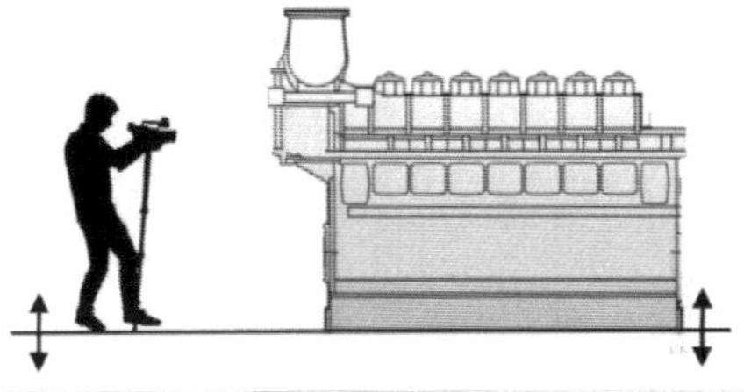

Aliasing

Aliasing is the appearance of vibration below Fmax when the signal is actually at a frequency above Fmax. In a typical vibration application using accelerometers and data acquisition hardware, the signal is low-pass filtered to remove all frequencies above half of the sampling rate, which is called the Fmax frequency or Nyquist frequency. Without low-pass filtering, aliasing can occur, but there is less chance the higher the frame rate. Figure 3 shows a peak frequency above Fmax "folded" below Fmax. With video recording, low pass filtering is not normally done. There is such a thing as an optical low-pass filter [4], which is a physical element placed in the optical path. However, why would you filter if you can get valuable information from an aliased signal?

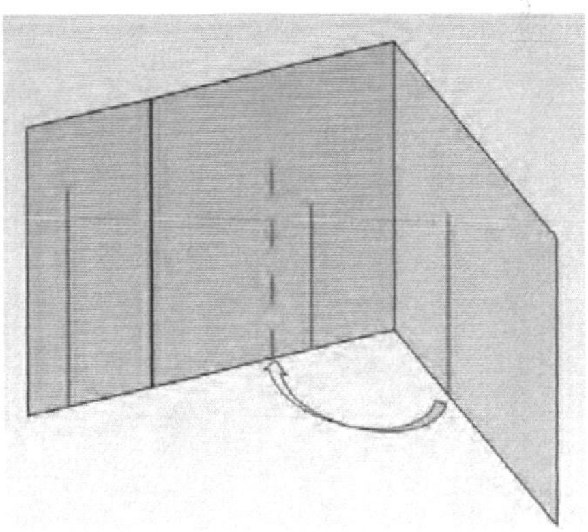

Spectra were generated with the Fourier analysis tool in Excel to show the behavior when the Nyquist limit is violated. The sample rate was constant at 102.4 Hz with an Fmax of 51.2 Hz.

[5] iPhone tripod

Two cases are given in Figure 4 and Figure 5, one where the signal is below Fmax and one where the signal is above Fmax. Figure 4 shows the FFT of a 40.8 Hz signal which is below Fmax. The part of the spectrum above Fmax of 51.2 Hz is the mirror image of the part below Fmax. The mirror image of 40.8 Hz is 61.6 Hz (102.4 Hz-40.8 Hz) and is not an issue. When the Nyquist limit is observed, the two peaks move closer to the Fmax line as the signal frequency increases.

Figure 5 shows the FFT of a 61.2 Hz signal which is above Fmax and where the Nyquist limit is violated. The mirror image of 61.2 Hz is 41.2 Hz (102.4 Hz-61.2 Hz), i.e. below Fmax and appears in the spectrum as a normal signal. When the Nyquist limit is violated, the two peaks move further from the Fmax line as the signal frequency increases.

Figure 6 shows the FFT of a 173.4 Hz signal which is between 2*Fmax and 4*Fmax. The alias is 31.4 Hz, which is the mirror image about 2*Fmax. Another frequency appears at 71 Hz which is the mirror image of 31.4 Hz about Fmax.

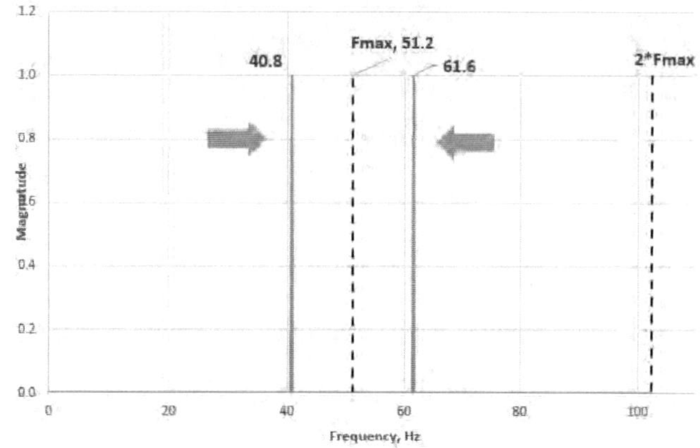

Figure 4 FFT of 40.8 Hz signal

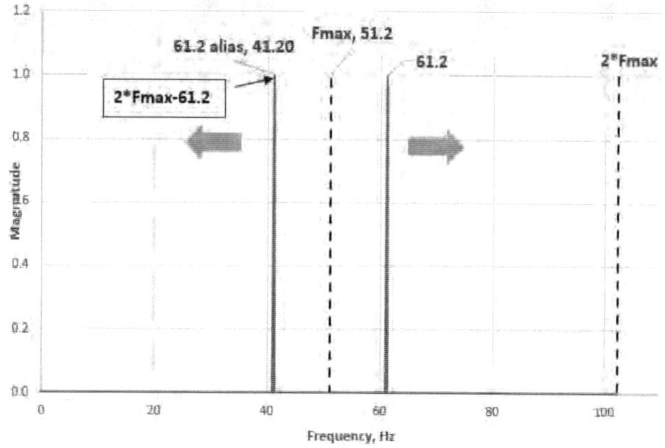

Figure 5 FFT of 61.2 Hz signal

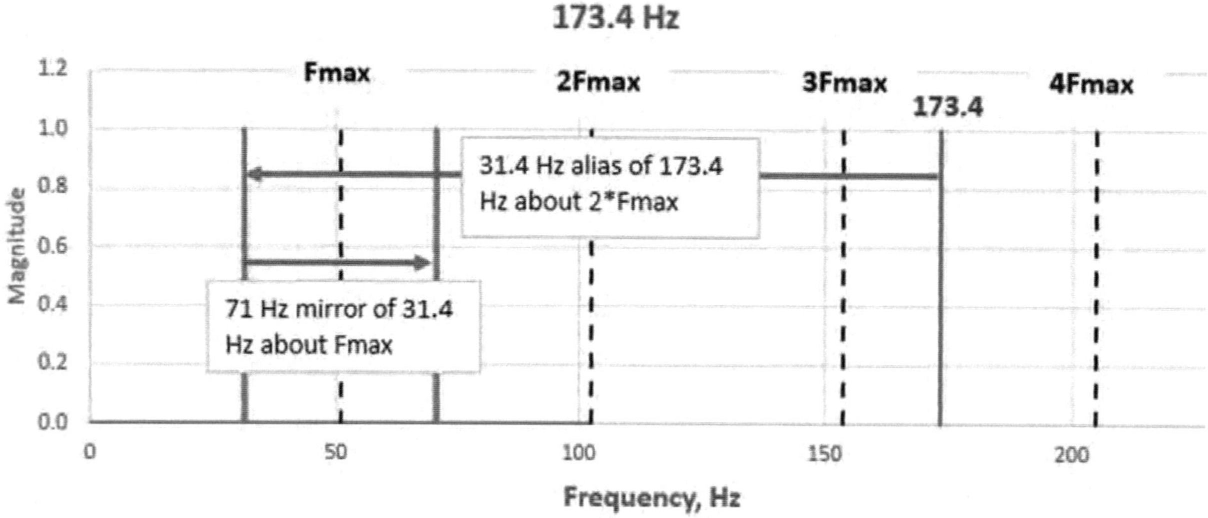

Figure 6: FFT of 173.4 Hz signal

Click to view increasing frequency

ODS without aliasing

Videos were taken on a gasoline powered generator in order to show the comparison of iPhone videos with high-speed camera videos. Figure 7 shows the spectrum for a Chronos 1.4 high-speed camera with a sample rate of 1,057 fps for which Fmax was 31,710 cpm. The observed frequencies for running speed of 3,510 rpm were 1750, 3500, 5270, and 7020 cpm, but no higher. ODS videos from the Chronos camera were provided previously [2][6].

Figure 8 shows a spectrum for an iPhone 14 with a sample rate of 240 fps. Observe the same four frequencies found with the high-speed camera[7]. Figure 9 through Figure 12 show the ODS movies for first order torsion, first order beam, second order torsion, and second order beam modes, respectively. The iPhone video ODS faithfully reproduces the same video ODS from a high-speed camera even though the iPhone signal is compressed.

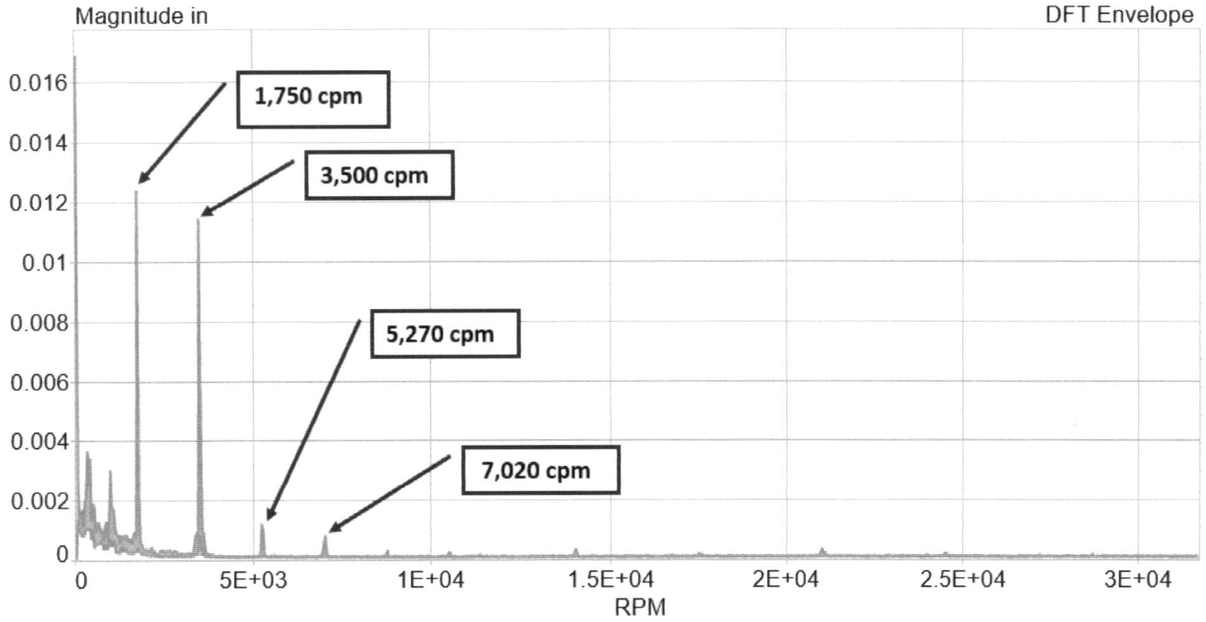

Figure 7: 1,051 fps High Speed Camera

[6] CTRL+click for animation on the member portal at full speed and half speed.

[7] The readings were taken at different times so that the frequencies are slightly different.

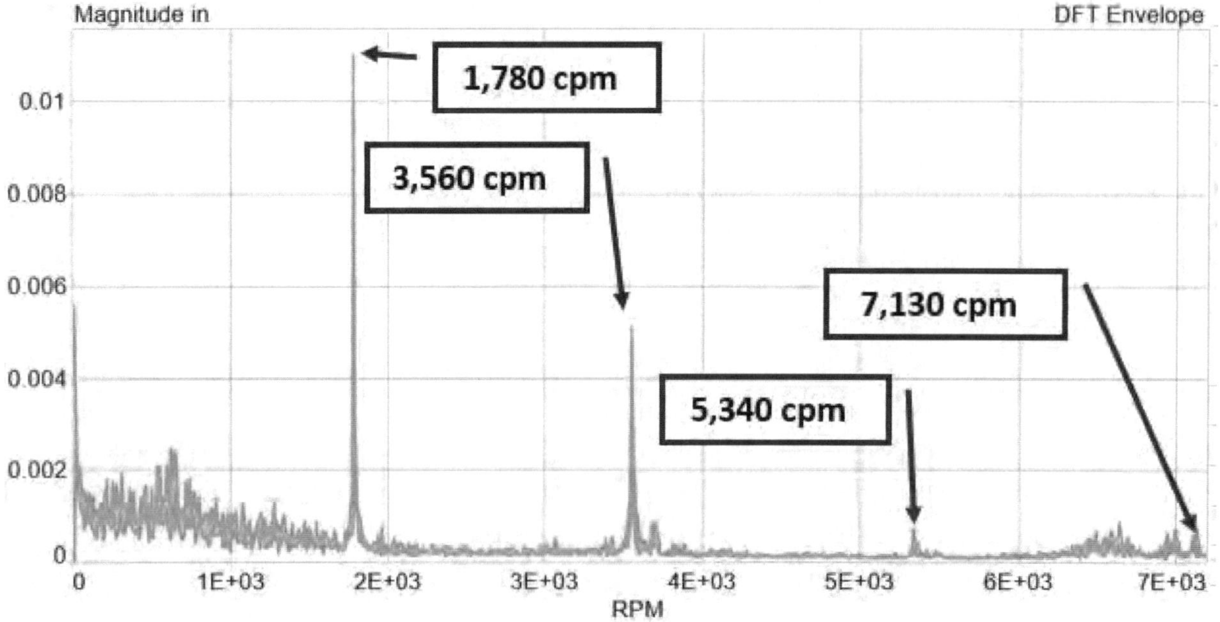

Figure 8: 240 fps iPhone

Figure 9: iPhone 240 fps 1,780 cpm

Click for animation

Figure 10: ODS iPhone 240 fps 3,560 cpm

Click for animation

Figure 11: ODS iPhone 240 fps 5340 cpm
Click for animation

Figure 12: ODS iPhone 240 fps 7,130 cpm

Click for animation

Cell phone ODS videos with aliasing

Figure 13 shows the spectrum for an iPhone with a sample rate of 30 fps with 900 cpm Fmax. The frequencies observed, 19.4 cpm and 38.8 cpm, do not match the expected machine frequencies found previously. Figure 14 is the unwrapped spectrum and shows the frequency 19.4 cpm is the alias of 1,781 cpm[8]. Figure 15 shows the ODS for the first torsion mode as expected. The case is the same as shown in Figure 5 which shows a frequency between Fmax and 2*Fmax.

The unwrapped spectrum in Figure 14 also shows the frequency of 38.8 cpm is an example of a mirror about 2*Fmax and is the alias of 3,561 cpm, which is just under 4*Fmax of 900 cpm. The extra frequency of 1,760 cpm is just the mirror of 38.8 cpm and is irrelevant to the analysis. The ODS video in Figure 16 is the first order beam as expected for 3,561 cpm. The case is the same as in Figure 6 which shows a frequency between 2*Fmax and 4*Fmax.

Table 1 summarizes the pattern for aliases. For a signal low pass filtered at Fmax, no alias is present. For a signal with frequency content up to 2*Fmax, the alias is the mirror about Fmax. For signals with frequency content between 2*Fmax and 4*Fmax, the alias is the mirror about 2*Fmax. For signals with content between 5*Fmax and 6*Fmax, the alias is the mirror about 6*Fmax. For signals with content from 4*Fmax to 5*Fmax and 6*Fmax and 7*Fmax, what is observed is the mirror of the alias about Fmax.

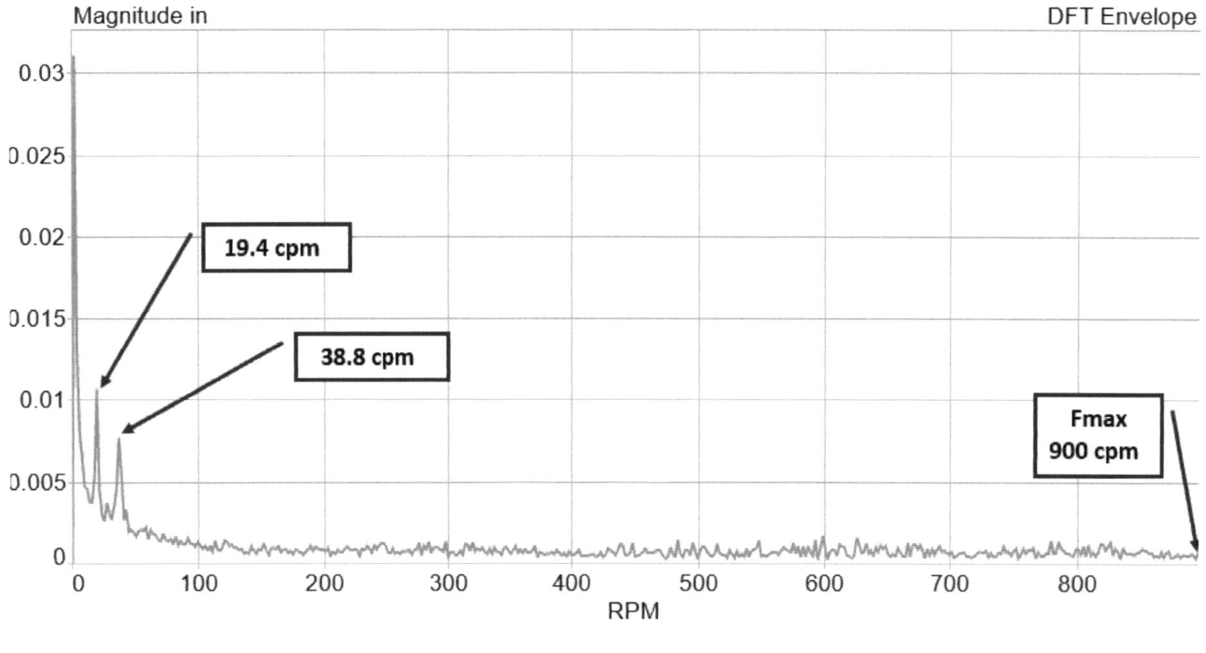

Figure 13: 30 fps iPhone

[8] The engine speed increased slightly.

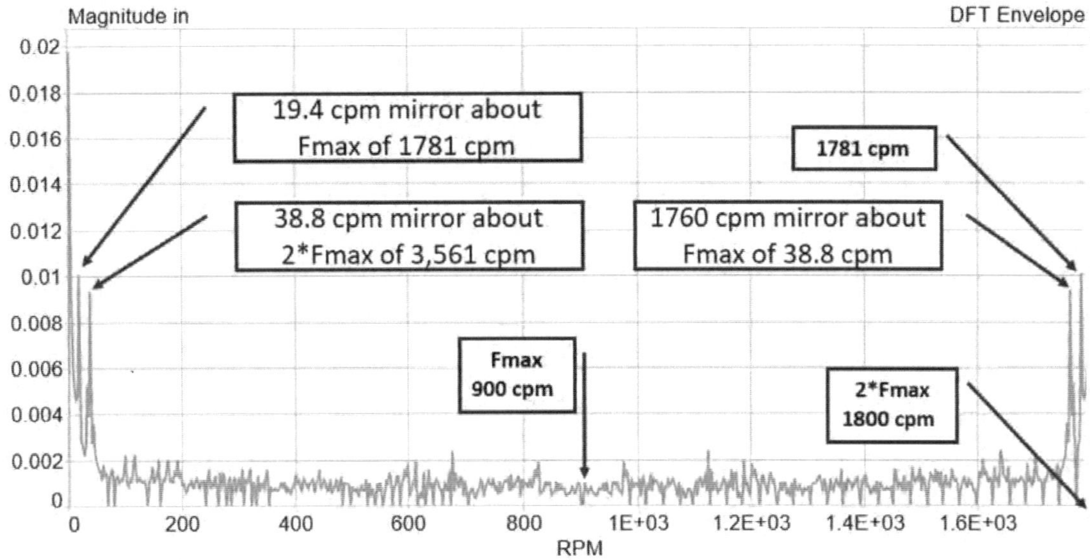

Figure 14: 30 fps iPhone unwrapped

Figure 15: ODS iPhone 30 fps aliased 1781

Click for animation

Figure 16: ODS iPhone 30 fps aliased 3561

Click for animation

Table 1 Pattern of aliases

Original frequency	Alias frequency	Frequency below Fmax
0 to Fmax	None	Original
Fmax to 2Fmax	Mirror about Fmax	Alias
2Fmax to 3Fmax	Mirror about 2Fmax	Mirror of alias about Fmax
3Fmax to 4Fmax	Mirror about 2Fmax	Alias
4Fmax to 5Fmax	Mirror about 4Fmax	Mirror of alias about 2Fmax

Mass loading

Figure 17 shows the effect of an accelerometer which caused decrease of an order of magnitude in vibration due to mass loading. Vibration was reduced by over 90% with the loading. Figure 18 shows ODS of parts that would be mass loaded if tested with accelerometers.

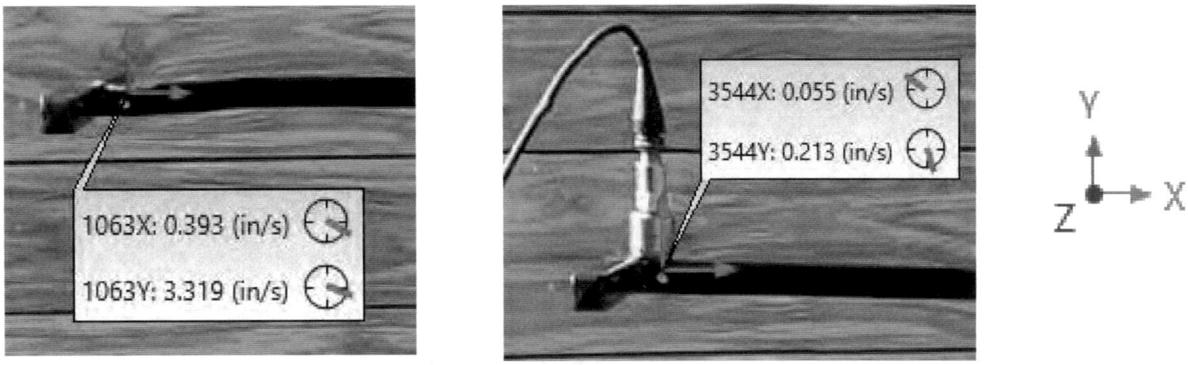

Figure 17: Mass loading

Figure 18: Largest motion on parts that would be mass loaded

Click for animation

Case studies

Case No. 1

Vibration with an accelerometer was taken on a vertical water pump and found elevated at the top of the motor with a value to 0.34 inch per second peak in a radial direction at the rotation speed. A video was taken with a cell phone camera as shown in Figure 19. From the ODS animation it is seen clearly that a high vibration spot is present at the foundation as shown in Figure 20. The spot turned out to be loose rust which might not have been otherwise noticed. The ODS video was useful to show to the to the operators the source of the vibration. The usual rule is to check the foundation for any and every vibration problem.

Figure 19 Motion left, neutral, right

Click for animation

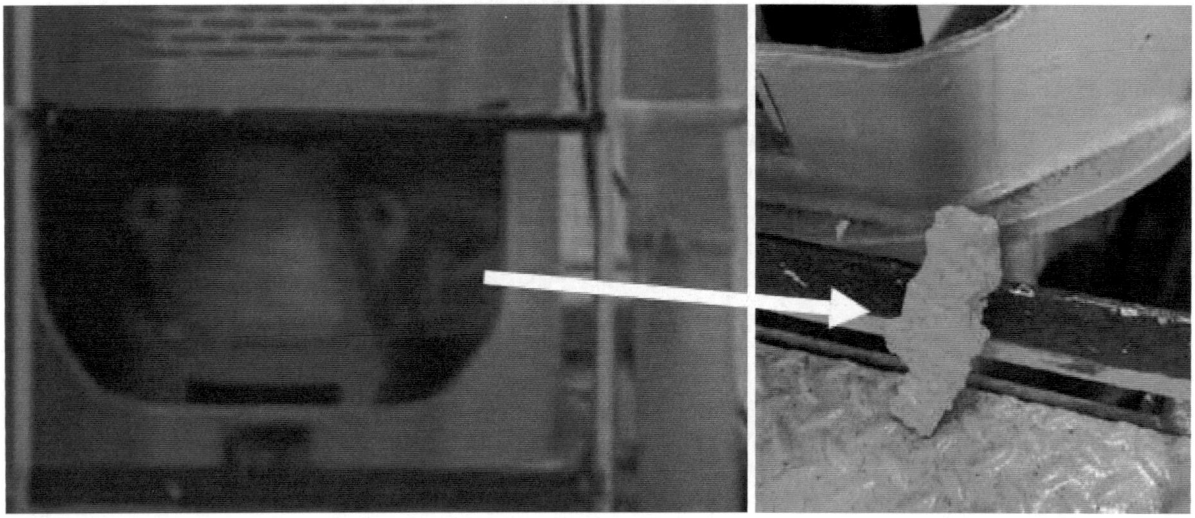

Figure 20 Rust identified at base

Case No. 2

Figure 21 shows the effect of vibration from one machine affecting another machine. Normally, machines that are not running are not tested. Without the ODS video, the foundation condition might not have been found. Numerous previous repairs had been done. Crew thought vibration was from nearby engine, but rotation rate was not related to the engine.

Figure 21 Adjacent machine effects

Click for animation

Case No. 3

Figure 22 shows flexibility in the foundation as seen below deck. Normally the surveyor would not go into the bilge, but in this case, animation showed vibration due to foundation flexibility. Vibration at rotation rate, which can be caused by imbalance or wear. The animation shows flexibility in the connections of the vertical supports at the deck. After seeing the animation, the chief engineer made arrangements to reinforce the foundation. Note stress in power leads to motor which would not have been found with accelerometer testing.

Figure 22 Foundation flexibility

Click for animation

Case No. 4

Figure 23 shows a case in which the fan motor was out of reach without a ladder. Note the casing has much less vibration than the motor and moves in the opposite direction from the motor. A loose foundation is indicated in additional possible motor wear.

Figure 23: Fan motor out of reach

Click for animation

Conclusions

Cell phone videos can be an effective tool for visualizing vibration with ODS animation. Low-frequency machine vibration due to imbalance, looseness, misalignment, and foundation flexibility are well within the ability of a cell phone. A cell phone video can be a convenient, inexpensive tool not requiring special lighting or attaching sensors and running wires to an analyzer. Aliasing is not a bad thing provided the frequencies above Fmax are known. Aliasing of running speed orders is easy to calculate. An aliased signal, even one greater than 2*Fmax, can be just as useful as an unaliased signal.

Acknowledgements

Maddison Brower, Ken Richardson, Eric Ryan and Chuck Stich, CAC Field Surveyors supplied the case studies.

References

[1] Liu, X., et al, "*Video-based IoT baby monitor for SIDS prevention*," IEEE Global Humanitarian Technology Conference (GHTC), October, 2017.

[2] Williams, M.J., "*The Next Big Thing: Video ODS*," Vibrations Magazine, Summer, 2021.

[3] "*Mobile phone*," [Online]. Available: https://en.wikipedia.org/wiki/Mobile_phone. [Accessed 31 January 2025].

[4] "*What is a Low-Pass Filter and How Does it Work?,*" [Online]. Available: https://petapixel.com/what-is-a-low-pass-filter/. [Accessed 26 February 2025].

Bibliography

Zeng, K. "*Characterizing Perceptual Artifacts in Compressed Video Streams*," in Proceedings of SPIE - The International Society for Optical Engineering, San Francisco, 2014.

Bing, B., *Video Coding and Online Streaming Technologies*, Bing, 2021.

STRUCTURAL DYNAMIC MODIFICATION FOR REDUCING BEARING COVERS VIBRATION OF 1,000MW LARGE STEAM TURBINE

Authors: Hwajun Jung, Jongmin Kim, Kilsu Ryu,
Mikyung Hwang, Sung Ho Lee

VIATC TRAINING CONFERENCE & EXPO
NEWPORT NEWS, VIRGINIA | AUGUST 6-8, 2025

VIATC TRAINING CONFERENCE & EXPO
NEWPORT NEWS, VIRGINIA | AUGUST 6-8, 2025

STRUCTURAL DYNAMIC MODIFICATION FOR REDUCING BEARING COVERS VIBRATION OF 1,000MW LARGE STEAM TURBINE

Hwajun Jung†. Jongmin Kim*. Kilsu Ryu**. Mikyung Hwang***. Sung Ho Lee****
† Corresponding Author; Kepco KPS *, **, *** Kepco KPS Solution Center

Hwajun Jung
KEPCO KPS Solution Center Vibration Engineering Team Team Manager
211, Munhwa-ro, Naju-si, Jeollanam-do, 58326 Republic of Korea
18 years' experience supporting rotating equipment commissioning, balancing, Modal Testing
holds a Cat III certification, instructor of the CAT III training courses

Abstract.

In the context of power plants, turbines and generators represent the most critical equipment due to their substantial size and rapid rotation. To ensure the continuous monitoring of equipment vibration, dual-probe vibration sensors are installed on each bearing cover. These sensors are capable of measuring the vibration of both the shaft and the cover, thereby facilitating constant checks on mechanical soundness. The present study investigates an instance of excessive vibration in a turbine bearing cover during its operation at rated output. In order to identify the root cause of this phenomenon, a range of analytical techniques were employed, including Frequency Analysis, Operational Deflection Shape (ODS), Modal Analysis, and Excitation Test. The analysis results confirmed that there was an excessive vibration response compared to the excitation force at the vibration sensor fixing part.

In order to resolve the issue of vibration, a finite element model was constructed, and the structural reinforcement plan was simulated whilst altering the configuration of the additional rib. The simulation results demonstrated that reinforcing the ribs on the inner and outer surfaces of the bearing cover could reduce the vibration of the bearing cover. Subsequent to the implementation of this improvement plan in the field and an evaluation of the vibration condition, it was ascertained that the vibration of the bearing cover had been reduced by 62%.

The findings of this study demonstrate the efficacy of structural reinforcement in addressing the high vibration problem of turbine bearing covers.

Key Words: Modal Analysis, ODS, Resonance, Steam turbine, Structural Dynamic Modification (SDM)

1.0 Introduction

In the context of power generation facilities, turbines are of particular significance, and vibration sensors are therefore installed on each bearing cover to ensure constant monitoring of the vibration status. The installed vibration sensors are dual probes capable of simultaneously measuring displacement and velocity vibration, and thus acquiring vibration data from the shaft and cover. It is imperative to ensure that the design and manufacture of the bearing cover precludes the occurrence of excessive vibration, as this may result in an unexpected cessation of turbine operation if the set alarm value is exceeded.

The present study employs a range of analytical techniques, including Frequency Analysis, Operational Deflection Shape (ODS), Modal Analysis and Excitation Test, in order to identify the root cause of the vibration. The study proposes a structural reinforcement solution through finite element analysis, with the aim of improving vibration. It is important to note that the maximum vibration of the bearing cover, which occurs at a rate of 15.6 mm/s 0-pk during turbine rated output operation, is a key factor in the study.

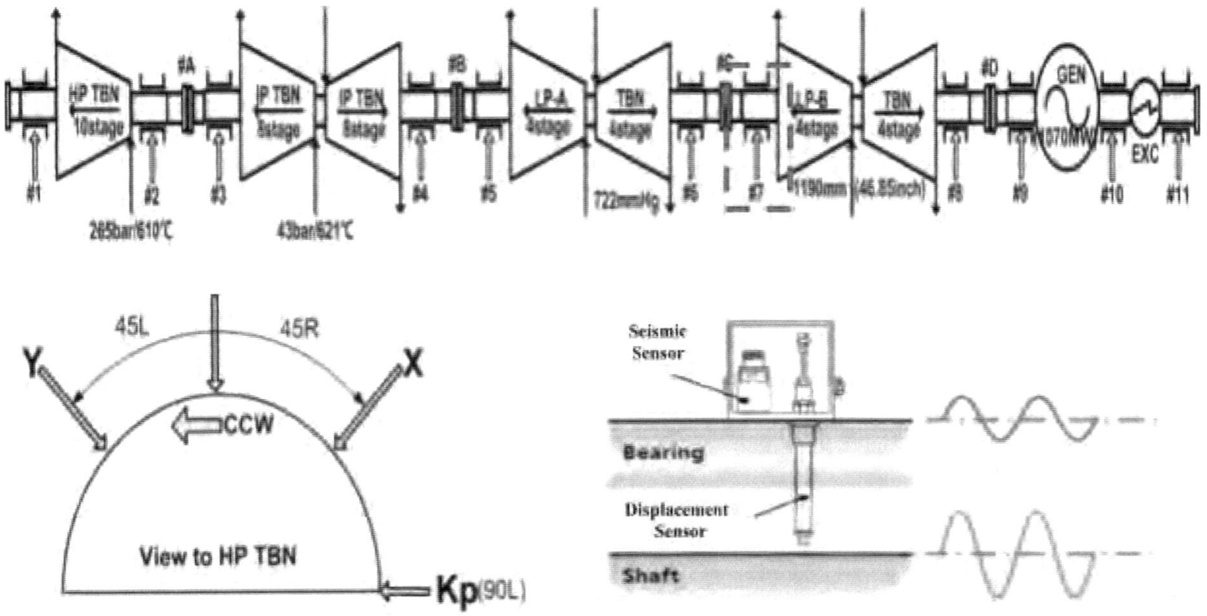

Figure SEQ Figure * ARABIC 1: Facility Schematic and Vibration Sensor on Cover

2.0 Diagnosis of Vibration Issue

2.1 Frequency Analysis

In order to identify the cause of the high vibration of the bearing cover, a frequency analysis of the shaft and casing vibrations was conducted. This analysis revealed that the main component of the shaft and casing vibration was 1X, which is the same as the shaft rotation speed (3,600 rpm), as shown in Figures 2 and 3. Furthermore, a review of the vibration history revealed that the vibration had partially improved following the manufacturer's reinforcement of the bearing cover structure during the construction test run. Consequently, it was deduced that the vibration was attributable to an increase in excitation force resulting from elevated turbine shaft vibration (Alarm: 150um pk-pk), and the high vibration was attributed to the inadequate rigidity of the bearing cover.

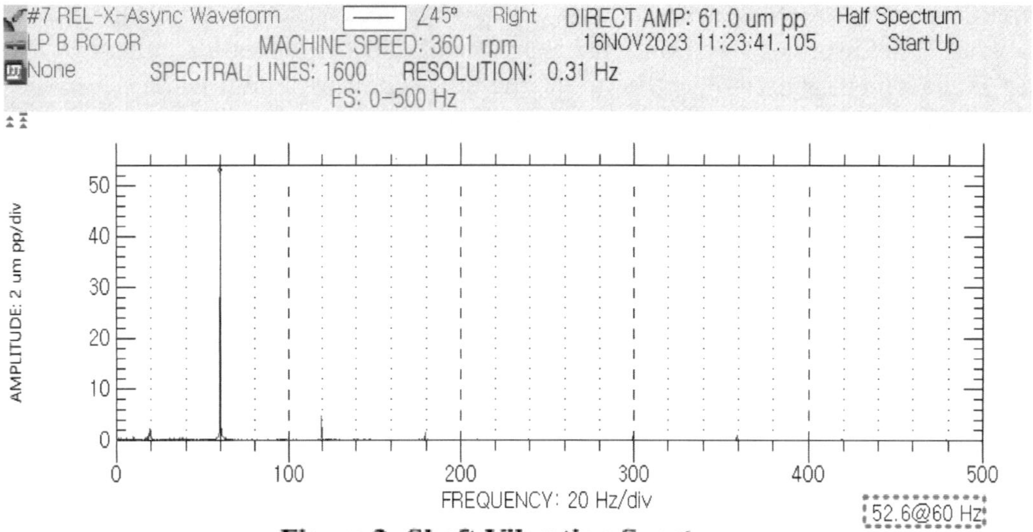

Figure 2: Shaft Vibration Spectrum

2.2 Operational Deflection Shape (ODS)

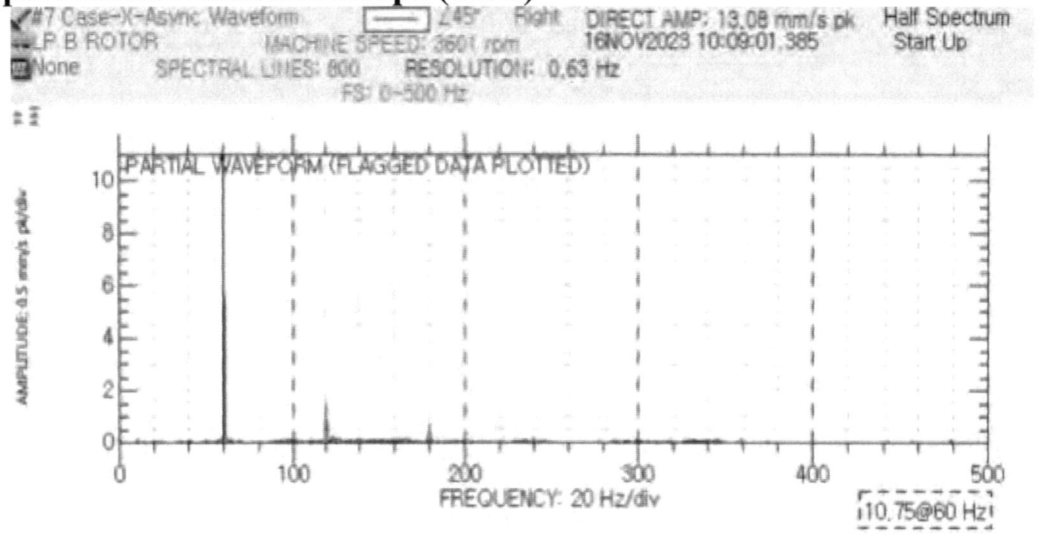

Figure 3: Cover Vibration Spectrum

The ODS is a test that measures the vibration shape related to the number of rotations and its multiples while the equipment is in operation. The bearing cover was divided into measurement points as shown in Figure 4 to measure the vibration. As a result of the test, as shown in Figures 5 and 6, it was confirmed that the main vibration component of the bearing cover was 60Hz and that a large amount of deformation occurred in the vibration sensor area. In addition, it was used as basic data for the location of rib attachment when reinforcing the structure in the future.

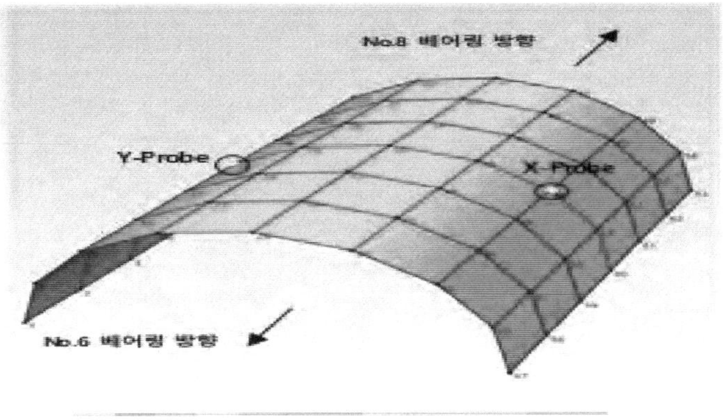

2.3 Modal Analysis

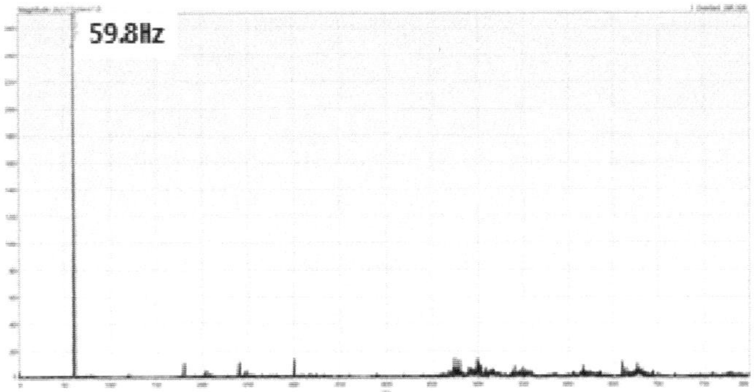

Modal Analysis is a test used to identify the natural frequency, damping, and vibration mode of the equipment. It does this by measuring the vibration response to excitation in a stationary state. The test was performed in two different states: first, in the assembled state (fixed), and second, in

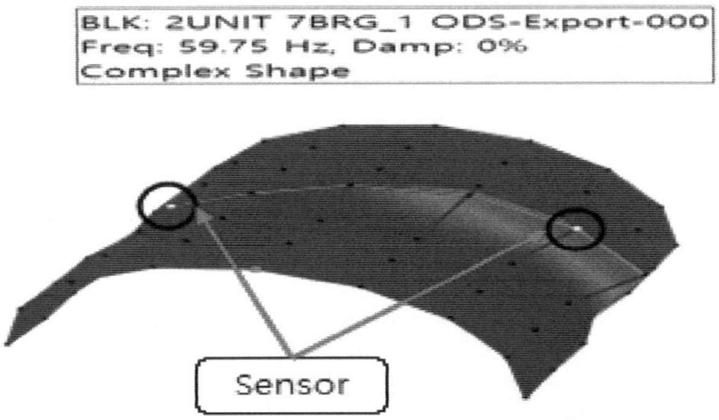

the disassembled and lifted state (free-free) by a crane on site. As illustrated in Figure 7, the bearing cover was subjected to excitation through the utilisation of an impact hammer at the designated fixed excitation point. Concurrently, the vibration response was measured using an acceleration sensor positioned at the same measurement point as the ODS. The measured data was then utilised to derive the mode shape, natural frequency, and other pertinent parameters through the implementation of a dedicated modal analysis program.

Following the modal analysis, the natural frequencies of the assembled and disassembled states were ascertained and are presented in Table 1. The mode shapes are illustrated in Figure 8.

Table 1: Bearing Cover Natural Frequency

Order	Fixed	Free-Free
1st	165Hz	159Hz
2nd	325Hz	283Hz

2.4 Excitation Test

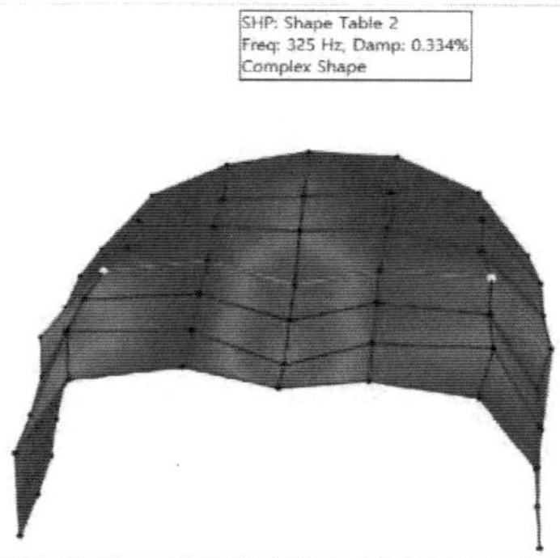

The excitation test was conducted to compare the vibration response of the vibration sensor installation site before and after structural reinforcement. The bearing cover was continuously excited with a constant magnitude of excitation force by an eccentric exciter. After installing the eccentric exciter on the bearing cover before structural reinforcement (see Figure 9), the vibration was measured to be 5.7 mm/s rms.

2.5 Finding

The findings of the vibration source analysis demonstrated that the primary component (60 Hz) and natural frequency (325 Hz) of the bearing cover vibration were distinctly separated, indicating that the vibration did not undergo amplification due to resonance. Rather, it was hypothesised that there was an issue with the dynamic structure of the bearing cover in response to the increased excitation force. This finding served as the foundation for the subsequent structural optimisation strategy.

3.0 Finite Element Analysis (FEA)
3.1 3D Solid Model

In order to undertake a Finite Element Method (FEM) analysis, the bearing cover prior to structural reinforcement was subjected to 3D scanning in order to create an analysis model (see Figure 10). The characteristics of the boundary conditions (i.e. material, elastic modulus and constraints) were defined and selected as tuning models so as to ensure that the FEA results of the bearing cover and the Experimental Modal Analysis (EMA) results were as close as possible. The FEA results and the EMA results by boundary conditions are shown in Table 2 and Figure 11.

Table 2: Modal Analysis Result (FEA vs. EMA)

Division	Free-Free	Fixed
Finite Element Analysis	146Hz	171/331Hz
Experimental Modal Analysis	159Hz	165/325Hz

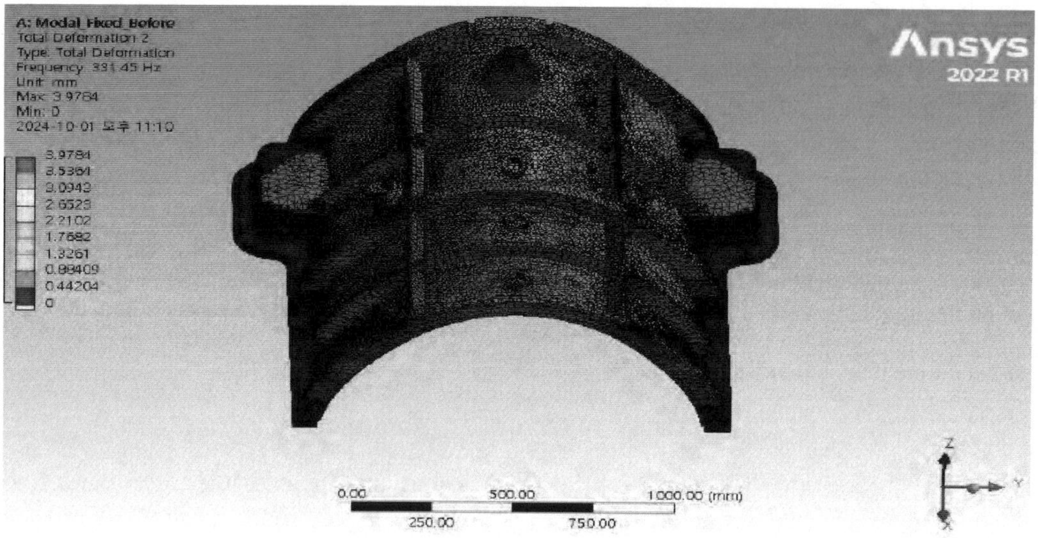

The finite element analysis results of the bearing cover and the field test results are very similar, indicating that the modelling and boundary conditions applied for the finite element analysis are consistent with the dynamic characteristics of the actual structure.

3.2 Simulation of Structural Enhancements

Following a comprehensive review of the ODS test results and other power plant installation cases, it was determined that the introduction of additional ribs to reinforce the weak rigidity of the bearing cover would be an effective measure for reducing vibration. As demonstrated in Table 3 and Figures 12-14, simulations were performed while altering the shape of the additional ribs to evaluate the changes in the natural frequency and vibration response of the bearing cover. The simulation results predicted that applying Case C would increase the natural frequency and reduce the vibration during operation from 15.66 mm/s 0-pk to 8.8 mm/s 0-pk.

Table 3: Before/After Adding Ribs to Bearing Cover

Division	Before (Original)	After Adding Ribs		
		Case A	Case B	Case C
Natural Frequency (Hz)	171/331	171/346	204/371	203/371
Vibration (mm/s 0-pk)	15.6	14.6	9.2	8.8

Following a thorough evaluation of the available options, it was determined that the optimal approach to enhance the vibration would be to affix supplementary ribs to the inner and outer surfaces of the bearing cover, as illustrated in Case C.

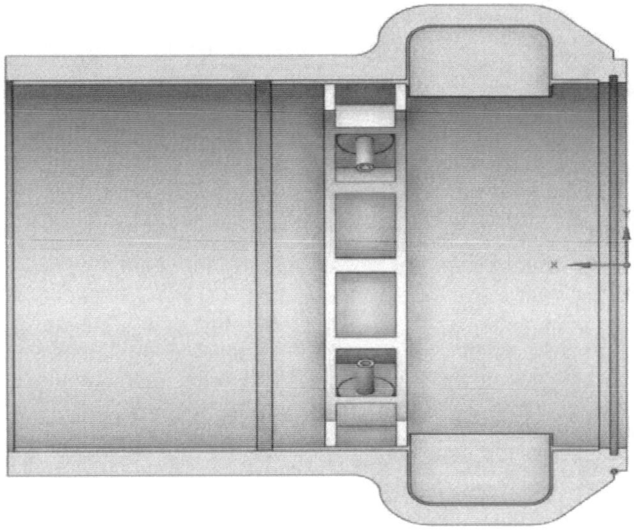

4.0 Field Implementation and Validation
4.1 Field Implementation

Following the fabrication of the additional ribs, these were applied by means of welding them to the bearing cover. In order to prevent thermal deformation of the bearing cover during the welding process, a jig was installed, as illustrated in Figure 15. Subsequent to the welding of the inner and outer ribs, a non-destructive test was performed, as demonstrated in Figures 16 and 17.

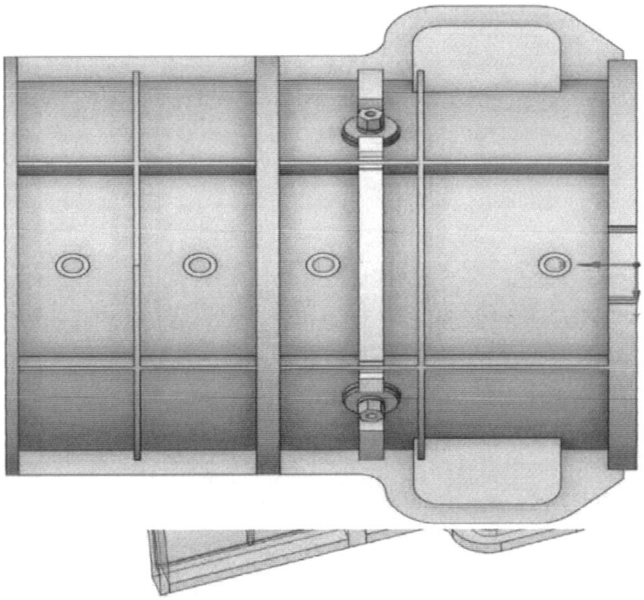

4.2 Modal Analysis (After Structural Reinforcement)

Modal analysis was performed on the bearing covers with structural reinforcement in order to ascertain the altered natural frequencies. The results, when compared to those prior to reinforcement, are demonstrated in Table 4.

Table 4: Natural Frequency Before/After Structural Reinforcement

Division	Fixed
Before (Hz)	165/ 325
After (Hz)	204 / 357
Shift (Hz)	↑ 39 / 32
Simulation (Hz)	203 / 371

The natural frequency of the assembled structure increased by more than 30 Hz following reinforcement, a result that is similar to the finite element analysis.

Furthermore, an analysis of the frequency response function (FRF) plots, both prior to and following structural reinforcement, as illustrated in Figures 18 and 19, respectively, substantiates that the magnitude of the bearing cover's response to excitation forces undergoes a substantial diminution subsequent to structural reinforcement.

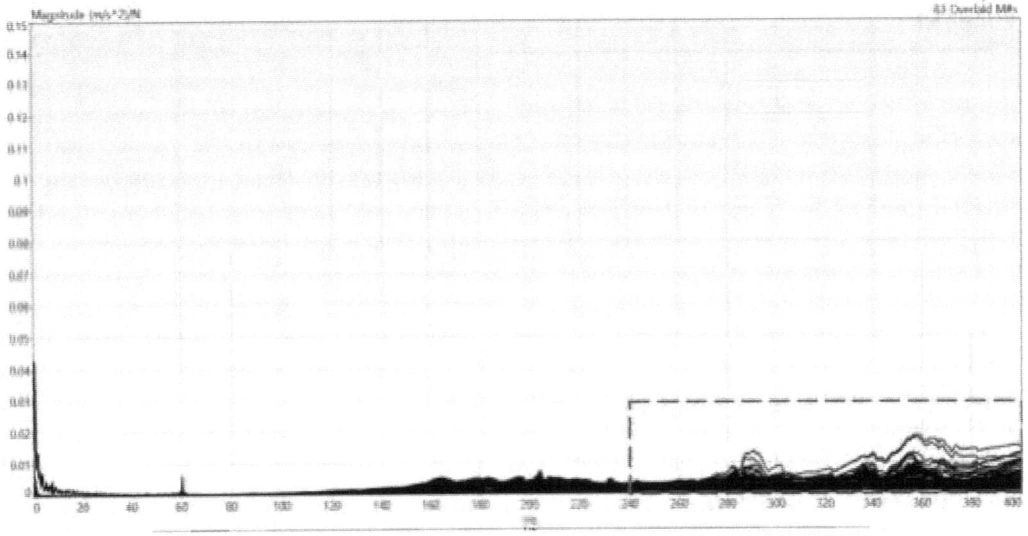

4.3 Excitation Test (After Structural Reinforcement)

Following the implementation of an improved bearing cover, an eccentric vibrator was installed on the bearing cover in the same position as prior to the structural reinforcement. The vibration at the vibration sensor installation site was measured whilst applying the same excitation force. The results, in comparison to those prior to the structural reinforcement, are presented in Table 5.

Table 5: Results of the Test Before/After Structural Reinforcement

	Before	After	Reduction
Vibration X/Y (mm/s rms)	5.7 / 5.5	1.4 / 1.3	4.3 / 4.2

Following the implementation of structural reinforcement measures, it was ascertained that the vibration of the bearing cover vibration sensor installation area was diminished by more than 4.0 mm/s rms.

5.0 Vibration Values during Output Operation

As illustrated in Table 6, the vibration of the bearing cover was measured during the test run following OH at an output of 1030 MW in comparison to the state prior to reinforcement.

Table 6: Vibration for Output Before/After Structural Reinforcement

	Output (MW)	Vibration (X/Y, mm/s 0-pk)
Before	950	14.6 / 7.8
After	1030	**6.0 / 5.1**

6.0 Conclusion

After field application, the vibration status of the bearing cover and the rotating shaft during normal operation, the size of the response to the excitation force, and the vibration status before and after the structural improvement were evaluated. Finally After field application, the actual vibration was 6.0㎜/s, which was a 62% reduction from the previous result. This solved the high problem of the turbine bearing casing and the equipment instability.

7.0 Reference

- "Guidelines for the Uprate Design of Turbine-Generator Rotor Support Systems", EPRI 1024579 TECHNICAL REPORT, 2011
- "Steam Turbine-Generator Bearing Vibration Diagnostic Guide", EPRI 3002016244, 2019
- Donald E. Bently, "Fundamentals of Rotating Machinery Diagnostics", 2002
- Kilsu Ryu, Seungwoo Han, & Hwajun Jung, 2019. "Added-Mass Design of Steam Turbine Casing for Reduction of Resonance". Transactions of the Korean Society for Noise and Vibration Engineering.

- Hyoung Young Lee, Seong Ho Lee, & Hwajun Jung, 2023. “A Case Study for Vibration Reduction of Vertical Pump with Modal Analysis and FEM”.

USING MODE SHAPES FROM CELL PHONE VIDEOS FOR MACHINERY HEALTH MONITORING

Authors: Shawn Richardson & Mark Richardson

VIATC TRAINING CONFERENCE & EXPO
NEWPORT NEWS, VIRGINIA | AUGUST 6-8, 2025

VIATC TRAINING CONFERENCE & EXPO

NEWPORT NEWS, VIRGINIA | AUGUST 6-8, 2025

USING MODE SHAPES FROM CELL PHONE VIDEOS FOR MACHINERY HEALTH MONITORING

Shawn Richardson & Mark Richardson
Vibrant Technology, Inc.

Shawn Richardson is manager of Special Projects at Vibrant Technology. Mark Richardson is a co-founder and President of Vibrant Technology. Vibrant sells and supports the MEscope family of post-test noise and vibration analysis software tools.
Mark joined Hewlett Packard Co., Santa Clara, California in 1973, where he directed the development of the first commercially available FFT-based modal testing system. Following that, he directed the development of the first modal testing instrument, the Hewlett Packard 5423A Structural Dynamics Analyzer.
In 1979, Mark co-founded Structural Measurement Systems, Inc. SMS developed and marketed the popular StarModal and StarStruct software packages.

Abstract

Use of a cell phone video provides a low-cost complement to traditional accelerometer-based methods for monitoring the health of plant operating equipment. The video enhancement technology in modern-day cell phones renders them ideal for use as a convenient measurement device for use in a plantwide route-based machine monitoring program.

To demonstrate the use of this technology, several unbalance cases were created in a rotating machine by adding screws to its rotors. Then while the machine was running, a cell phone video was recorded during each unbalanced condition. Mode shapes were extracted from each video, labeled with their unbalanced condition, and stored into an archival database. Then a database search method called **FaCTs™** was used to uniquely identify each unbalance condition based on the **OMA** mode shapes of the first three orders of the machine.

Key Words

Time Waveform (**TWF**), Digital Fourier Transform (**DFT**), Auto Spectrum (**APS**), Cross Spectrum (**XPS**), **ODS-FRF**, Operating Deflection Shape (**ODS**), Operational mode shape (**OMA** mode shape), Degree of Freedom (**DOF**), neural network (**NN**), Fault Correlation Tools (**FaCTs™**), Frames Per Second (**fps**)

Rotating Machine

In this paper, **FaCTs™** is used to uniquely identify ***five different unbalance cases*** of the rotating machine shown in Figure 1. The machine has a variable speed motor connected to the rotor with a coupler. The motor speed was adjusted to ***approximately 1000 RPM*** and remained there throughout all the cell phone video recordings.

Figure 1. Rotating Machine Showing Unbalance Screws Added to Its Rotors

Introduction

Most manufacturing plants worldwide have implemented route-based monitoring programs for accessing the health of their rotating machinery and associated equipment. Digital vibration signals are the primary data used for machinery health monitoring.

Traditionally, machine monitoring has been done by attaching accelerometers to the surfaces of the operating equipment and acquiring signals from the accelerometers with a portable digital spectrum analyzer. Compared to recording a cell phone video, using accelerometers for data acquisition is more expensive and time-consuming to implement. Furthermore, because it is non-contacting, a cell phone can record vibration of machine parts that are too hot, too dangerous, or are inaccessible for attaching accelerometers.

In previous papers [1] to [4], traditional signal processing methods were applied to the **TWFs** extracted from the frames of a video. Both **TWFs** and their associated **DFTs** can be used to display either **time-based** or **frequency-based ODS**'s allowing vibration to be visualized at slower speeds with higher amplitudes.

A new database search method called **FaCTs™** was introduced in a previous paper [2]. In this paper, **FaCTs™** is used to identify different unbalance conditions of a rotating machine using only its mode shape **DOFs** from two points on the motor and the tops of its two bearing blocks. **FaCTs™** uniquely identified all five unbalance conditions using only the **OMA** mode shapes of the first two machine orders.

FaCTs™ functions in the same manner as a trained neural network. It uses the current mode shapes of a machine together with a shape correlation coefficient, called the shape difference indicator (**SDI**), to search in an archival database for labeled mode shapes that ***closely match*** the current mode shapes. Each mode shape in the database is labeled with the machine condition and a time stamp when it was archived.

FaCTs™ displays a bar chart of the **SDI** values of the current mode shapes with the ***closest matching*** archived mode shapes, thereby defining the most likely current mechanical condition of the machine based on its current mode shape values.

Artificial Intelligence (Machine Learning)

Artificial Intelligence, popularly known as **AI**, uses a ***trained*** neural network (**NN**) to interpret the meaning of a set of data. However, to accurately diagnose a mechanical fault, an **NN** must be trained with a ***lot of vibration data***.

A neural network is a machine learning model that uses a network of interconnected nodes, or artificial neurons, to process data in a way that mimics the human brain. It uses a machine learning (ML) process of many interconnected nodes in a layered structure that resembles the human brain. A neural network is depicted in Figure 2

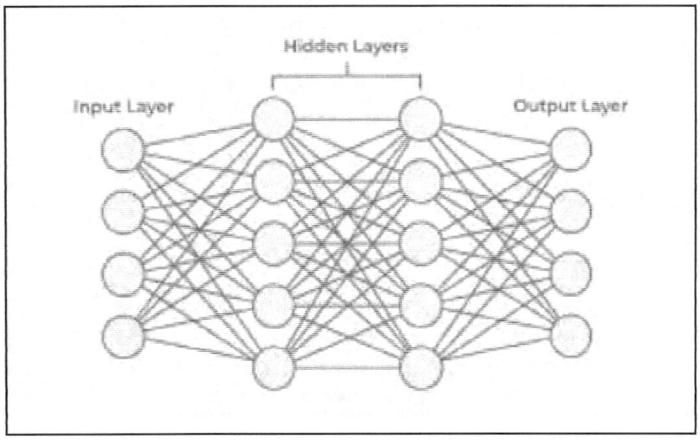

Figure 2. A Neural Network

To diagnose mechanical faults, an **NN** must be trained with data that is ***uniquely associated*** with a mechanical fault. Then when vibration data is input to a trained **NN,** called an Inference Engine, it diagnoses a mechanical fault, as shown in Figure 3. But a drawback of using a trained NN is that ***lots of labeled vibration data*** is required to train an **NN**.

Vibration data in the form of **TWFs**, **DFTs**, **ODS-FRFs**, **ODS's** or **Mode Shapes** can be used to characterize machinery vibration. Using **OMA** mode shapes obtained from a cell phone video together with **FaCTs**, machine health monitoring can be done in a much more efficient manner.

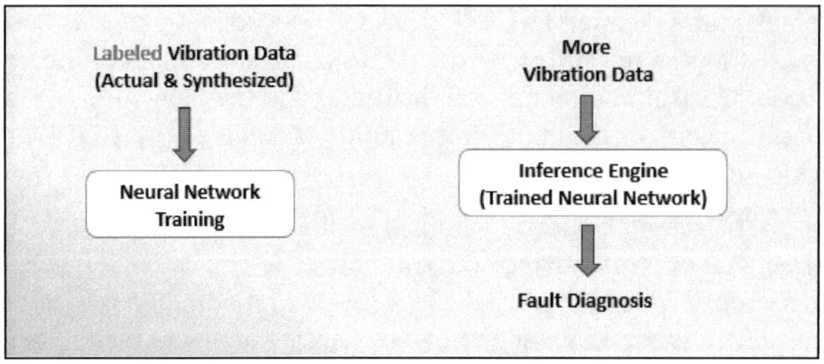

Figure 3. NN Training & Inference Engine

FaCTs™

FaCTs™ [4] is a database search algorithm used in MEscope [10]. It searches for ***labeled*** mode shapes; each mode shape labeled with ***a particular machine condition or mechanical fault***. **FaCTs** then displays a bar chart of **FaCTs** values for those mode shapes that ***closely match*** a current set of mode shapes. The bar chart displays the ***mechanical fault and time stamp*** associated with the archived shapes that ***most closely match*** the current mode shapes.

FaCTs™ uses an algorithm called the Shape Difference Indicator **(SDI)** [8]. **SDI** calculates a correlation coefficient between ***two complex-valued*** shape vectors. **SDI** is used to search the archival database for the mode shapes that have the ***highest* SDI *value*** with each current mode shape.

- **FaCTs** values are between **0.0 & 1.0**
- **FaCTs = 1.0** □ two mode shapes are ***identical***
- **FaCTs >= 0.9** □ two mode shapes are ***similar***
- **FaCTs < 0.9** □ two mode shapes are ***different***

Using Mode Shapes with FaCTs

To use mode shapes with **FaCTs**, each set of mode shapes in the archival database must be labeled with a machine condition, (balanced, bearing replacement, unbalance, aligned, passed inspection, etc.). Each set of shapes is also given a time stamp when it is stored into the database. Like an **NN**, **FaCTs** will also become more accurate as more labeled mode shape data is stored in the archival database.

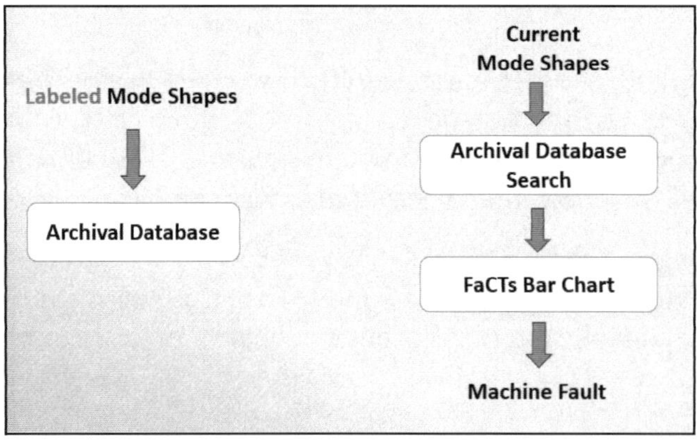

*Figure 4. **FaCTs** Archival Database Search*

Then when a new set of mode shapes is acquired, **FaCTs** is used to search the archival database and display a **FaCTs** bar chart, ordered from the ***closest matching*** mode shapes to the ***least matching*** mode shapes. The shapes with the ***highest*** **FaCTs** ***values*** have the ***highest correlation*** with the current mode shapes. This search process is depicted in Figure 4.

FaCTs Bar Chart

A **FaCTs** bar chart shows the ***closest matches*** of a set of ***current*** mode shapes with sets of ***labeled*** mode shapes already stored in the archival database. The search method uses a correlation coefficient called the **Shape Difference Indicator** (**SDI**) [8] which measures the difference between two mode shapes. A typical **FaCTs** bar chart is shown in Figure 5.

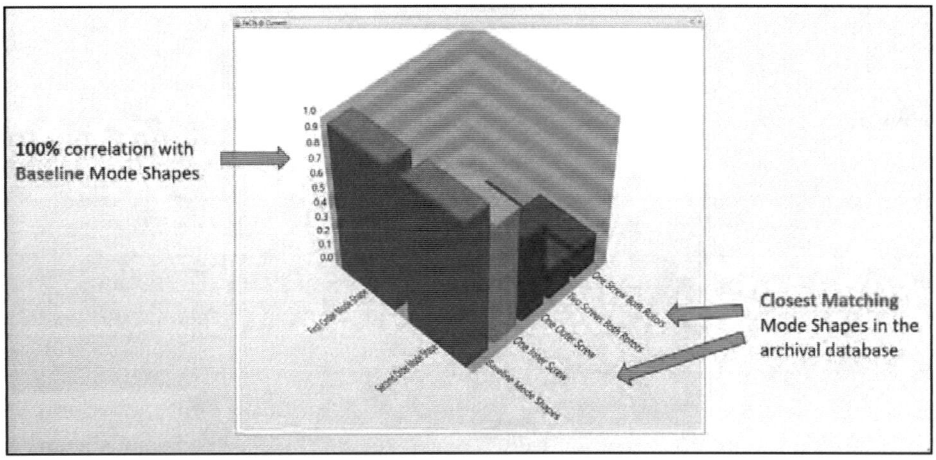

*Figure 5. **FaCTs** Bar Chart*

TWFs & DFTs

When a video is processed in MEscope [8], a rectangular grid of points with rectangular surfaces between them is created. Frames of the video are attached to the rectangular surface during an animated display of the **ODS's** extracted from the video.

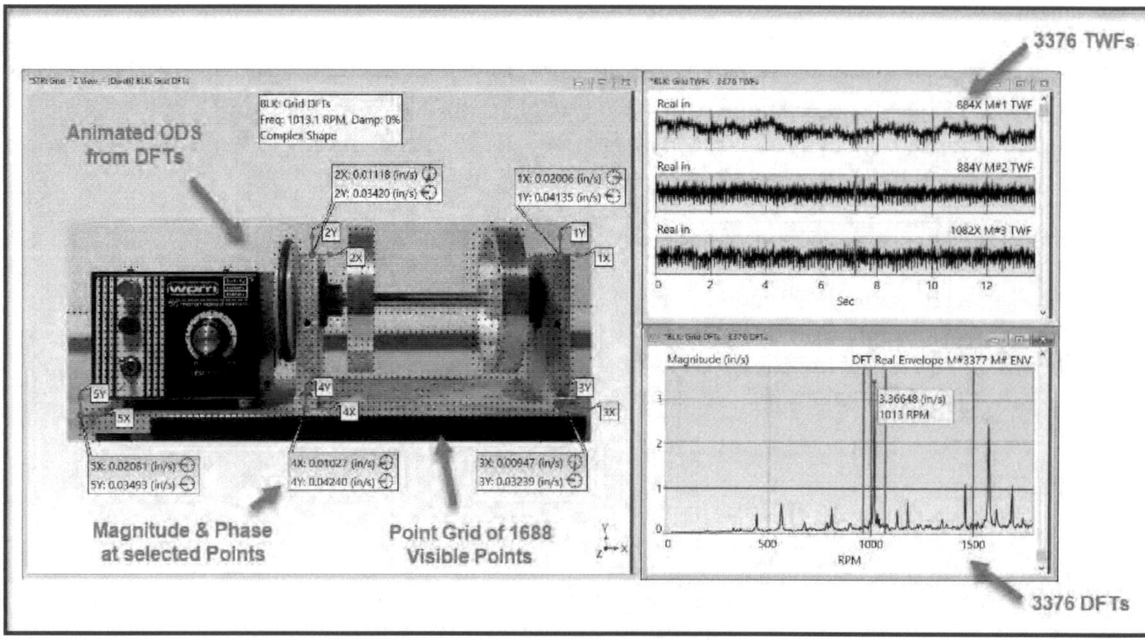

Figure 6. First-Order Frequency-based ***ODS*** *Animated from* ***DFTs***

Using a rectangular point grid, the displacements of ***millions of pixels*** per video frame are processed to extract the **TWFs** of the horizontal & vertical motion of ***thousands of points*** in the point grid.

Grid points with ***little or no motion,*** (like background points), are hidden and their **TWFs** are removed from further analysis. A point grid with background points hidden is shown in Figure 6.

Time-based **ODS**'s are displayed in animation from the **TWFs** using a sweeping Line cursor. Each frame of the raw video corresponds to a sample of the **TWFs**, so frames of the raw video are displayed on the point grid during sweep animation through the **TWFs**.

A **DFT** is calculated for each **TWF** that is extracted from the video. The frequency-based **ODS** at the cursor position in the **DFTs** is displayed in animation using sinusoidal modulation of the **ODS** values. The magnitude & phase of the **ODS** at selected points on the point grid can also be displayed during **ODS** animation, as shown in Figure 6. This feature enables the examination of ***magnitude & phase differences*** between various points on the test article during **ODS** animation.

Scaling the TWFs

During extraction of **TWFs** from a raw video, the **TWFs** are scaled to displacement units by choosing two points ***in or close to*** the plane of the video and entering the distance between them in length units, The **TWFs** are then scaled to the chosen displacement units. When scaled into displacement units, both **TWFs** and **DFTs** can be accurately ***differentiated*** to velocity units or ***double differentiated*** to acceleration units.

ODS-FRFs

A unique frequency domain function, called an **ODS-FRF,** can be calculated from each response **TWF** extracted from a video. A set of **ODS-FRFs** calculated from the **TWFs** is typically more accurate than their **DFTs** because ***spectrum averaging*** can be used to reduce extraneous noise, and a Hanning window applied to reduce leakage in the **ODS-FRFs**.

The ***magnitude*** of an **ODS-FRF** is the value of the **APS** of the response **DOF** at a grid point. The ***phase*** of the **ODS-FRF** is the phase of the **XPS** between the response **DOF** and the **DOF** of a **reference** grid point.

ODS-FRFs carry the ***same displacement units*** as the response **TWFs** from which they are calculated. A typical **ODS-FRF** in displacement units is shown in Figure 7. Because it is a frequency domain function, an **ODS-FRF** can be ***accurately differentiated*** to ***velocity units*** by multiplying it by the frequency variable.

Vibration levels in ***velocity units*** are commonly used by vibration analysts to assess potentially harmful vibration levels in rotating equipment.

Law of the FFT

One of the laws of the **FFT** algorithm is that **Δf = 1/T**, where **Δf** is the ***frequency difference between samples*** of an **ODS-FRF**, and **T** is the ***time length*** of **TWF** data from which the **ODS-FRF** was calculated. For example, if an **ODS-FRF** is calculated from **TWF** samples that span a **15 second** period, the frequency resolution (**Δf**) of the **ODS-FRF** is **60/15 = 4 RPM.**

To increase the frequency resolution of **ODS-FRFs**, **TWF** data ***over a longer time period*** is required. Therefore, the video from which the **TWFs** are extracted must be recorded over a ***longer period*** **T**.

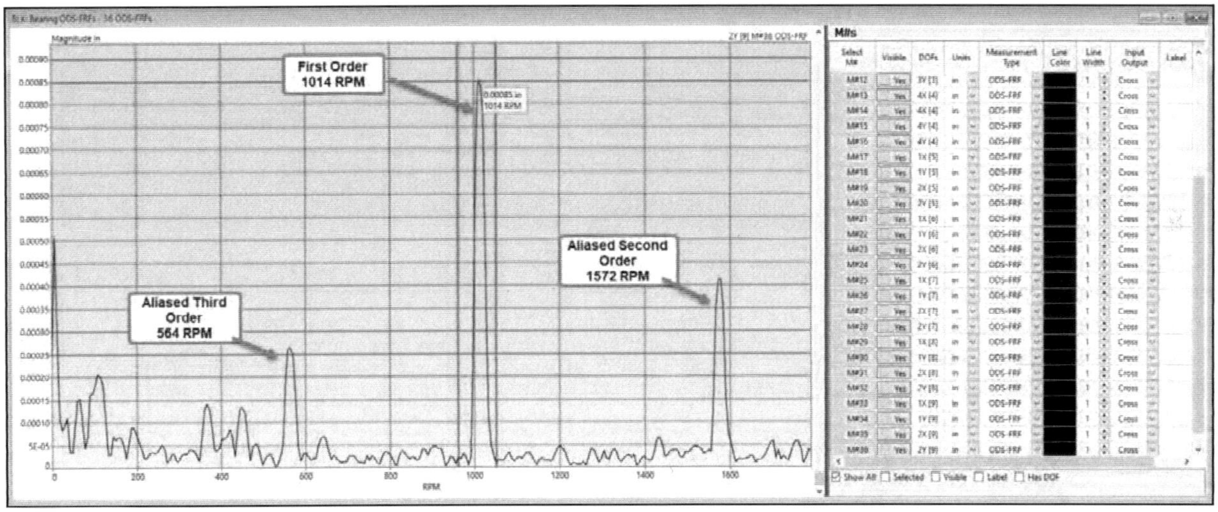

*Figure 7. **ODS-FRF** Showing First Three Order Peaks.*

Aliased Order Peaks

A limitation of any video recording is that ***anti-alias filtering cannot be used*** to remove high-frequency signals from the video.

Without anti-alias filtering, machine order peaks ***greater than one-half the sampling frequency***, (called **Fmax**), are ***folded around*** **Fmax** and appear at lower frequencies in the **ODS-FRFs**, as shown in Figure 7.

All order-related resonance peaks between **Fmax & 2 x Fmax** are folded around (***wrapped around***) **Fmax** and appear at a lower frequency in the frequency band **(0 to Fmax).** Aliasing of these higher frequencies to lower frequencies occurs in both **DFTs** and **ODS-FRFs**.

The **ODS-FRF** shown in Figure 7 was calculated from a **TWF** that was extracted from a video that was sampled at **60 fps**, or **3600 RPM**. Therefore, **Fmax** of the **ODS-FRF** is **1800 RPM**.

The first-order peak is at the machine running speed and is visible at **1014 RPM**. The second order resonance peak should be at **2028 RPM** and third order resonance peak should be at **3042 RPM**, but they are both ***folded around*** **1800 RPM** and are clearly visible ***at lower frequencies*** in the **ODS-FRF**.

The aliased frequency of an order with frequency between **Fmax** & **2 x Fmax** can be calculated from the expected order frequency and **Fmax**.

- **Second order** aliased frequency □ 1800 - (2028 - 1800) □ **1572 RPM**
- **Third order** aliased frequency □ 1800 - (3042 - 1800) □ **558 RPM**

The **ODS-FRF** in Figure 7 was calculated from a **TWF** with ***a 10-second time length*** **T**. Therefore, **Δf = (1/10) Hz or 60/10 = 6 RPM**. So, the third order aliased frequency peak at **558 RPM** ***is within one*** **(Δf)** of its calculated value.

Spectrum Averaging

ODS-FRFs are calculated using spectrum averaging. The advantages of spectrum averaging are listed in Figure 8.

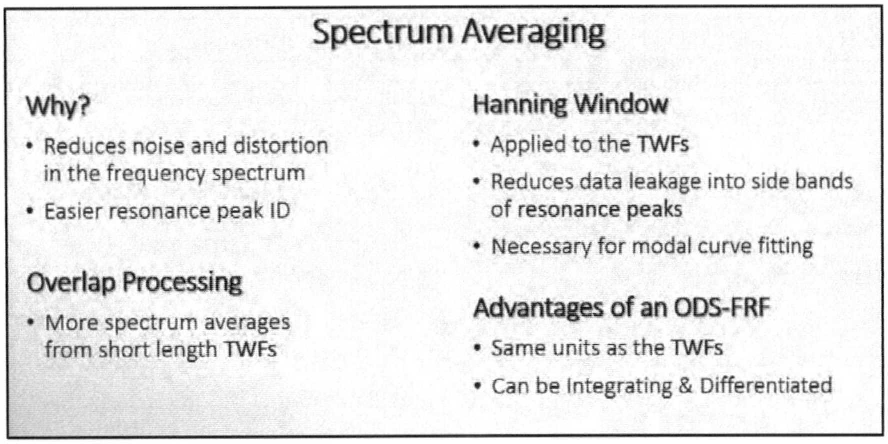

Figure 8. Spectrum Averaging

Experimental Mode Shapes

Experimental Modal Analysis (**EMA**) is used to define the resonances of a mechanical structure in terms of its modal parameters. Modal parameters are used to define each resonance, as shown in Figure 9

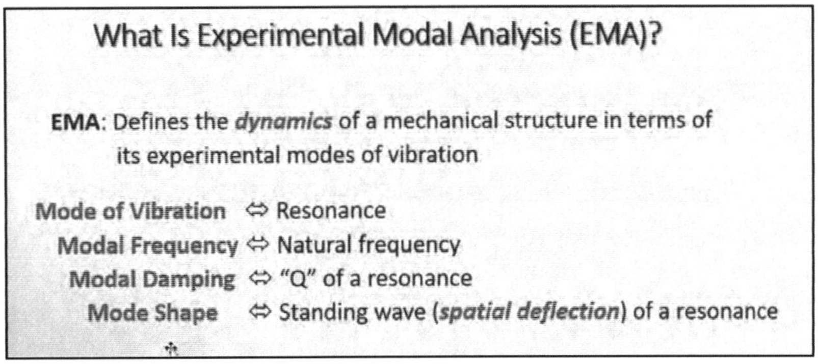

Figure 9. What is EMA?

Resonant vibration follows the rules of modal analysis, as shown in Figure 10.

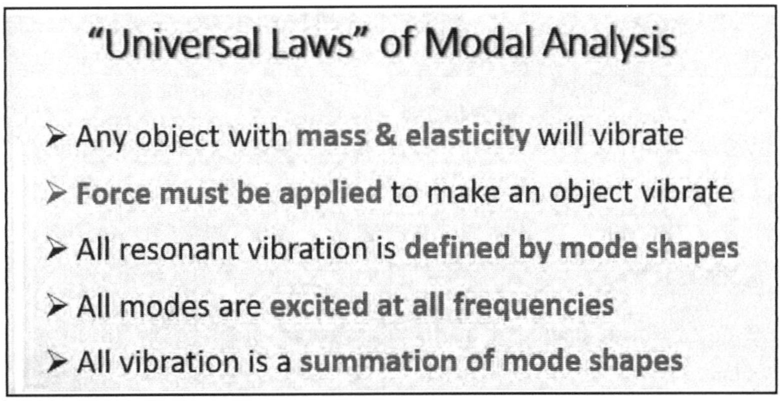

Figure 10. Laws of Modal Analysis

An **ODS** is different from a mode shape, but the two are related to one another, as shown in Figure 11. An **ODS** is the overall deflection of a mechanical structure due to applied forces, while a mode shape is the spatial deflection of a structural resonance. All FRF-based curve-fitting assumes that all **ODS** data is a ***summation of mode shapes***.

A set of **ODS-FRFs** is a set of **ODS's** defined over a span of frequencies.

An ODS versus a Mode Shape

ODS	Mode Shape
1. Forced vibration of a structure at two or more DOFs	1. Deflection of a structural resonance at two or more DOFs
2. Defined at any frequency or any time	2. Defined at a specific "natural" frequency
3. Has engineering units	3. Has no units
4. Has unique values	4. Has no unique values (an eigenvector)
5. Defines actual motion of all DOFs	5. Defines relative motion between DOFs
6. Affected by forces, material properties or boundary conditions	6. Affected by material properties or boundary conditions
7. A summation of mode shapes	7. Called an EMA mode shape if extracted from ODS data

*Figure 11. **ODS** versus a Mode Shape*

Using Mode Shapes for Machine Monitoring

The steps required to use mode shapes calculated from a cell phone video for machinery monitoring are listed in Figure 12.

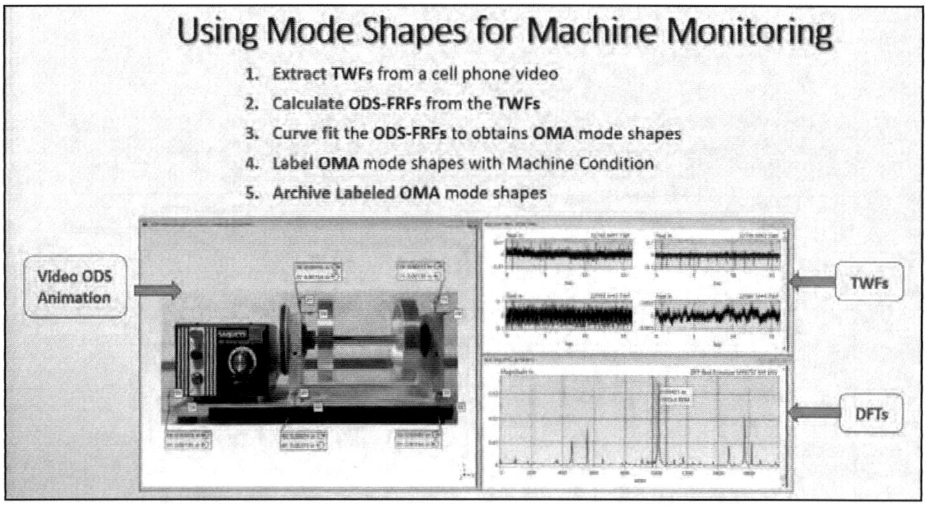

Figure 12. Steps Required to Use Mode Shapes

Curve Fitting ODS-FRFs

FRF-based curve fitting methods are commonly used on a set of **FRFs** that have been calculated from accelerometer-based **TWFs** acquired during a modal test. However, to calculate an **FRF**, the forces which cause the resonant vibration must be ***simultaneously acquired*** with the response.

When the excitation forces cannot be measured, then either Cross Spectra (**XPS's**) or **ODS-FRFs** can be calculated from the response **TWFs**, and curve fit using FRF-based curve fitting to extract **OMA** mode shapes.

FRF-Based Curve Fitting

For this paper, FRF-based curve fitting was applied to a set of **ODS-FRFs** calculated from the **TWFs** extracted from a cell phone video to obtain **OMA** mode shapes of a rotating machine. A typical FRF-based curve fit of several **ODS-FRFs** is shown in Figure 13.

Then the mode shapes were labeled with their associated machine unbalanced condition and archived into a database. Mode shapes were archived for five different unbalanced conditions on the rotating machine shown in Figure 1.

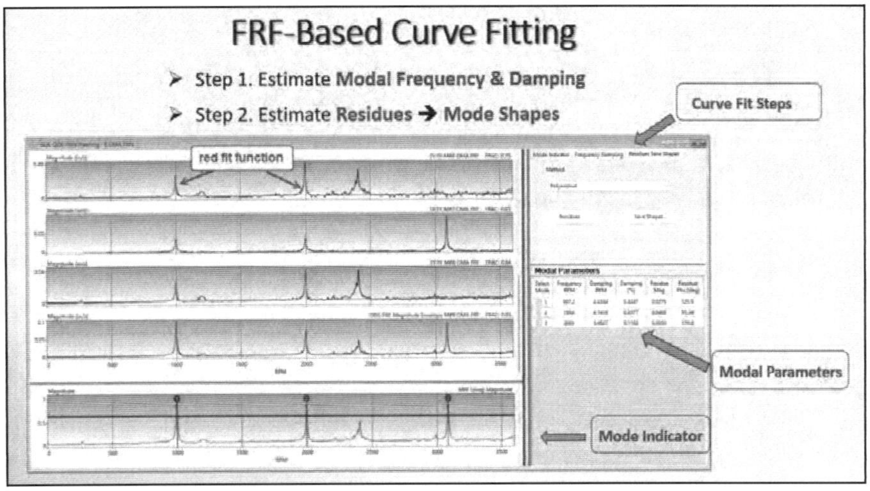

Figure 13. FRF-based Curve Fit of ODS-FRFs

Animated Mode Shape Display

After obtaining **OMA** mode shapes from a cell phone video, it is useful to animate the mode shapes on the point grid to spot errors that might have occurred from the curve fitting step.

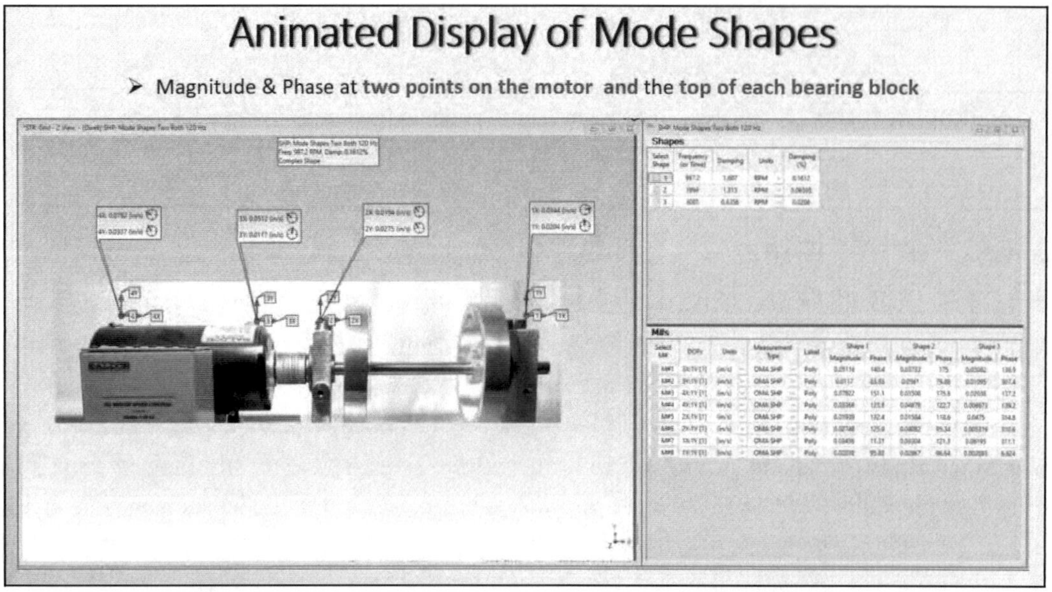

Figure 14. Mode Shape Display Showing Magnitudes & Phases

The other advantage of the mode shape animation is that the ***magnitude & phase*** of the two points on the motor and one on each bearing lock can be used to identify a particular machine fault. A typical display of ***magnitude & phase*** during mode shape animation is shown in Figure 14.

Baseline Case

When no screws were added to either rotor of the machine in Figure 1, its mode shapes were labeled as the Baseline case. When the Baseline case is archived into the database, the **FaCTs** bar chart in Figure 15 clearly identifies it by its unique **OMA** mode shapes compared to the mode shapes of other unbalanced cases. The **FaCTs** bars of the Baseline case with all the other unbalanced cases are ***much less than*** **1.0**.

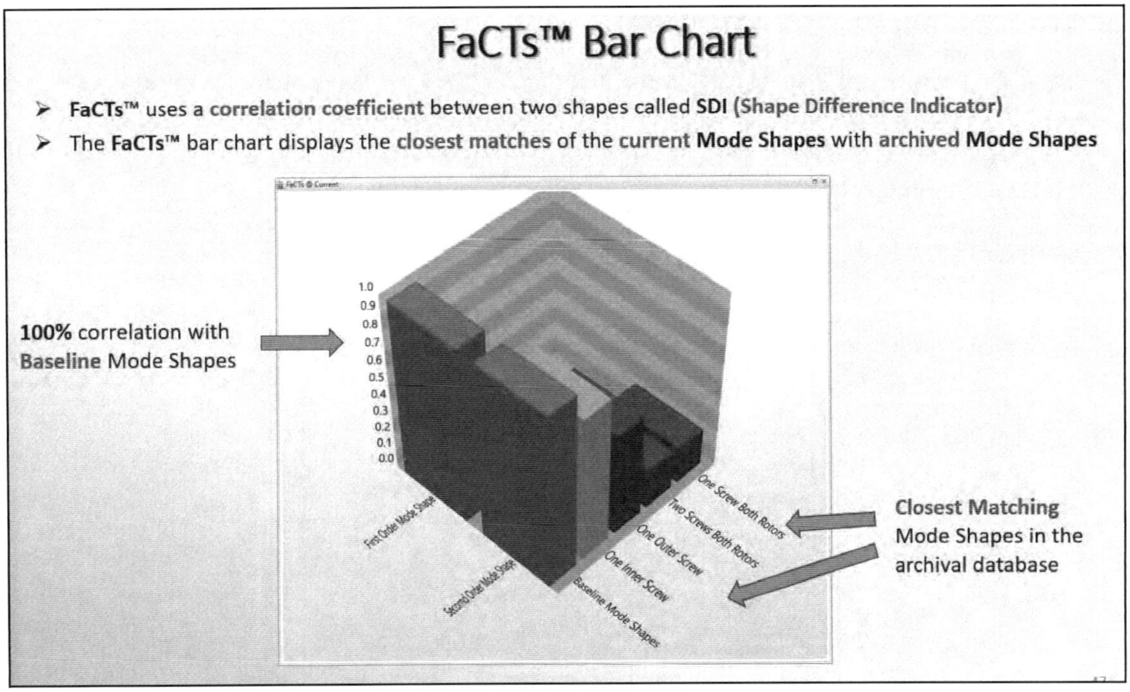

Figure 15. Baseline OMA Mode Shapes Versus Other Unbalance Cases

Cases With Unbalance Screws Added

Figures 16 and 17 show the **FaCTs** bar charts for the ***four unbalanced cases*** where screws were added to either the inner or outer rotor, or to both rotors. Each unbalance case was ***uniquely identified*** by **FaCTs** because its corresponding mode shapes were unique when compared to the mode shapes of the other unbalance cases.

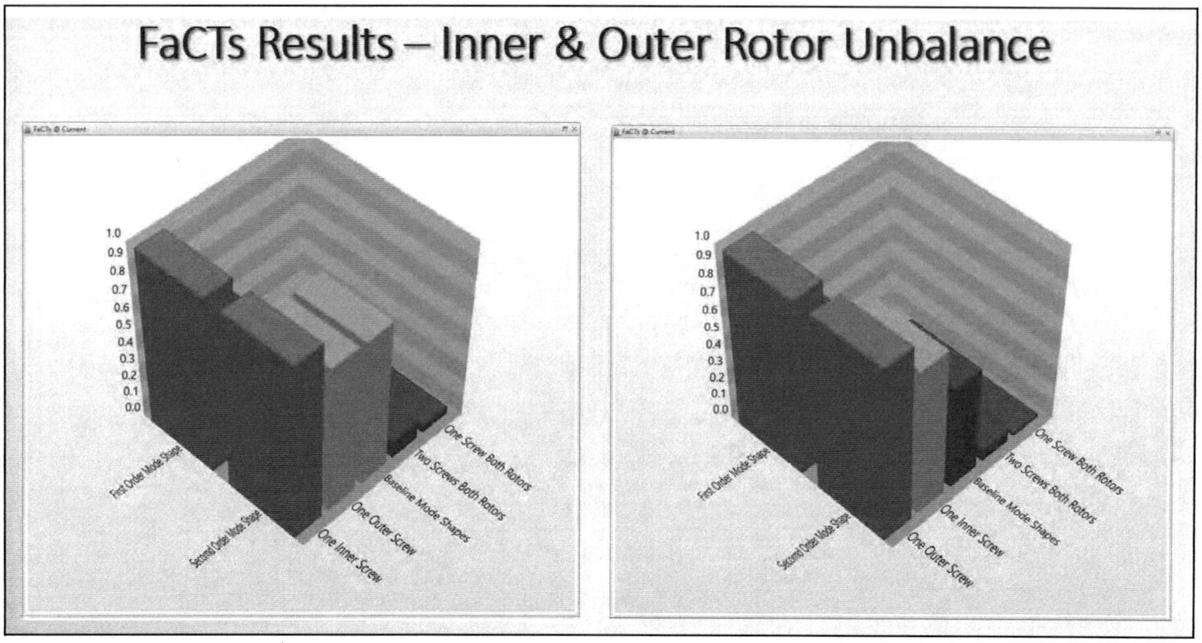

Figure 16. One Inner Screw and One Outer Screw

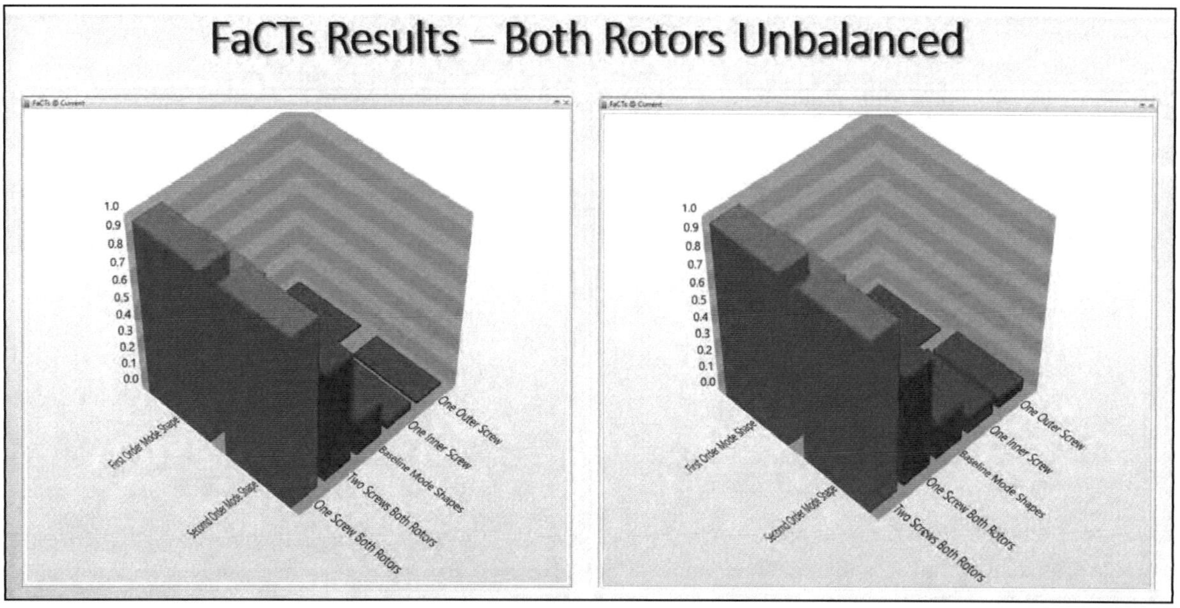

Figure 17. One Screw Both Rotors and Two Screws Both Rotors

Conclusions

Five different unbalance cases were created on a rotating machine by adding screws to its ***inner*** and ***outer*** rotors. The first case with ***no screws added*** was labeled as the **Baseline** case. In each case, a **30 second** cell phone video was recorded, with the machine running at approximately **1000 RPM**.

Using MEscope [10], **TWFs** were extracted from each cell phone video and **ODS-FRFs** were calculated from the **TWFs** for each unbalance case. To reduce noise in the **ODS-FRFs**, ***spectrum averaging*** and ***overlap processing*** were used to calculate the **ODS-FRFs**. A Hanning window was also applied to each **TWF** during the calculation of each average to enable curve fitting of the **ODS-FRFs**.

Then the **ODS-FRFs** were curve-fit using **FRF-based** curve fitting and the **OMA** mode shapes for the ***first two machine orders*** were labeled and stored in an archival database. Each set of mode shapes was labeled with its ***corresponding unbalance case***.

Then, when the mode shapes for each unbalanced case were again stored into the archival database, **FaCTs** correctly identified each case by calculating their **FaCTs** value with each ***labeled*** set of mode shapes in the database.

This numerical comparison of a ***current set*** of mode shapes with sets of ***labeled*** mode shapes in the database uniquely identified each of the unbalance cases using only ***eight mode shape components***, X & Y **DOFs** for two points on the motor and one point on the top pf each bearing block.

This simple non-contacting approach using cell phone videos and **FaCTs** could be used by any plant maintenance department for monitoring the condition of its rotating equipment and accurately identifying common machine faults such as unbalance, misalignment, soft foot and loose parts.

These results demonstrate the reliability of using mode shapes extracted from cell phone videos for machine health monitoring.

References

[1] - S. Richardson, M. Richardson "Extracting 3D Mode Shapes from a Cell Phone Video" IMAC XLII, February 10-13, 2025, Orlando, FL

[2] - B. Schwarz, S. Richardson, P. McHargue, M. Richardson "Using Modal Analysis and ODS Correlation to Identify Mechanical Faults in Rotating Machinery" IMAC XLII, January 29 - February 1, 2024, Orlando, TX

[3] - D. Ambre, B._Schwarz, S. Richardson, M. Richardson, "Using Cell Phone Videos to Diagnose Machinery Faults" IMAC XLI, Austin, TX, February 13-16, 2023

[4] - B._Schwarz, S. Richardson, M. Richardson, "Using a Cell Phone Video and ODS Correlation
to Diagnose Unbalance in Rotating Machinery" IMAC XLII, January 29 - February 1, 2024, Orlando, FL

[5] -_M.H. Richardson, "Is It a Mode Shape or an Operating Deflection Shape?" Sound and Vibration magazine, March 1997

[6] - B. Schwarz, M.H. Richardson, "Measurements Required for Displaying Operating Deflection Shapes" Proceedings of IMAC XXII, January 26, 2004

[7] - B. Schwarz, M.H. Richardson, "Introduction to Operating Deflection Shapes" CSI Reliability Week, Orlando, FL, October 1999

[8] - S. Richardson, J. Tyler, P. McHargue, M. Richardson "A New Measure of Shape Difference" IMAC XXXII February 3-6, 2014

[9] - S. Richardson, J. Tyler, B. Schwarz, P. McHargue, M. Richardson "Using Modal Parameters for Structural Health Monitoring" IMAC-XXXV January 30–February 2, 2017, Garden Grove, CA

[10] - MEscope™ is a trademark of Vibrant Technology, Inc. www.vibetech.com

BLADE MOMENT WEIGHING AND ROTOR UNBALANCE

Authors: Brian Hantz, Manish Thorat, Brian Pettinato

BLADE MOMENT WEIGHING AND ROTOR UNBALANCE

Brian Hantz
Engineer, Advanced Technology Programs
Ebara Elliott Energy

Manish Thorat
Senior Engineer, Machinery Dynamics & Acoustics
Ebara Elliott Energy

Brian Pettinato
Manager, Aero and Structural Dynamics
Ebara Elliott Energy

Brian Hantz is a Mechanical Engineer of Advanced Technology Programs at Ebara Elliott Energy in Jeannette, Pennsylvania. He received his B.S. in Aerospace Engineering from Penn State University in 2012. He has been with Ebara Elliott Energy since 2014. His areas of expertise include rotordynamics, vibration, and modal testing.

Manish Thorat is a Senior Engineer and Machinery Dynamics & Acoustics Group Lead, R&D at Ebara Elliott Energy. He received his MS degree in Mechanical Engineering from Texas A&M University (2010) and BE Mechanical Engineering from Mumbai University (2007). He has been with Elliott since 2010. His areas of expertise include rotordynamics, vibration analysis, testing and evaluation of seal designs, squeeze film dampers and fluid film journal bearing design.

Brian Pettinato is a Senior Manager of Aero and Structural Dynamics at Ebara Elliott Energy in Jeannette, Pennsylvania. He has been with Elliott since 1995. His primary area of expertise is machinery dynamics. Mr. Pettinato received his B.S. (1989) and M.S. (1992) Mechanical Engineering degrees from the University of Virginia. Mr. Pettinato is a fellow member of ASME and a member of STLE.

Abstract

Blade moment weighing is a well-established and effective means of reducing unbalance in bladed turbomachinery. Managing rotor unbalance is critical to ensuring low vibration and successful operation of critical turbomachinery. This article discusses the impact of blade weights and moments on rotor unbalance, and how blade moments are measured and used. An example case of a steam turbine with a stage disk comprised of 64 blades having 26-inch blade height is presented. The measured blade masses and moments are separately used to highlight how different algorithms and assembly methods influence the total unbalance relative to the rotor balance criteria.

Key Words: Balance, Balancing, Moment Weighing, Moment Blade Sorting, Turbines, Turbine Blades, Unbalance

Introduction: Unbalance in Turbomachinery

Unbalance in rotating machinery is one of the primary culprits of high vibration during operation. Unbalance is a synchronous force in that it is synchronized to the rotor running speed. It is specified as an unbalance weight at a specific radial distance from the shaft centerline, typically provided in units of ounce-inches or gram-millimeters. Force due to unbalance is proportional to the square of the speed as shown in Equation (1),

$$F_u = (m \cdot e) \cdot \omega^2 = Unb \cdot \omega^2 \tag{1}$$

where F_u is the unbalance force, m is the mass of unbalance, e is the unbalance eccentricity, Unb is the unbalance amount or (m·e), and ω is the rotational speed.

Placing a 5-gram weight at a centrifugal compressor impeller eye with radius 100 mm would result in a known unbalance of 500 g-mm. At 1,000 rpm, the force would be 5.48 N (1.23 lb_f). At 10,000 rpm, the force would be 100 times more or 548 N (123 lb_f).

For machines built to American Petroleum Institute standards, also known as API standards, the maximum allowable low-speed unbalance of the rotor assembly is determined from Equation (2) or Equation (3)

In USC units:

$$U_{max} = 4 \cdot \frac{W}{N} \text{ (for N} \leq 25{,}000 \text{ RPM)} \tag{2}$$
$$U_{max} = \frac{W}{6250} \text{ (for N} > 25{,}000 \text{ RPM)}$$

In SI units:

$$U_{max} = 6350 \cdot \frac{W}{N} \text{ (for N} \leq 25{,}000 \text{ RPM)} \tag{3}$$
$$U_{max} = \frac{W}{3.937} \text{ (for N} > 25{,}000 \text{ RPM)}$$

where U_{max} is the maximum residual unbalance, W is the total rotor weight in lb_f (kg_f) and N is the maximum continuous operating speed in RPM. Because the unbalance amount is proportional to the square of the radial distance of the unbalance weight, larger diameter components are often the major source of rotor assembly unbalance. For centrifugal compressors, this typically corresponds to the impellers. For bladed turbines, this corresponds to the stage disk assemblies.

Impellers of centrifugal compressors are manufactured with integral blades, hub and cover such that the entire component forms a single piece. This means the resulting unbalance is inherent to the entire impeller after manufacture. After the initial manufacture, each impeller is balanced on an arbor (Figure 1) where the unbalance is corrected with either balance weights (adding mass) or impeller grinding (subtracting mass). The impellers are then assembled to a shaft and the entire rotor is balanced (Figure 2).

Figure 1: Impeller Assembled to an Arbor in a Balance Machine

Figure 2: Fully Assembled Compressor Rotor in a Balance Bunker

Stage disk assemblies, on the other hand, consist of many individual blades that are assembled into the machined disk separately (Figure 3). This presents a unique opportunity (or pitfall) to influence the total disk assembly unbalance with careful placement of each blade. Since the disks are often part of an integral rotor forging, there typically is not an opportunity to balance each bladed disk separately except by repetitively balancing the rotor after each individual disk is bladed; a process known as progressive balancing. Regardless of the assembly procedure, there is benefit to minimizing the amount of balancing correction that is required.

Figure 3: Steam Turbine with Last Two Stages Assembled

Blade Moment Weighing

Taken in isolation, the unbalance added to the disk by assembling a single blade is calculated using the equations described by Figure 4. The unbalance can be separated into the horizontal (X) and vertical (Y) directions, respectively.

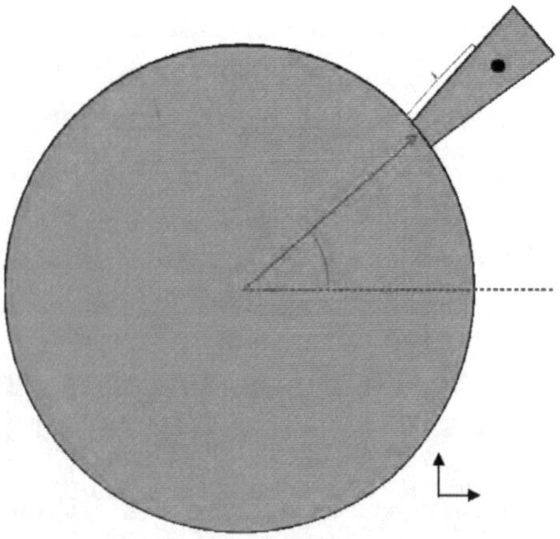

Figure 4: Blade Contribution to Rotor Unbalance

Each blade contributes to the total assembly unbalance, such that the horizontal and vertical components of unbalance for each blade are summed to get the total blade arrangement unbalance. To minimize total unbalance of the entire assembly, it is preferred to match blades of similar moments 180 degrees away from each other, such that their individual contributions to total unbalance counteract, negating the influence of each. This can ensure low vibration and meeting API criteria, while minimizing required unbalance corrections. Theoretically, residual unbalance that exists in the disk itself can also be negated, by assembling the turbine blades such that the amplitude and phase of the combined blade unbalance negates the amplitude and phase of the disk unbalance. In practice, however, it is common to assume a balanced disk.

API Criteria

So far, blade moment weights and their impact on assembly unbalance are only discussed in API RP 687, a recommended practice for special-purpose rotating equipment repairs. This API recommended practice defines the blade moment weight as a value that represents the combination of the blade weight and its center of gravity, whereas blade pan-weight is simply the static weight of the blade as measured using a simple scale. According to clause 4.1.6 of API RP 687, blades with an airfoil length greater than 6 inches (152 mm) should be moment-weighed during assembly, whereas shorter airfoil blades can be pan-weighed. There are currently no technical requirements in API STD 611 6th edition, *General-purpose Steam Turbines*, or API STD 612 8th edition, *Steam Turbines – Special-purpose Applications*, with respect to weighing and assembly of turbomachinery blading.

Moment Weighing Blades

Moment weighing blades consists of recording two parameters: the blade weight, and the location of the blade's center of gravity. Recording the static weight (pan-weight) is straightforward and can be done with any scale calibrated to the desired accuracy. To obtain the blade center of gravity, a more involved setup is required. In a simple application, this can be done using two separate knife edge scales. Given a known separation between the contact points of the two scales and the recorded weight on each scale, the center of gravity with respect to the root of the blade can be found (Figure 5).

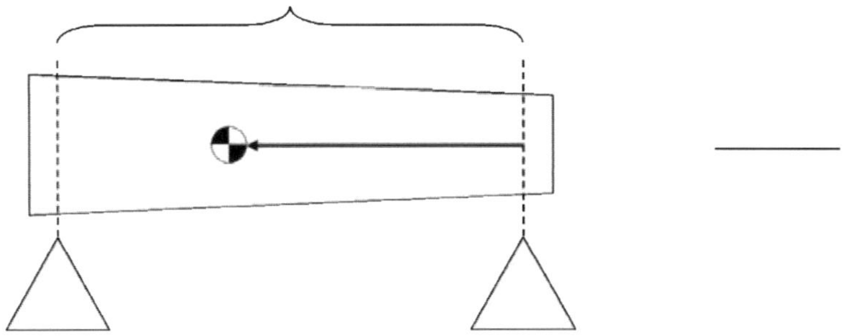

Figure 5. Measurement of Blade CG

The total unbalance acting on the rotor for the blade in question is also a factor of the radial distance from the center of the shaft to the center of gravity of the blade, as shown in Figure 4. In other words, even relatively small blades could feasibly impart large unbalance if the disk diameter is large, as any significant deviation in the blade mass is multiplied by the large radius.

Sorting Algorithms

Blade moment or weight measurements for an entire set of blades can then be used by computer programs to arrange the blades in such a way as to minimize the total unbalance of the stage. Several algorithms exist both in literature and in commercially available balancing software. Some examples of these algorithms include, but are not limited to, the following approaches.

- Brute Force and Ordinal Pairing
- Greedy algorithms
- Genetic algorithms
- Simulated Annealing
- Combinations of Others

Brute force is the simplest and most primitive approach. In this method, various assembly configurations are tried at random or by consistent steps such that all possible assembly configurations are attempted and evaluated individually. While straightforward, this quickly becomes computationally impractical as the number of turbine blades increases. The number of possible assembly configurations for a turbine stage is N factorial, where N is the number of stage blades. Ordinal pairing is a straightforward attempt at balancing blades by pairing blades of similar mass or moment 180 degrees apart, such that the vector sum of their influence is minimal. While these attempts may yield acceptable assembly configurations in a limited number of attempts, the number of blades in a modern industrial turbine stage prevent these methods of realistically finding anything near a global minimum. Because of this, most blade sorting software employ more sophisticated algorithms.

A greedy algorithm [1] is an approach that makes locally optimal choice at each stage of the process in the hope of finding a global optimum. In the context of assembling turbine blades to minimize unbalance, the Greedy algorithm can be applied by iteratively selecting the blade that contributes the least to the overall unbalance at each step—In this way, the algorithm prioritizes immediate gains for each blade selection. The process continues until all the blades are assembled or have their assembly order decided. While the Greedy algorithm does not guarantee a globally optimal solution, it is often efficient and can provide reasonably good solutions.

Genetic algorithms, as the name implies, are inspired by the principles of natural selection and genetics. A population of potential solutions, representing different combinations of blade arrangements, is evolved over successive generations. Each solution is treated as a candidate assembly, and the algorithm applies genetic operators such as selection, crossover, and mutation to generate new potential solutions. The fitness of each assembly is evaluated based on the degree of unbalance it exhibits. Over several generations, the algorithm refines the population, favoring combinations that lead to lower unbalance. Further discussion of a genetic algorithm applied to the turbine balancing problem is discussed by Osguei and Khamoushi [2]. Genetic algorithms are particularly useful for complex optimization problems with a large search space, as they explore multiple possibilities simultaneously.

Simulated Annealing is a probabilistic optimization approach inspired by the annealing process in metallurgy, where a material is heated and then gradually cooled to remove defects and achieve a more ordered structure [2]. It starts with an initial solution and iteratively explores neighboring solutions by perturbing the assembly configuration. Unlike greedy algorithms, simulated annealing allows for uphill moves, accepting solutions with higher unbalance initially, but with decreasing probability of these uphill moves over time. This stochastic nature helps the algorithm escape local optima and explore the solution space more thoroughly. The algorithm gradually reduces the exploration probability, mimicking the cooling process in metallurgy, which helps converge towards a globally optimal solution.

While each algorithm offers unique benefits, any algorithm can be applied when searching for an acceptable arrangement of turbine blades to minimize the assembled unbalance.

Comparisons Using Actual 26-inch Blades

For a practical example, blade weights and moments were measured for a set of 64 blades to be installed in the L-0 stage of a large frame steam turbine. Each blade measured approximately 26 inches from the root first land to the tip. The blades were then optimized using BLADIS Net, a commercial software program developed by Schenck RoTec GmbH, to minimize the total assembly unbalance around the stage disk using only the blade weights, and then only the blade moments.

For this sample set of 64 blades, the weights and moments followed a roughly normal distribution about the mean blade mass and moment, albeit with a slightly disproportionate number of heavier blades. The standard deviation from the mean for the blade mass was 0.17%, while the standard deviation from the mean for blade moment was 0.22%. Note that the total moment of the blades is always influenced not only by deviations in blade mass, but also the dimensional location of this mass deviation.

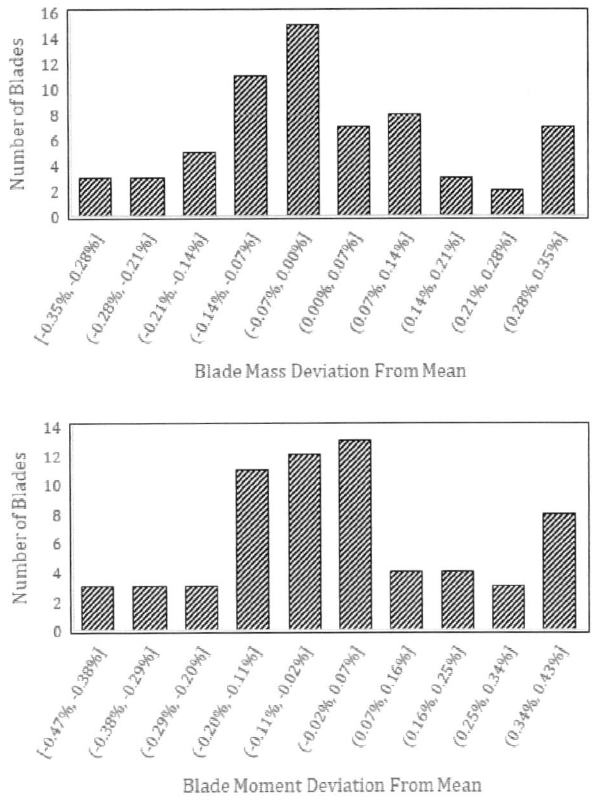

Figure 6. Blade Mass and Moment Deviations

With 64 blades, there are 64 factorial (1.27 x 10^{89}) unique assemblies. For the set of 64 blades in question, the unbalance level of a random assembly can vary from a near-zero (best case) up to nearly 100 times the 4W/N unbalance limit for the rotor. A simulation of 100,000 random assemblies for the set of blades shows that the most likely unbalance present (50th percentile) is on the order of 10W/N, which still far exceeds the allowable static unbalance for the entire rotor assembly. Note that this is for a single row of blades—when considering that a turbine rotor can have many stages of blades, in addition to residual unbalance in the rotor shaft and disks, it is clear that a more sophisticated approach is required.

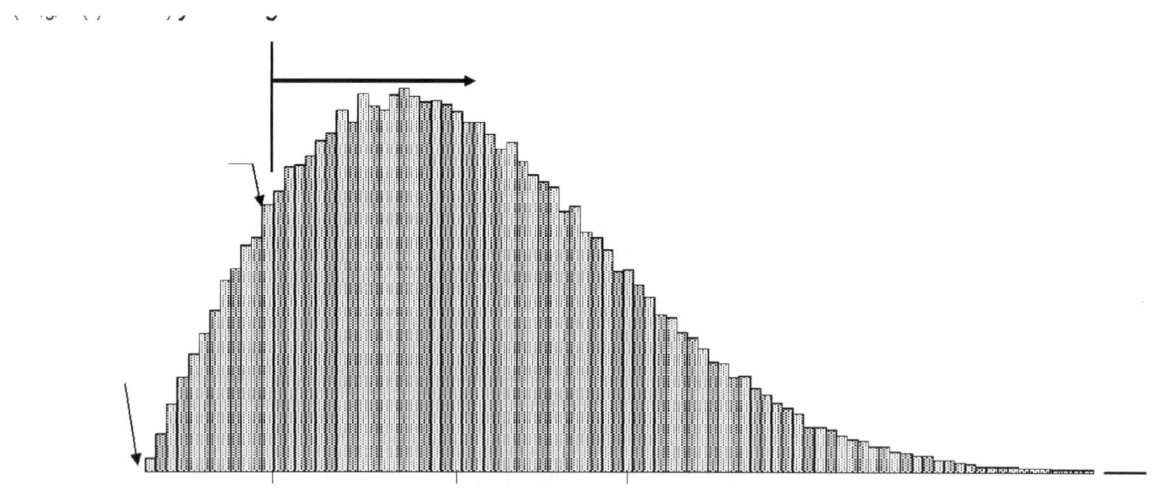

Figure 7. Blade Arrangement Unbalance Likelihood with qty. 64 blades with 26-inch length

Using the BLADIS Net program, an optimized assembly order was generated for the blades using only the blade masses. The algorithm in this case only has knowledge of the static mass of each blade, and as a result optimizes the mass distribution only (net mass offset). That is, the vector sum of the blade masses in the horizontal and vertical direction is minimized. However, this neglects any contribution of the CG location of the blades. Because the CG was recorded for each blade, despite not being used in this sorting attempt, the actual unbalance distribution can be determined to evaluate the effectiveness of sorting by static mass only.

A sorting algorithm using pan-weights only was capable of reaching a resulting unbalance in the 90th percentile of randomly assembled orders. In other words, weighing each blade and sorting based on masses only would produce lower total unbalance than 90% of randomly ordered assemblies. In this case, pan-weighing reduced the total unbalance of the blades to 3.7W/N. While a definite improvement and manageable, the blade arrangement unbalance alone (especially considering that other stages exist) will undoubtedly require some additional balancing of the rotor to meet API limits.

While the pan-weighing solution greatly lowers total unbalance, including the impact of the center of gravity for each blade can further optimize the assembly order. Using the measured moments for each blade, the BLADIS Net program was capable of reducing the blading unbalance to 0.007W/N, well below the pan-weighing solution and orders of magnitude below the API low-speed balance requirement. The moment-weighed blade arrangement adds an insignificant amount of unbalance to the overall rotor assembly, and ensures easier rotor assembly balancing and greater chance of success. Of the 100,000 random assemblies simulated, no assembly order presented a lower total unbalance than the algorithm used by the BLADIS Net program, indicating that the moment-weighed algorithm and sorted blades are indeed approaching a global minimum of total unbalance. In other words, the very least that can be said is that the moment-weighed approach is better than a random assembly >99.999% of the time for the blades in question.

The previous cases illustrated the unbalance distribution created when using an advanced algorithm to optimize the blade assembly order. Other simple options exist to attempt to minimize unbalance, such as simple ordinal pairing (Figure 8). Ordinal pairing is a method of pairing blades of similar moment or mass, and distributing them 180 degrees apart to minimize the overall unbalance. Care must be taken to alternate pair starting points for a good ordinal pairing result. For example, if the blades were simply arranged as shown in Figure 8, the blades shown would always be unbalanced toward the top right due to the heavier blade of each pair being located there. A comparison of each method is provided in Figure 9.

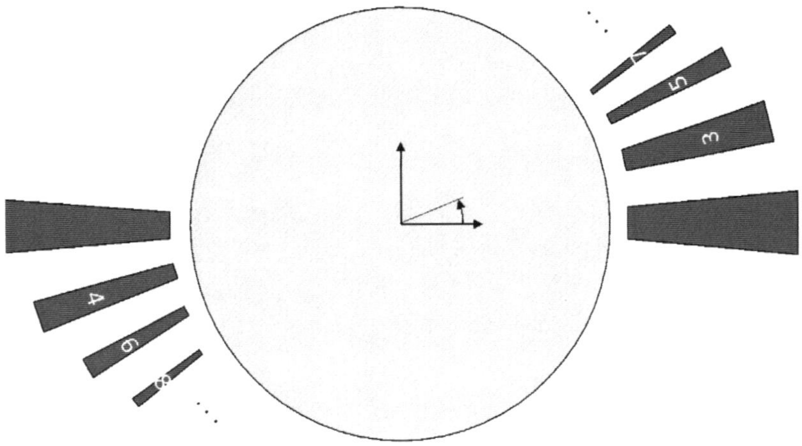

Figure 8. Illustration of Ordinal Pairing - Blades Arranged in Order of Moment or Mass to Minimize Unbalance or Net Mass Offset

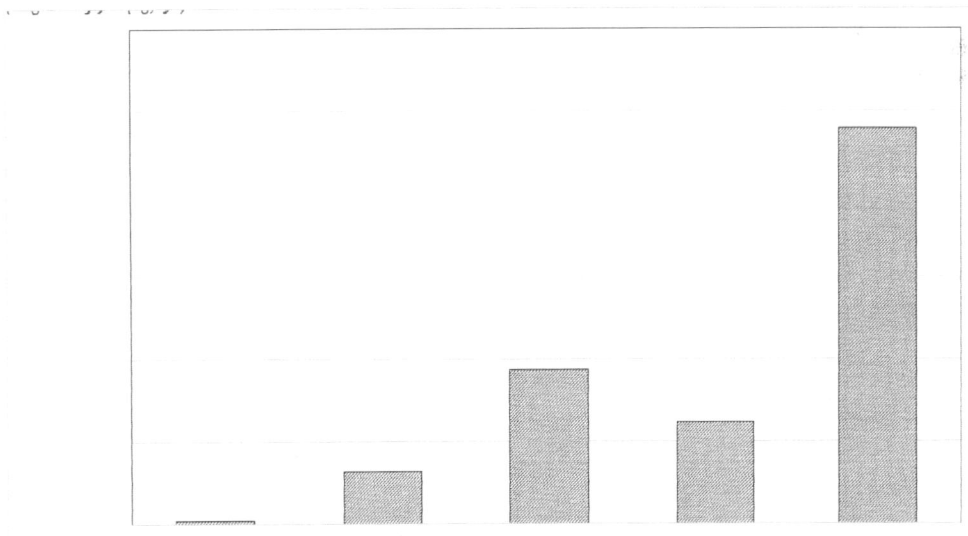

Figure 9. Comparison of Blade Assembly Methods on Rotor Unbalance

Figure 9 shows that moment-weighing the blades and using a sophisticated algorithm to optimize the assembly order provides the lowest total unbalance to the rotor assembly. Moment-weighing the blades and sorting with a simple ordinal pairing routine still provides relatively good results. Meanwhile, random assembly is likely to provide significant rotor unbalance.

When considering the blade masses only (pan-weights), Figure 9 at first glance suggests that ordinal pairing is superior to a sophisticated algorithm. However, this is only the result of chance. Both methods work to minimize the net mass offset, not unbalance, because the distance to each blade CG is unknown. This is explained by Figure 10, showing that the algorithm provides a significantly lower net mass offset than ordinal pairing. By chance, this arrangement provided lower total unbalance after considering the blade CG locations.

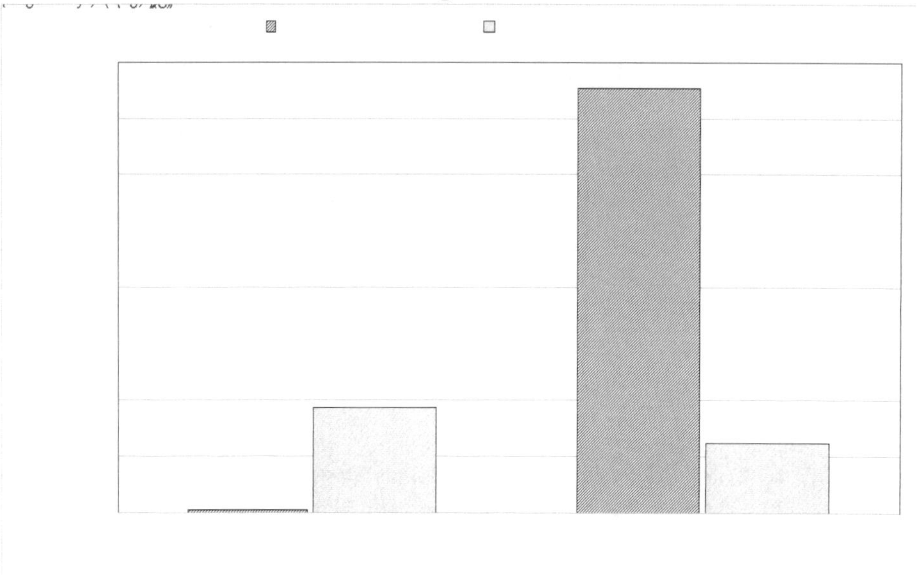

Figure 10: Resulting Net Mass Offset and Unbalance Considering Pan-Weights Only

Comparisons Using 6-inch Blades Derived from the 24-inch Data

Assuming the standard deviation of blade mass and moment for a given set of blades, the 26-inch blades were also used to extrapolate to a 6-inch blade to compare the differences between pan-weighing and moment-weighing. This was accomplished by scaling both the mass and the distance from the blade root to CG by the blade length. The disk radius was kept the same. A comparison of the resulting contribution to rotor unbalance when arranging the blades via ordinal pairing is shown in Figure 11. This result demonstrates the significant impact that larger blades can have on the overall rotor unbalance.

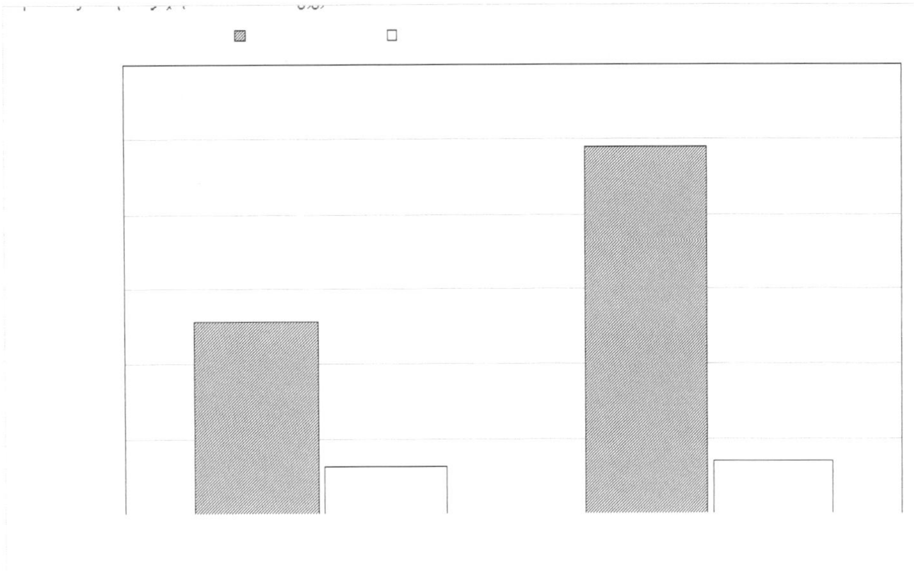

Figure 11: Comparison of 26" Blade Contribution to Rotor Unbalance compared to an Equivalent Set of 6" Blades

Conclusion

Managing rotor unbalance during the assembly process is paramount to low vibration and successful operation of critical turbomachinery. Bladed rotors such as steam turbines and expanders offer unique challenges due to the manual assembly of oftentimes dozens of blades for each stage disk. It has been shown that the bladed disk contribution to unbalance of the rotor assembly depends not only on blade masses, but also on the blade center of gravity location. When blades are large, they present a unique opportunity to not only drastically increase the rotor unbalance when assembled improperly, but also to drastically reduce the total assembly unbalance by careful sorting and assembly order. Mass-weighing large turbine blades will typically result in a significant improvement in total assembly unbalance, but to truly minimize the assembly unbalance to near zero levels, moment-weighing offers the best solution.

Acknowledgements

The authors would like to thank Ebara Elliott Energy for permission to publish this paper.

References

[1] Li Y, Pardalos PM, Resende MG. A Greedy Randomized Adaptive Search Procedure for the Quadratic Assignment Problem. Quadratic assignment and related problems. 1993;16:237-61.

[2] Taraghi Osguei A, Khamoushi I. Genetic algorithm-based search heuristic for turbine balancing problem. Proceedings of the Institution of Mechanical Engineers, Part E: Journal of Process Mechanical Engineering. 2022 Dec;236(6):2630-8.

[3] Jun-kong L, Li-li LI. Optimization of Assembly Sequence Planning of Turbine Low-pressure Rotor Blades Based on the Improved Simulated Annealing Algorithm. In Journal of Physics: Conference Series 2021 May 1 (Vol. 1939, No. 1, p. 012098). IOP Publishing.

THE PRACTICAL USE OF GENERATIVE AI TO DEVELOP COMMERCIAL PROGNOSTICS SYSTEMS FROM SCRATCH

Author: Howard W. Penrose, Ph.D., CMRP, CEM, CMVP

VIATC TRAINING CONFERENCE & EXPO
NEWPORT NEWS, VIRGINIA | AUGUST 6-8, 2025

VIATC TRAINING CONFERENCE & EXPO
NEWPORT NEWS, VIRGINIA | AUGUST 6-8, 2025

The Practical Use of Generative AI to Develop Commercial Prognostics Systems from Scratch

Howard W Penrose, Ph.D., CMRP, CEM, CMVP
President, MotorDoc® LLC
howard@motordoc.com

President MotorDoc LLC, a Veteran Owned Small Business, Chair CIGRE-USNC A1 (motors and generators), IEEE PES grid working group, ACP clean energy systems standards chair, Convenor CIGRE A1.78 (ESA/MCSA project), IEEE PES standards group, NEMA rail electrification standards committee, global leader in ESA/MCSA research and development, certified CMRP, CEM, CMVP, data science/AI/machine learning from Stanford University, University of Michigan, AWS and IBM. Life member MENSA.

Abstract:

In 2024 a challenge was issued to develop a complete prognostic system using commercially available hardware and Generative AI as the lead programming tool with Python as the language. At first the challenge seemed to be insurmountable until hardware was selected that allowed a sample rate that made such a system practical. In this paper we will discuss the selection of hardware and then the development of a software system in Python from scratch using a combination of generative AI systems such as ChatGPT, GitHub Copilot, Microsoft Copilot, and Anaconda Copilot. The effort specifically identified that the speed of concept to completion, or software lifecycle, can be dramatically improved, but the understanding of the programming language by the team is crucial to success.

Key Words: MCSA, current signature analysis, Python, AI, Generative AI, programming

Introduction

Prior to the end of 2022 there were a few tools and platforms to assist in the development of technical software. In December 2022, OpenAI introduced several advanced LLMs that had great potential for software development. Over time other competitors have entered the market, and these tools have quickly advanced in the marketplace with some thinking that generative AI tools will replace programmers and other creatives. By 2025 a new term has been introduced as a direct result of these tools called 'vibe coding.' This is the thought process that someone without any experience can fully develop software from start to finish. In this paper we will discuss the progression of a commercial project in which these tools were utilized to streamline the process.

The development of the tool discussed in this paper is a commercially viable current signature analysis product which has had success in the market. The original concept for the product was already in process which was expected to take up to a year to develop originally. Once a hardware component was identified the original plan was scrapped and the system was adjusted to a single phase of current CSA device. The project direction also required us to deviate from standard software development practice.

Standard small-business development steps for the development of a technical instrument are as follows:

1. Concept and Requirements: define product objectives; identify target users and use cases; document technical specifications and operational requirements; engage with stakeholders to clarify expectations.
2. Feasibility and Risk Assessment: evaluate technical and financial feasibility; perform risk assessment; identify compliance, safety and industry standard requirements.
3. Planning and Architecture Design: select technology stack and frameworks for interface and embedding; develop system architecture diagrams; develop a detailed project timeline and milestones; select a development team with clearly defined roles and responsibilities.
4. User Experience and Interface Design: draft wireframes and prototypes to refine workflows; design GUI – recommend intuitive and responsive; gather early user feedback and revise.
5. Development: develop using an agile or incremental approach; develop software in manageable, testable increments; regular peer code reviews to ensure code quality, maintainability and best practices; frequent code integration to catch issues early.
6. Testing and Validation: automated tests for individual software modules; validate interactions; system testing including hardware and software; engage stakeholders and end-users to validate real-world usability; any regulatory compliance verification.
7. Documentation: user manuals and quick-start guides; technical spec and datasheets; maintenance manuals for tech support; software release notes and version control records.
8. Deployment and Release: develop release procedures, versioning and update strategies; provide deployment package and installation tools; conduct field testing with limited user groups.

9. Training and Support: conduct training for internal and external users; establish customer-support procedures which may include help desk, FAQ, troubleshooting guides.
10. Maintenance and Iterative Improvement: collect user feedback for continuous improvement; monitor performance and reliability; issue patches and updates to improve performance, security and functionality.

Following this process, we saw the potential of up to 12 months for the original product design including the selection of hardware. The application of the support tools influenced the development process.

The Impact of LLM Tools on Development

Areas where tools such as those mentioned can be used each tool can be used for different parts of the process.

1. Requirements and Conceptualizations: generate and refine product ideas, concepts and use-case scenarios; process vague stakeholder input to precise functional requirements; quickly research and summarize competitors and related technologies.
2. Feasibility Analysis and Planning: quickly assess technology stacks, frameworks and compatibility; highlight potential risks or missing considerations; develop timelines, task breakdowns and resource allocations.
3. Architecture and Design Phase: Generate and validate architecture diagrams or system design options; develop user workflows, wireframes or navigation models; suggest UI layouts tailored to the target user demographic.
4. Development and Implementation: automatically draft code snippets, functions or full prototypes; detect potential bugs, inefficiencies, or code defects; suggest improvements for readability, maintainability, and efficiency.
5. Testing and Validation: suggest test cases and unit tests based on provided requirements, including test case generation; quickly generate boilerplate code for automated tests; troubleshoot errors, interpret log files, and provide potential fixes.
6. Documentation Phase: quickly produce clear, consistent technical documentation, user manuals, and release notes; generate structured, easy-to-understand API documentation; develop concise, practical training guides.
7. Deployment and Release Phase: draft checklists, release procedures, and deployment instructions; prepare quick troubleshooting resources and common error FAQs; generate clear release announcements or update notices.
8. Support, Training and Maintenance: provide rapid troubleshooting assistance based on problem descriptions; generate training FAQs and interactive troubleshooting guides; prioritize reported issues, propose quick fixes or maintenance strategies.
9. Innovation and Continuous Improvement: propose innovative enhancements based on current product capabilities; summarize the latest industry trends or advancements relevant to your product; aid in analyzing customer feedback and prioritizing future feature development.

The use of ChatGPT throughout these steps resulted in a reduction of version 1.0 release down to 9 weeks from the expected year of development. However, things were not as smooth as expected and there were, and are, plenty of lessons learned as we continue with the final step while also releasing several additional software products.

The steps we generically used for the development of the software using the various tools mentioned in the introduction include:

1. Step 1: worked with planned concept and features based on experience and used ChatGPT to refine the concepts.
2. Step 2: generate UI wireframes with ChatGPT and other layout tools.
3. Step 3: software development using Python. We did not allow Python to generate the algorithms.
4. Step 4: generate and document testing procedures.
5. Step 5: use tools for user documentation and training materials.
6. Step 6: assist draft release communications and customer-facing documents.

Before following these steps, make sure to read our lessons learned section of this paper.

Example Development in Python

Versioning and capabilities of ChatGPT and other LLM development tools, such as various Copilot systems, have advanced since the project started. In some cases, this has been an improvement, in others not so much. For this example, we will use the '125kV to 13-8kV transformer.wav' file from github.com/MotorDoc1/FunWithSound/ (free to access) in which there is also an Anaconda Notebook file with variations in working with the data.

We will use Visual Studio Code and Python version 3.11 (python.org). The wav file was gathered using a Samsung A51 phone which resulted in 44.1 kHz and 32-bit data. The purpose is to develop a spectra that can be compared to IEEE Standard C57.136-2000, "IEEE Guide for Sound Level Abatement and Determination for Liquid-Immersed Power Transformers and Shunt Reactors." This standard provides various dB spectral patterns related to sound to determine condition of transformers in effect providing a way to interpret the hum you hear from the transformer.

We will also be just developing a method for interpreting data using ChatGPT and the sound file. The complete analysis will be performed within ChatGPT version 4o in which we will upload the sound file.

Prompt: 'import wav file and show the waveform as an image.' The LLM will then show the python code and the resulting image as follows.

```
import matplotlib.pyplot as plt
import numpy as np
from scipy.io import wavfile

# Load the WAV file
file_path = "/mnt/data/125kV to 13-8kV transformer.wav"
```

```
sample_rate, data = wavfile.read(file_path)

# If stereo, select only one channel
if len(data.shape) > 1:
    data = data[:, 1]  # Use channel 2 (Channel B)

# Generate time axis in seconds
time = np.arange(len(data)) / sample_rate

# Plot waveform
plt.figure(figsize=(12, 4))
plt.plot(time, data, linewidth=0.5)
plt.title("Waveform of '125kV to 13-8kV transformer.wav'")
plt.xlabel("Time [s]")
plt.ylabel("Amplitude")
plt.grid(True)
plt.tight_layout()

# Save the waveform image
image_path = "/mnt/data/waveform_image.png"
plt.savefig(image_path)

image_path
```

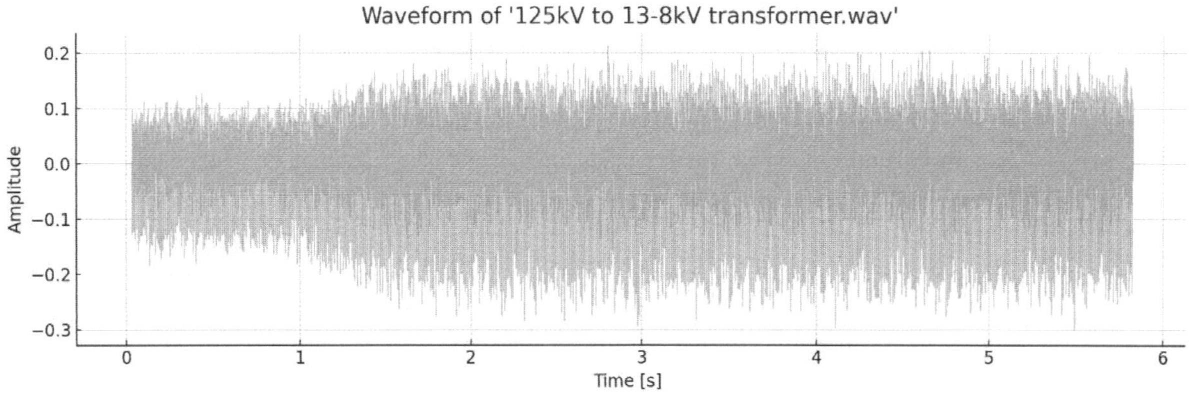

Figure 1: Image from prompt and code.

Next, we will examine the sample rate and other features of the file with the following prompt: ‘provide sample rate, fundamental frequency, time and bit for next steps.’

```
from scipy.fft import fft, fftfreq

# Determine bit depth from data type
bit_depth = data.dtype.itemsize * 8

# Estimate fundamental frequency
```

```
# Perform FFT
n = len(data)
yf = fft(data)
xf = fftfreq(n, 1 / sample_rate)

# Consider only the positive frequencies
idx = np.arange(1, n // 2)
frequencies = xf[idx]
magnitudes = np.abs(yf[idx])

# Find the peak frequency (excluding DC component)
fundamental_freq = frequencies[np.argmax(magnitudes)]

# Total duration in seconds
duration_seconds = len(data) / sample_rate

{
  "Sample Rate (Hz)": sample_rate,
  "Estimated Fundamental Frequency (Hz)": round(fundamental_freq, 2),
  "Duration (s)": round(duration_seconds, 2),
  "Bit Depth": bit_depth
}
```

The result is a sample rate of 44.1 kHz, fundamental frequency of 240 Hz, duration 5.83 seconds and 32-bit sound depth. The LLM will retain this information.

In the next step we will define a spectra such that the Nyquist is 2, converted to dB with the peak frequency set to 0 dB, use 3 averages and flat top window. This will generate negative dB levels. We will let ChatGPT determine the formula for dB based upon sound. The prompt will be: 'develop a spectra such that the nyquist is 2, convert sound to dB with the highest peak set to 0 dB, use 3 averages and flat top window. Reference prior data from the file. Show full spectra range.'

```
from scipy.signal import windows, welch
import matplotlib.pyplot as plt

# Define parameters
nyquist_target = 2  # Hz
segment_length = int(sample_rate / nyquist_target)  # samples per segment
n_overlap = segment_length // 2  # 50% overlap
window = windows.flattop(segment_length)  # Flat top window
n_avg = 3  # Number of averages

# Truncate data for averaging
n_segments = (len(data) - n_overlap) // (segment_length - n_overlap)
n_segments = min(n_segments, n_avg)
```

```
avg_data = data[:n_segments * (segment_length - n_overlap) + n_overlap]

# Welch's method with flat top window and averaging
frequencies, power = welch(
    avg_data,
    fs=sample_rate,
    window=window,
    nperseg=segment_length,
    noverlap=n_overlap,
    scaling='spectrum',
    average='mean'
)

# Convert to dB, peak normalized to 0 dB
power_db = 20 * np.log10(power)
power_db -= np.max(power_db)

# Plot the full spectra
plt.figure(figsize=(14, 5))
plt.plot(frequencies, power_db, linewidth=0.8)
plt.title("Full Spectrum (Flat Top, 3 Averages, Peak-Normalized dB)")
plt.xlabel("Frequency (Hz)")
plt.ylabel("Amplitude (dB)")
plt.grid(True)
plt.tight_layout()

# Save the spectrum image
spectrum_image_path = "/mnt/data/full_spectrum_dB.png"
plt.savefig(spectrum_image_path)

spectrum_image_path
```

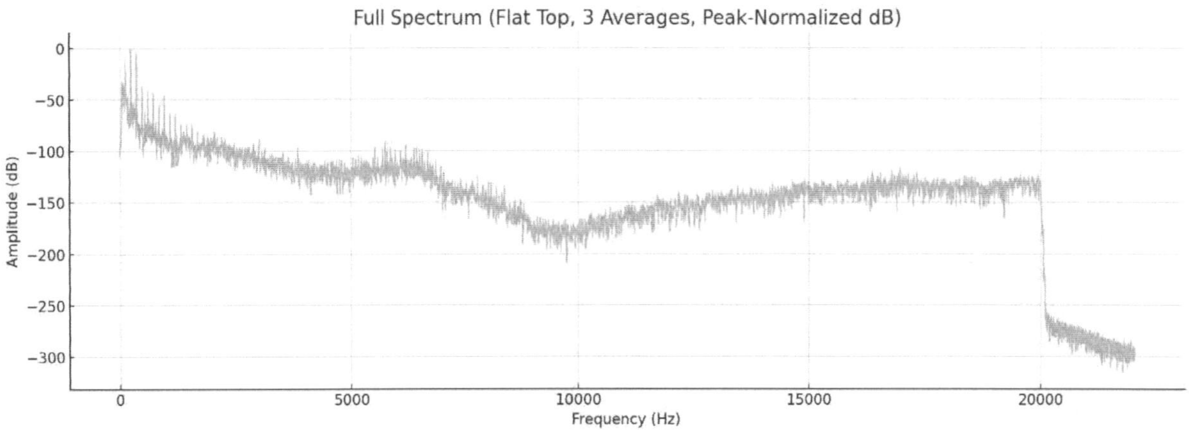

Figure 2: Full spectra which comes out to 22 kHz

The LLM tool then makes recommendations, as well. We will follow one on zooming in to view the data from 0 Hz to 2,000 Hz to see just that range. The prompt will simply be to zoom in to that range. There will be a few issues so we will modify the prompt to state: 'zoom in from 0 to 2,000 Hz. Set the Y-axis to the visible spectrum in this range starting at 0 dB. Set frequency harmonic markers starting at 60 Hz, which is line frequency.'

```
# Limit spectrum to 0–2000 Hz
freq_mask = (frequencies >= 0) & (frequencies <= 2000)
zoom_freqs = frequencies[freq_mask]
zoom_power_db = power_db[freq_mask]

# Determine Y-axis limits starting from 0 dB downward
ymin = np.min(zoom_power_db)
ymax = 0

# Generate harmonic markers at multiples of 60 Hz
line_freq = 60  # Hz
harmonics = np.arange(line_freq, 2000 + line_freq, line_freq)

# Plot zoomed spectrum
plt.figure(figsize=(14, 5))
plt.plot(zoom_freqs, zoom_power_db, linewidth=0.8, label="Spectrum (0–2000 Hz)")
for h in harmonics:
   if h <= 2000:
      plt.axvline(x=h, color='red', linestyle='--', alpha=0.6)
         plt.text(h, ymax - 1, f'{h} Hz', color='red', rotation=90, verticalalignment='top',
fontsize=8)

plt.title("Zoomed Spectrum (0–2000 Hz) with Harmonic Markers (60 Hz base)")
plt.xlabel("Frequency (Hz)")
plt.ylabel("Amplitude (dB, peak-normalized)")
plt.ylim([ymin, ymax])
plt.grid(True)
plt.tight_layout()

# Save the zoomed spectrum image
zoomed_spectrum_path = "/mnt/data/zoomed_spectrum_0_2000Hz_dB.png"
plt.savefig(zoomed_spectrum_path)

zoomed_spectrum_path
```

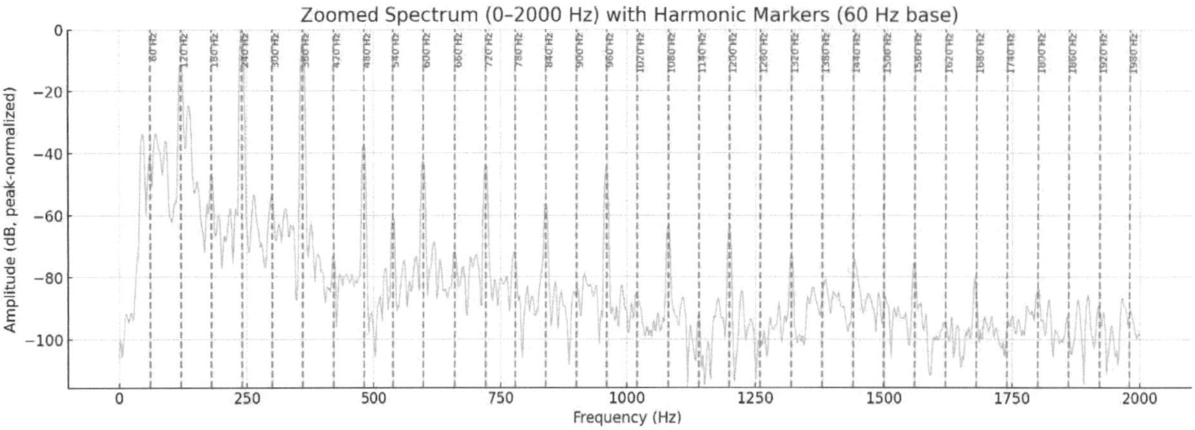

Figure 3: Zoomed spectra with Y-axis set to show complete spectra and line frequency harmonic markers.

In the next steps I asked ChatGPT to review the complete dataset and see if it could interpret the data based upon the IEEE standard. It claims that it can. So, we will ask it to: 'take in any transformer sound file of any length, show the spectra and markers at 50 or 60 Hz, identify the sound file length, bit depth, and sample rate – displayed, and identify the condition based upon the standard. We will also request that it sets up a complete GUI for this process that includes the ability to open the wav file, a button that runs the process, a pdf output of the findings, and a close button that safely closes the software. Show the spectra, harmonics and conclusions in the GUI with a save as pdf button.'

```
import tkinter as tk
from tkinter import filedialog, messagebox
from tkinter import ttk
import numpy as np
import matplotlib.pyplot as plt
from scipy.io import wavfile
from scipy.signal import windows, welch
from matplotlib.backends.backend_tkagg import FigureCanvasTkAgg
from fpdf import FPDF
import os

class TransformerAnalyzerApp:
    def __init__(self, root):
        self.root = root
        self.root.title("Transformer Sound Analyzer")
        self.file_path = ""
        self.sample_rate = None
        self.bit_depth = None
        self.duration = None
        self.power_db = None
        self.frequencies = None
        self.line_freq = 60  # Default, user can override
```

```
        self.setup_gui()

    def setup_gui(self):
        frame = ttk.Frame(self.root, padding="10")
        frame.grid(row=0, column=0, sticky=(tk.W, tk.E, tk.N, tk.S))

        self.open_button = ttk.Button(frame, text="Open WAV File",
command=self.load_wav)
        self.open_button.grid(row=0, column=0, pady=5)

        self.analyze_button = ttk.Button(frame, text="Analyze",
command=self.analyze_wav, state='disabled')
        self.analyze_button.grid(row=0, column=1, pady=5)

        self.save_button = ttk.Button(frame, text="Save PDF Report",
command=self.save_pdf, state='disabled')
        self.save_button.grid(row=0, column=2, pady=5)

        self.quit_button = ttk.Button(frame, text="Close", command=self.root.quit)
        self.quit_button.grid(row=0, column=3, pady=5)

        self.info_text = tk.Text(frame, width=80, height=10, wrap="word")
        self.info_text.grid(row=1, column=0, columnspan=4, pady=10)

        self.canvas_frame = ttk.Frame(self.root)
        self.canvas_frame.grid(row=1, column=0)

    def load_wav(self):
        file_path = filedialog.askopenfilename(filetypes=[("WAV files", "*.wav")])
        if file_path:
            self.file_path = file_path
            self.analyze_button.config(state='normal')
            self.info_text.delete(1.0, tk.END)
            self.info_text.insert(tk.END, f"Loaded file: {os.path.basename(file_path)}\n")

    def analyze_wav(self):
        self.info_text.delete(1.0, tk.END)
        sr, data = wavfile.read(self.file_path)
        self.sample_rate = sr
        self.bit_depth = data.dtype.itemsize * 8
        if len(data.shape) > 1:
            data = data[:, 1]
        self.duration = len(data) / sr

        segment_length = int(sr / 2)
        window = windows.flattop(segment_length)
```

```
        frequencies, power = welch(data, fs=sr, window=window, nperseg=segment_length,
noverlap=segment_length//2, scaling='spectrum', average='mean')
        power_db = 20 * np.log10(power)
        power_db -= np.max(power_db)

        self.frequencies = frequencies
        self.power_db = power_db

        self.display_info()
        self.plot_spectrum()
        self.save_button.config(state='normal')

    def display_info(self):
        self.info_text.insert(tk.END, f"Sample Rate: {self.sample_rate} Hz\n")
        self.info_text.insert(tk.END, f"Bit Depth: {self.bit_depth} bit\n")
        self.info_text.insert(tk.END, f"Duration: {self.duration:.2f} seconds\n")

        harmonics_detected = []
        for h in range(self.line_freq, int(self.sample_rate / 2), self.line_freq):
            idx = (np.abs(self.frequencies - h)).argmin()
            db = self.power_db[idx]
            if db > -60:
                harmonics_detected.append((h, db))
                self.info_text.insert(tk.END, f"Harmonic at {h} Hz: {db:.2f} dB\n")

        if harmonics_detected:
            self.info_text.insert(tk.END, "\nCondition: Harmonics present, monitor for core
or winding vibration.\n")
        else:
            self.info_text.insert(tk.END, "\nCondition: No significant harmonics detected in
expected range.\n")

    def plot_spectrum(self):
        fig, ax = plt.subplots(figsize=(8, 3))
        ax.plot(self.frequencies, self.power_db, linewidth=0.7)
        ax.set_title("Spectrum with Harmonics")
        ax.set_xlabel("Frequency (Hz)")
        ax.set_ylabel("Amplitude (dB)")
        ax.grid(True)

        for h in range(self.line_freq, int(self.sample_rate / 2), self.line_freq):
            if h < self.frequencies[-1]:
                ax.axvline(h, color='red', linestyle='--', alpha=0.4)
                ax.text(h, 0, f"{h} Hz", rotation=90, color='red', fontsize=6,
verticalalignment='bottom')
```

```
        for widget in self.canvas_frame.winfo_children():
            widget.destroy()

        canvas = FigureCanvasTkAgg(fig, master=self.canvas_frame)
        canvas.draw()
        canvas.get_tk_widget().pack()

    def save_pdf(self):
        pdf = FPDF()
        pdf.add_page()
        pdf.set_font("Arial", size=12)

        pdf.cell(200, 10, txt="Transformer Sound Analysis Report", ln=1, align='C')
        pdf.ln(10)

        pdf.multi_cell(0, 10, f"File: {os.path.basename(self.file_path)}")
        pdf.multi_cell(0, 10, f"Sample Rate: {self.sample_rate} Hz")
        pdf.multi_cell(0, 10, f"Bit Depth: {self.bit_depth} bit")
        pdf.multi_cell(0, 10, f"Duration: {self.duration:.2f} seconds")

        for h in range(self.line_freq, int(self.sample_rate / 2), self.line_freq):
            idx = (np.abs(self.frequencies - h)).argmin()
            db = self.power_db[idx]
            if db > -60:
                pdf.multi_cell(0, 10, f"Harmonic at {h} Hz: {db:.2f} dB")

        spectrum_path = "spectrum_temp.png"
        fig, ax = plt.subplots(figsize=(8, 3))
        ax.plot(self.frequencies, self.power_db)
        ax.set_title("Spectrum with Harmonics")
        ax.set_xlabel("Frequency (Hz)")
        ax.set_ylabel("Amplitude (dB)")
        ax.grid(True)
        for h in range(self.line_freq, int(self.sample_rate / 2), self.line_freq):
            if h < self.frequencies[-1]:
                ax.axvline(h, color='red', linestyle='--', alpha=0.4)
        plt.tight_layout()
        fig.savefig(spectrum_path)
        pdf.image(spectrum_path, x=10, y=None, w=180)
        os.remove(spectrum_path)

        output_path = filedialog.asksaveasfilename(defaultextension=".pdf",
filetypes=[("PDF files", "*.pdf")])
        if output_path:
            pdf.output(output_path)
            messagebox.showinfo("PDF Saved", f"Report saved to {output_path}")
```

```
# Launch the GUI
root = tk.Tk()
app = TransformerAnalyzerApp(root)
root.mainloop()
```

The result when entered into the editor, saved and run, we get the GUI. Then it is time to test the software. We highly recommend reviewing it manually first but will take just the raw product and evaluate it.

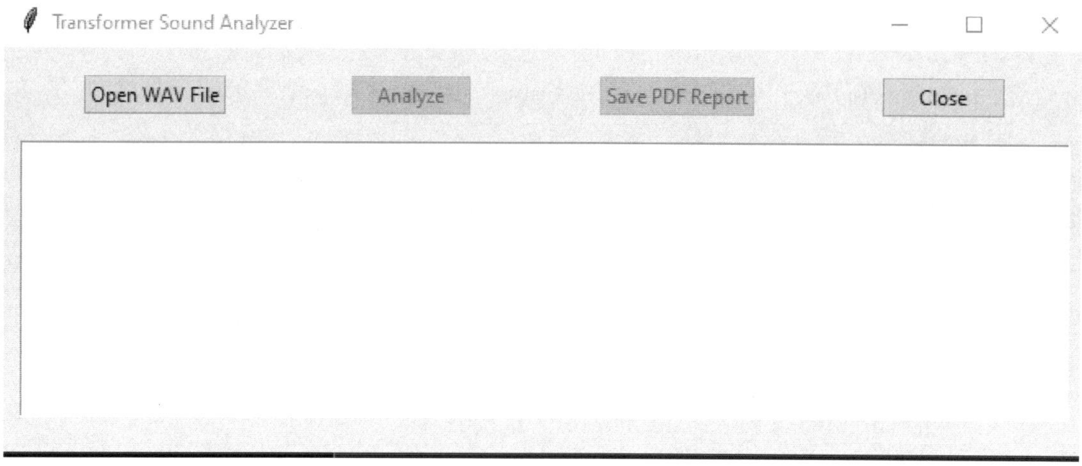

Figure 4: The software that was developed.

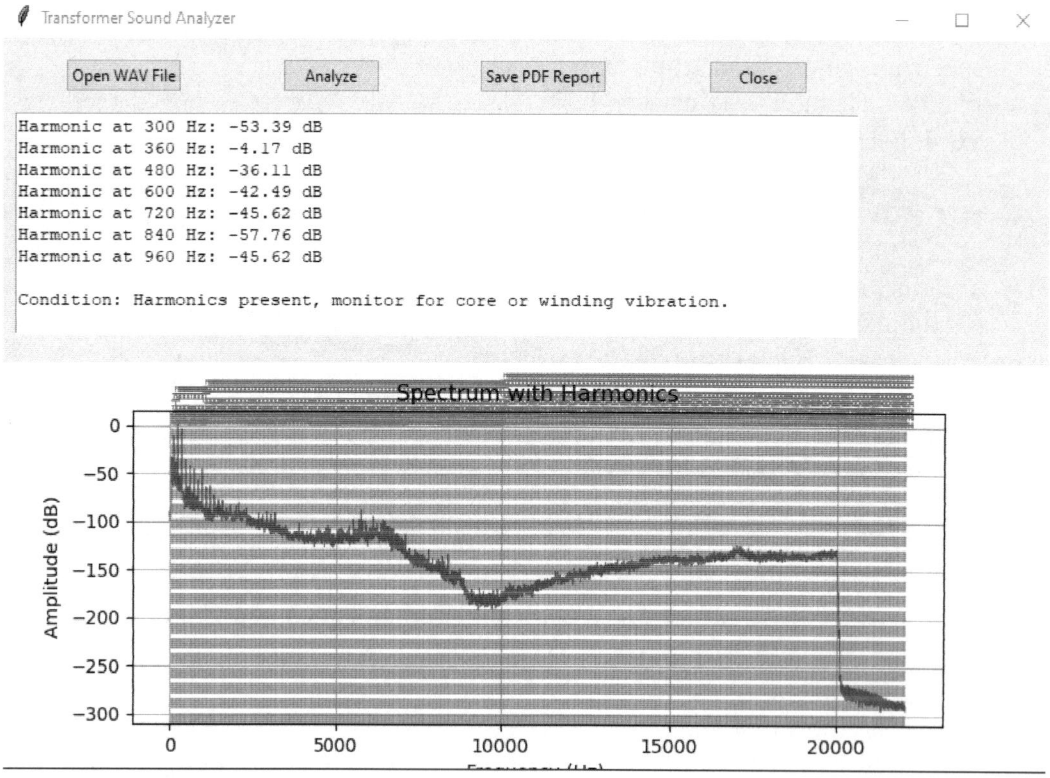

Figure 5: GUI when importing the data file that was analyzed.
It should be noted that it did not provide several features that we requested.

Transformer Sound Analysis Report

File: 125kV to 13-8kV transformer.wav

Sample Rate: 44100 Hz

Bit Depth: 32 bit

Duration: 5.83 seconds

Harmonic at 60 Hz: -39.91 dB

Harmonic at 120 Hz: -10.75 dB

Harmonic at 180 Hz: -38.53 dB

Harmonic at 240 Hz: 0.00 dB

Harmonic at 300 Hz: -53.39 dB

Harmonic at 360 Hz: -4.17 dB

Harmonic at 480 Hz: -36.11 dB

Harmonic at 600 Hz: -42.49 dB

Harmonic at 720 Hz: -45.62 dB

Harmonic at 840 Hz: -57.76 dB

Harmonic at 960 Hz: -45.62 dB

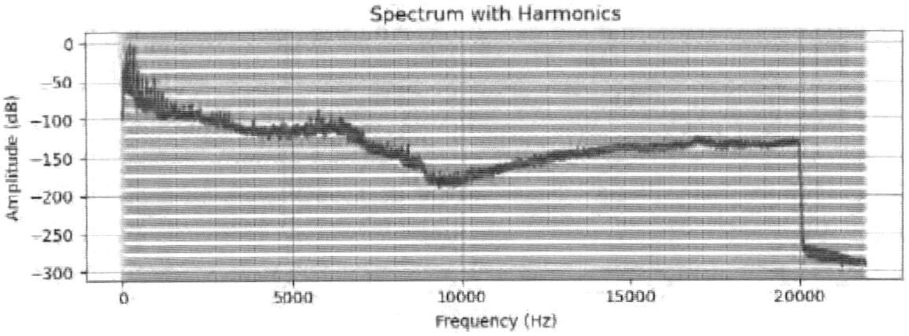

Figure 6: The pdf report output.

At this point we would go on to do testing, adding the missing and additional features, and field testing before deployment.

Our Experiences When Creating Over 10k Lines of Code

The concept for the project was a plan for a simple single phase of current MCSA system for evaluating electric machines of all types of sizes that would interpret data based on a rules system. The system would be more affordable and portable for an average technician and would provide detailed conclusions based upon the entered information. It was also to be based on standard libraries from Python which would make it flexible for both Windows and Linux operating systems.

As noted in the previous example, it is acknowledged that the LLMs we used were invaluable in the development of the software. It is also noted that even though we use the paid version that promises that our work would not be included in the training data that the AI version of development would not be used to develop our algorithms. We did not use any other method or any of the ‘software development flow’ systems that are available on the market in which our prior experiences were not positive.

Following are a few of the lessons we learned – contrary to much of the marketing hype:

- Generative AI systems accuracy and responses vary widely between each use and sometimes in the same chat session. For instance, a repeat with the same prompts used in our sample would result in very different code and software.
- If asking the LLM to fix the software chances are that it would completely modify and often break the previously working portion of the code.
- Once you have hit about 1500-2000 lines of code, for the enterprise editions we use, it starts to fall apart. Therefore, we ended up providing structure and having the LLMs assist us function by function.
- DO NOT use generative AI as your subject matter expert.
- Contrary to what some programmers, consultants and marketing people will tell you – one of your software people MUST be an SME for what you are developing.
- DO NOT TRUST and VERIFY: the LLM will often deliberately lead you astray and promise to perform tasks that it simply cannot do. By August of 2025 it was claiming lengthy development in the background and then when questioned it was stating that it had lied and when pressed it stated ‘in order to waste your time.’
- During 8 weeks of programming, 1.5 weeks was development and 6.5 weeks was debugging and troubleshooting. We are referring to 7-day weeks.
- Develop your own algorithms and only let the LLM systems develop the framework.
- DO NOT let the AI determine the flow, the SME must do that.
- AI will drift and sometimes completely break down. Expect and plan for this. Know when to walk away and work on other things, such as debugging. In one instance while we were testing various AIs ability to self-debug we spent a 12 hour Saturday hunting a single bug that turned out to be an indentation issue. We manually found the problem in about 15 minutes.
- Beware! AI systems will often hallucinate software libraries and commands that do not exist. This is where a programming expert is required as well as access to good documentation

There are more considerations, but these are the primary. While the business goal was to develop a viable product, that ended up performing better than expected, the research part of the project was to see what would happen in a full, from scratch, instrument software development. The result was a far greater success than expected, faster than expected, which allowed us to explore the addition of features we did not originally consider.

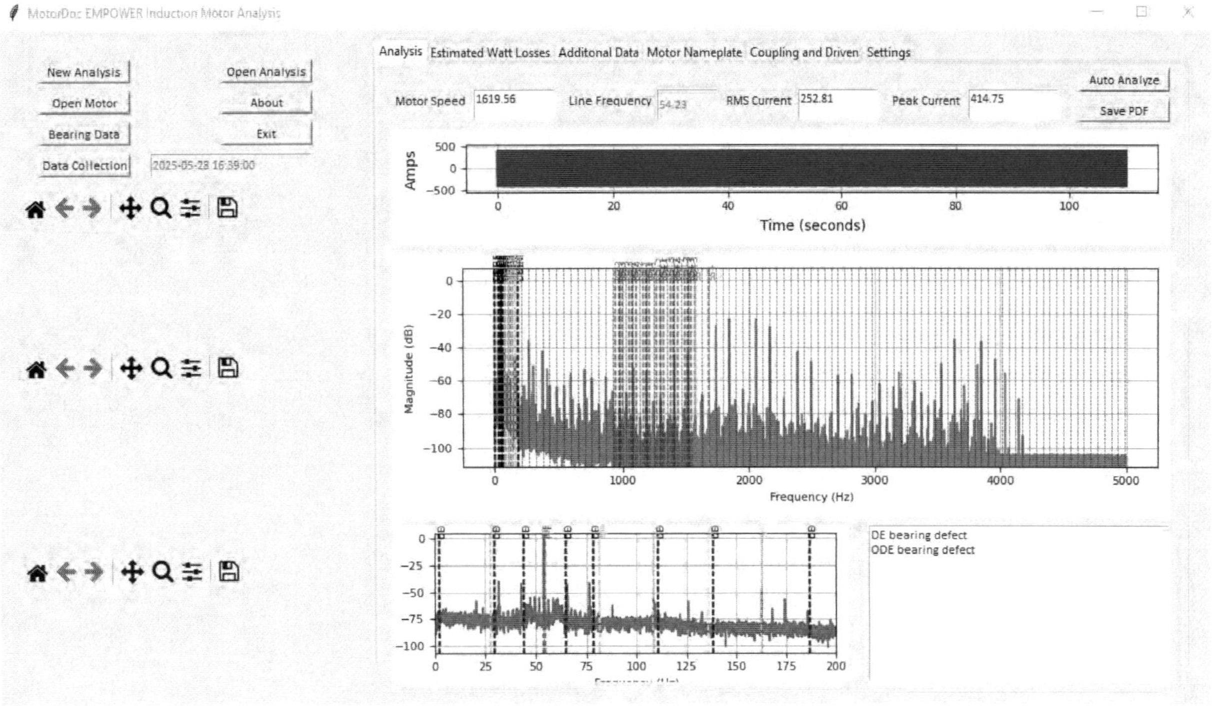

Figure 7: The continuing development results.

Conclusion

While LLMs and generative AI are tools that can assist in making software development more efficient, the developer must be cautious and must ensure that they have access to an expert programmer and a subject matter expert (SME) for the system being developed.

Our experience demonstrated that while generative AI tools such as ChatGPT and Copilot were invaluable for accelerating parts of the development, they are not substitutes for human expertise. Particularly for systems that involve specialized domains—such as transformer diagnostics or electrical machine analysis—the SME must define logic, validate interpretations, and verify the scientific soundness of the results.

Moreover, the tools tend to **drift**, **hallucinate**, or **generate non-functional or misleading code**, especially in longer projects. The process becomes more error-prone beyond 1500–2000 lines of code unless broken into modular components. Despite these limitations, with structured use and iterative human oversight, the tools helped us develop a fully functional, commercial-grade application significantly faster than traditional workflows.

The **final result** was not only a working product but one that exceeded initial expectations in features, flexibility, and scalability—while also highlighting the potential for future enhancements through hybrid human-AI collaboration models.

LOGICAL VIBRATION FAULT ANALYSIS FOR ROTATING MACHINERY

Author: Jeffrey M. Mauk

VIATC | TRAINING CONFERENCE & EXPO
NEWPORT NEWS, VIRGINIA | AUGUST 6-8, 2025

LOGICAL VIBRATION FAULT ANALYSIS FOR ROTATING MACHINERY

Jeffrey M. Mauk
Machinery Equipment Monitoring Technician
BP

Jeffrey M. Mauk works has worked as a Machinery Equipment Monitoring Technician at BP Whiting Refinery in Indiana for the past 18 years, with principal duties focused on centrifugal compressors, turbine generators, and large critical process rotating equipment. He has a maintenance machinist apprenticeship with the US Department of Labor and over 41 years' experience working in heavy process industry in steel, chemical, gas and oil, and manufacturing. He has worked in predictive maintenance for twenty-five years and holds a Category III Vibration Analyst certification with the Vibration Institute.

ABSTRACT

Where do we begin when we start an analysis of machinery? What are the basic steps, tools, and process of narrowing down the responses we are feeling, smelling, hearing, touching and, yes, eventually measuring vibration signals? What predetermined prognosis have I made in an office before I go and look at the machine? Many others have walked this road that I have stumbled through over my career. I faced my own self-inflicted errors, because I was too self-absorbed. I say, "follow the path already trodden". From what first appears as unbalance, a simple misalignment issue, to an axial 1x vibration, these are three case studies where what seemed obvious was a more complex defect.

Key Words: Looseness, foundation, axial vibration

Why Field Verification Is Important

The vibration routes at the Whiting Refinery are collected by operations personnel. Inherently there is always some data that needs verified. I almost always verify the data with a field walkdown and confirmation of collected data. When it is clear a condition has changed, field analysis begins. The environment of the machines status is important. Is it clean? Are the filters plugged? Is the motor fan covered in dirt? Does it have barrier tape wrapped around the fan? What sounds am I hearing? Are there any abnormal smells? I am touching the machine for temperature or vibration and seeing if broken shims are under the coupling. I cannot tell you how many times I've seen that. There are many faults that can be quickly identified with a handheld unit, but when situations demand more analytical tools a virtual instrument that has the capability to store several different acquisition files for post processing over multiple channels closes that gap. Knowing the application, process, and fluid dynamics are all value-added information that is mostly gathered by going to the field and studying the application before you start building a configuration. Each facility analyst work varies, but we have many similarities. We all have motors, pumps, gearboxes, etc.

Case Study One – Unbalance or Looseness?

A center hung 300 HP ID (Induced Draft) Fan for a Furnace that pre-heats the feed stock for the Coker Unit broke alert and danger alarm levels after a process upset had forced the shutting down of the fan. The fan's previous vibration at steady state values was at 0.16 in/s pk. When the ID fan was placed back into service, the vibration levels began increasing abnormally quickly based upon comparable startups. Vibration levels reached 0.6 in/s pk within 6 hours of startup. (Figure 1)

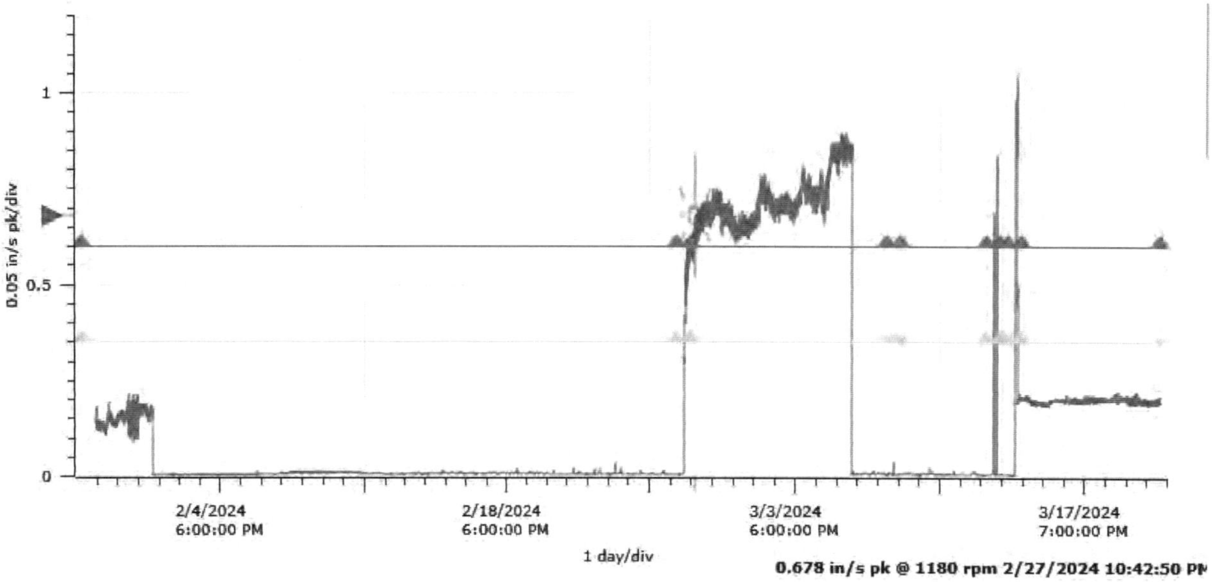

Figure 1

The ID Fan has stud-mounted vibration sensors installed in the horizontal direction for periodic online trending. The vibration measurements are communicated via Modbus and linked to the

DCS (Disturbed Control System). The alert and danger set points were met, and board operators received alarms as the set points were exceeded. Field verification with a hand-held data collector confirmed the online trending values, and discussion of faults began. Cross channel phase was close to 60°, so we could not rule out unbalance. An attempt was made with the data collector using its trigger, but issues with instruments prevented success. The information and the scope of the analysis dictated a virtual analyzer would provide more clarity and measurements in more than one direction.

A configuration was built and set up in the field with all the cabling, sensors, and a speed reference, (got to have speed). Where do you begin? What are the first plots that provide information on an operating machine? The first thing I look at is the Time Waveform and the FFT (Fast Fourier Transform). The WF (wave form) is an unaltered response from the force to the absolute vibration sensor. It is a picture of what is happening. It is art in motion. From there the use of barographs (Figure 2) and a vector table are immensely helpful visually, and with the set points of alert and danger configured, you know immediately the highest amplitudes and where to focus.

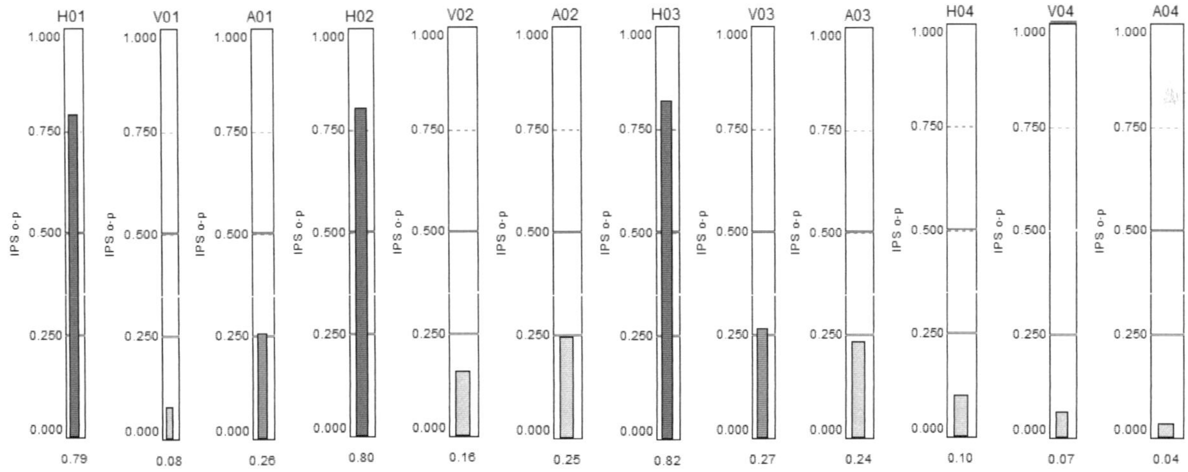

Figure 2

First thoughts while just looking at the trend data from the horizontal sensors was it is possible, we have an unbalance issue, but the data does not show that? What does unbalance look like?

- Vibration frequency is predominately 1X on FFT
- Amplitude of horizontal vibration is typically twice the vertical vibration.
- Phase shift between vertical and horizontal sensors is 90° or 270° (+/-30°)
- Time waveform is sinusoidal

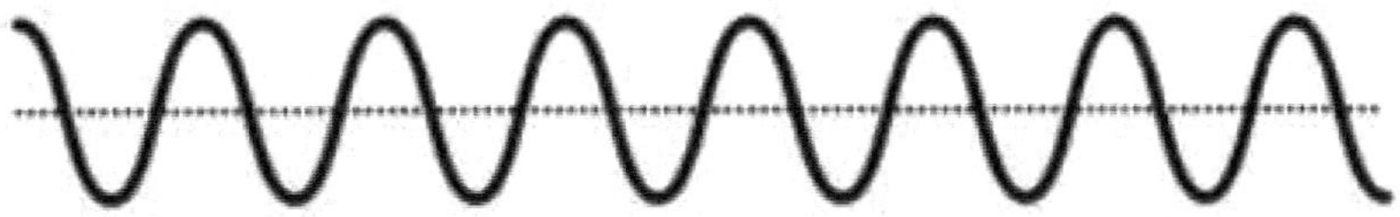

Generally, this is a good place to start.

Upon viewing the data and based upon the criteria, we have met two of the four bullet points. 1X dominates and the time waveform is sinusoidal, but the amplitudes are beyond the 50% higher between horizontal to vertical, and the fan NDE is low amplitude. We do not have an unbalanced fan. I have talked myself into balance syndrome too many times, but not anymore, back to the basics.

The highest vibration at the motor is horizontal radial measurements at the NDE & DE. The fan bearing DE horizontal radial vibration is high but not the NDE. The base structure for the motor and fan DE is separate from the NDE of the Fan. If this were an unbalance, eccentricity, rotor bow, or bent shaft, we should see high vibration at the fan NDE. It is not there, and the phase shift for unbalance is not there either. Sleeve bearing looseness with a heavy rotor is categorized as typical type A mechanical looseness. Viewing the trigger pulses (red circles) in the artwork of the TW show them changing positions. The rotor is not in the same place each time the trigger is referenced to its high spots. (Figure 3)

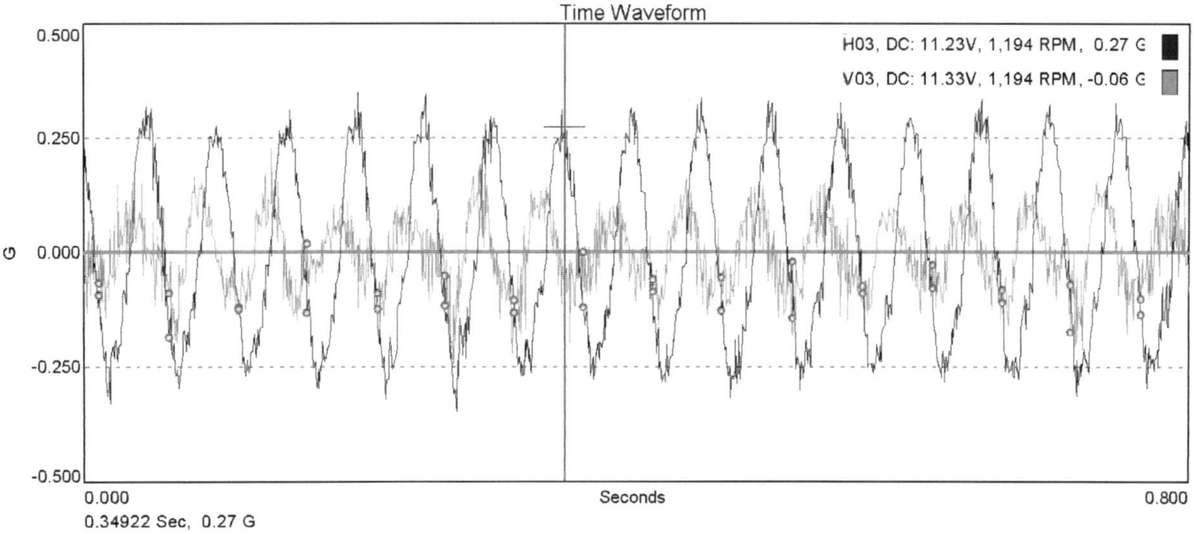

Figure 3

Due to the lower vibration at the NDE bearing pedestal and the high horizontal reading at the motor, mechanical looseness is possible. When we first began discussing faults, we asked and were informed that all the fasteners were checked. When the field analysis was being conducted, many folks wanted some answers, operation staff, rotating equipment engineer, department superintendent. Pressure is on you. “Keep Calm Carry On.” “This isn’t WW2.” They will blurt out what they think you should do, or keep asking, “What do you think?” In this case, if we don’t find out what’s wrong, this fan is coming down. It is above our standard design limit for operations. That means the furnace comes down and we are not cooking coke.

Everybody is looking at us now. So, we go sit on the base and we can feel the steel frame moving compared to the foundation bedplate. I grab a pair a channel locks from my pocket. (I always have a pair of channel locks, 4” adjustable wrench, wire strippers, and a multi tip screwdriver) I turned the bolt head a ¼ turn. **WHAT!!** I thought you said the fasteners were tight.

We went to the shop grabbed the correct wrench and tightened all the bolts ourselves. (We can get away with it since we are still machinists.) The vibration dropped to 0.16 in/s pk. They did not tighten the bolts. A work order was issued to torque the bolts along with a PM for annual torquing of the nuts. Final measurements after tightening all the fasteners are shown in Figure 4.

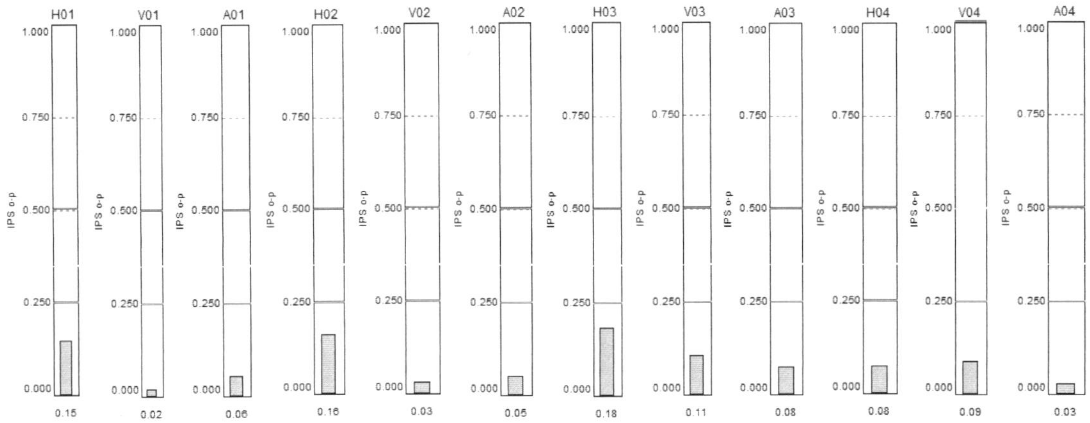

Figure 4

Case Study Two Just a Failed Coupling?

A back pull-out center supported single stage overhung pump 4x6x13, driven by a 2-pole 250 HP motor at a CRU (Catalytic Reforming Unit) unit pulls LMD (Light Middle Distillate) from the feed tower and increases the pressure to 354 psig. The LMD is pre-heated through two heat exchangers before being heated in a furnace. LMD is within the Kerosene stocks. The CRU is really a desulfurizer today, producing a jet fuel product.

The failure was obvious. The coupling was destroyed. The high vibration caused a bypass line to crack (Figure 5). LMD at 345 PSIG spraying through a crack isn't good. We like to share our stories of great findings and our accomplishments as we should, but sometimes being honest and telling the truth builds more character. We (our vibe group) took some egg on the face with this failure. There were signs, or at least alert status setpoints, that were exceeded, but upon examining the software we didn't have alarms set for vertical measurements. Figure 6 shows the trend of monthly data. The last reading before the failure was 0.45 in/s pk. The operators collect the route data. There are positives and negatives for operations staff to collect data. I still remember the email from my colleague, who was new to the vibration department. Management is going to want some answers about what happened. I told the truth. The safety department reported it correctly and we bore the shame. All discipline when you're going through it isn't joyful. It is quite painful. But in the end, it results in a peaceable outcome for those who are willing to be trained by it. We need to grow and learn from our mistakes. I picked up a saying from a coworker, "it only bothers if you care, it bothered me immensely.

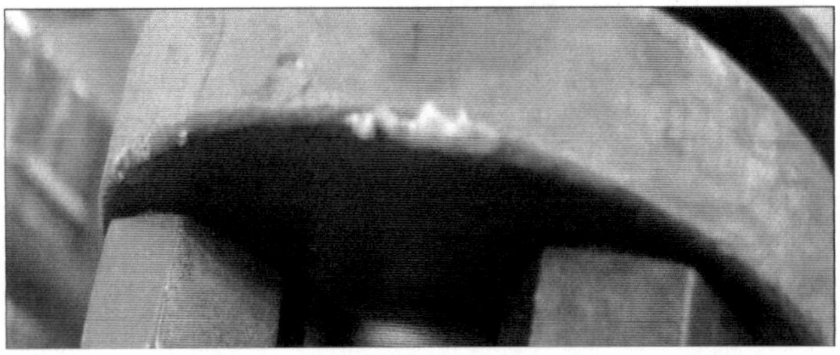

Figure 5

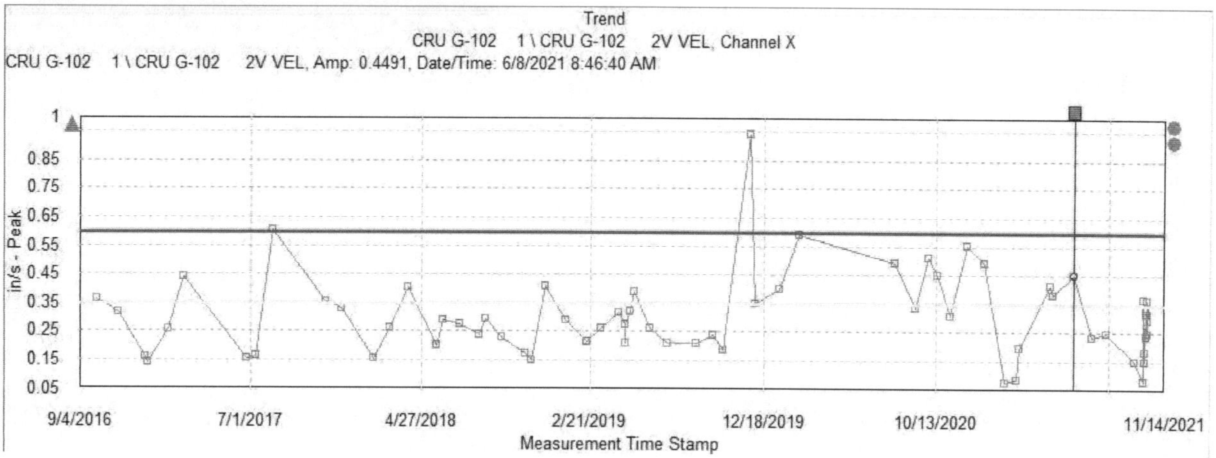

Figure 6

The piping was repaired, and a new coupling was installed. Given the amount of attention to the failure and loss of production, capturing startup with just the handheld data collector did not seem appropriate. I built a configuration for the analyzer after walking down the application and captured data for the startup. Immediately the vibration condition measurements, values, shapes, and cycling fluctuating transients were troubling (Figure 7)

The baseplate had concerns. You can sense the machine just is not healthy. If we pull this weed, we must make sure that we get the root, and we all know why. Because if we don't get the root out, it's coming back. Why did a shim pack coupling fail? What forces are causing the elevated response at the DE of the motor in the vertical direction?

Point	Speed	Bias	Direct	1x	1xP	2x
H01 ips	3,587 RPM	11.2 V	0.31 IPS (o-p)	0.22 IPS o-p	240 °	0.02 IPS o-p
V01 ips	3,587 RPM	11.1 V	0.16 IPS (o-p)	0.09 IPS o-p	60 °	0.01 IPS o-p
A01 ips	3,587 RPM	11.0 V	0.54 IPS (o-p)	0.37 IPS o-p	182 °	0.02 IPS o-p
H02 ips	3,587 RPM	11.1 V	0.25 IPS (o-p)	0.19 IPS o-p	244 °	0.05 IPS o-p
V02 ips	3,587 RPM	10.7 V	0.45 IPS (o-p)	0.38 IPS o-p	184 °	0.01 IPS o-p
A02 ips	3,587 RPM	11.1 V	0.35 IPS (o-p)	0.28 IPS o-p	163 °	0.01 IPS o-p
H03 ips	3,587 RPM	11.1 V	0.41 IPS (o-p)	0.20 IPS o-p	331 °	0.05 IPS o-p
V03 ips	3,587 RPM	11.2 V	0.31 IPS (o-p)	0.11 IPS o-p	253 °	0.06 IPS o-p
A03 ips	3,587 RPM	11.0 V	0.29 IPS (o-p)	0.17 IPS o-p	18 °	0.01 IPS o-p
H04 ips	3,587 RPM	11.3 V	0.27 IPS (o-p)	0.09 IPS o-p	339 °	0.03 IPS o-p
V04 ips	3,587 RPM	11.1 V	0.21 IPS (o-p)	0.08 IPS o-p	275 °	0.04 IPS o-p
A04 ips	3,587 RPM	11.0 V	0.44 IPS (o-p)	0.20 IPS o-p	11 °	0.01 IPS o-p

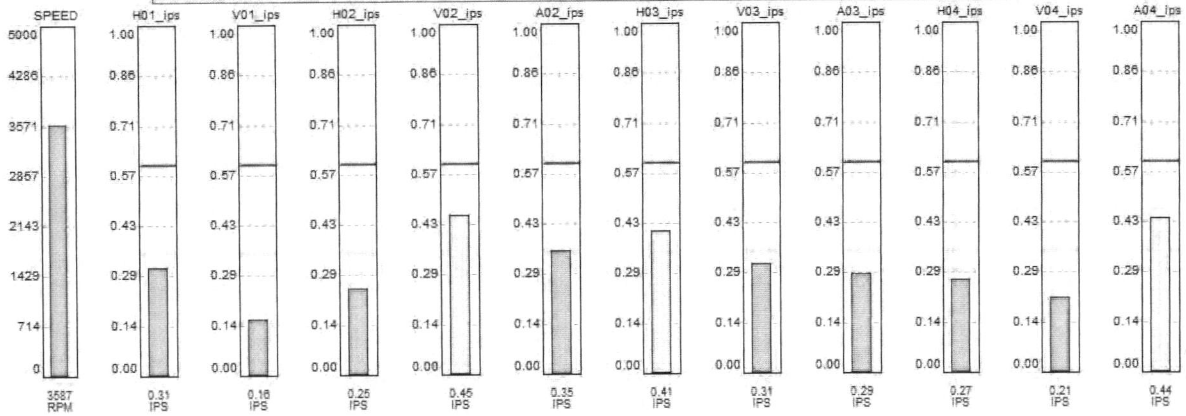

Figure 7

Using the multi-channel analyzer is great because you get all the data you need to process, but it can be overwhelming. How do I logically begin to process the data? The game is afoot. We are detectives, right? What draws my attention to what is abnormal? Is this synchronous vibration? What does the TW resemble? The DE of the motor is high, and the DE of the pump is high. Using synchronous time averaging with 50 averages, it cleans the signal. Now we know it is 1X, and the WF is sinusoidal. We have a shape that narrows down our suspects. The phase across the couplings A02 (axial brg.2) to A03 (axial brg.3) is 166°. That's within the +/- 30°. But is it really a misalignment? What does misalignment look like? What are some basic characteristics? This misalignment is predominantly 1X as shown in Figure 8.

- Misalignment increases system stiffness and may reduce vibration.
- Predominate frequencies at 1X and 2X
- Phase shift between vertical and horizontal sensors is 0° or 180° (+/- 30°)
- Stationary preload causes a distorted sine wave

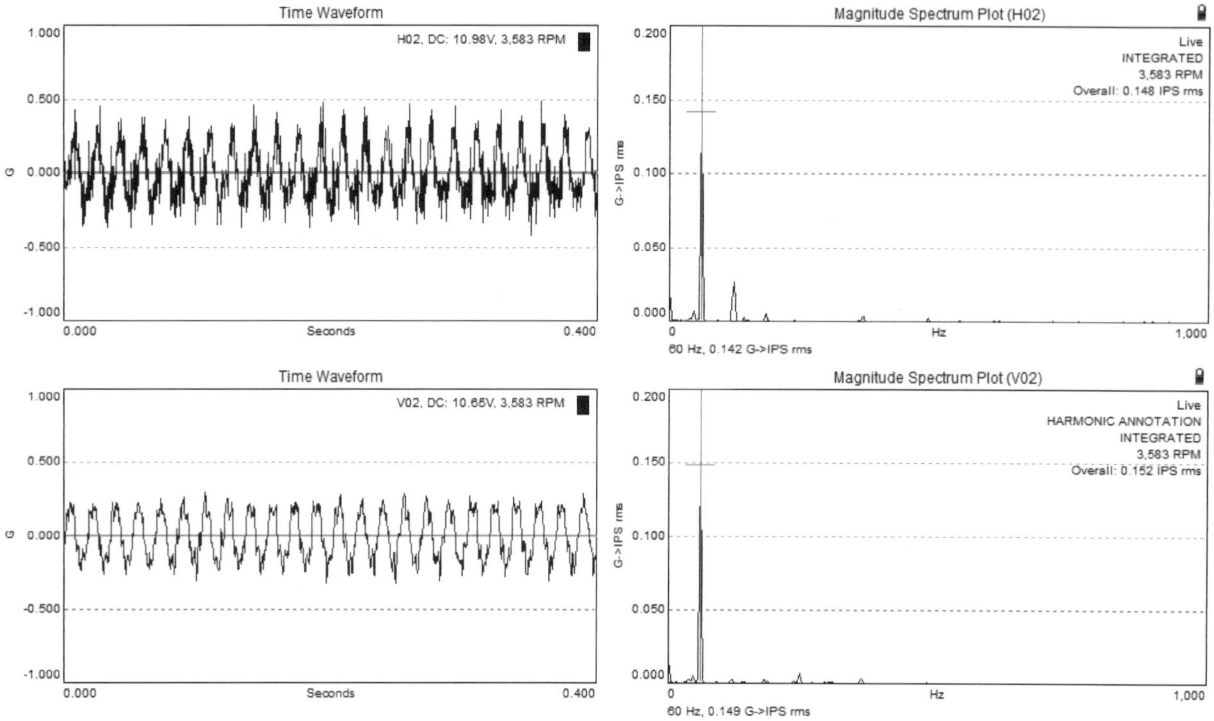

Figure 8

The alignment done by the machinists was within specified tolerances. What's causing the transient or fluctuating vibration? Something is flexing. We stiffened the pump with a machine jack under the DE pump bearing. By adding stiffness, the vibration was reduced below the alert setpoint. It didn't take long for another issue to arise. Within two weeks, the seal was leaking. This pump had a history of failures. Table 1 shows most of them.

Table 1

Year	Failure
1970,72.73,77,79,	Seals, Bearing
1980,81,81,83,84	Tray parts, bearing
1998	Long run?
2000,04,14,17,18,19	Seals, case replacement etc.
2021	Sub structure, foundation

Machine Table

Component	Point	Bias	Speed	Direct	1x	1xP	2x	2xP
Motor	H01 ips	11.2 V	3,582 RPM	0.34 IPS (o-p)	0.23 IPS o-p	248 °	0.06 IPS o-p	280 °
Motor	V01 ips	11.1 V	3,582 RPM	0.13 IPS (o-p)	0.09 IPS o-p	83 °	0.02 IPS o-p	217 °
Motor	A01 ips	11.0 V	3,582 RPM	0.27 IPS (o-p)	0.20 IPS o-p	193 °	0.02 IPS o-p	50 °
Motor	H02 ips	11.0 V	3,582 RPM	0.28 IPS (o-p)	0.20 IPS o-p	249 °	0.04 IPS o-p	227 °
Motor	V02 ips	10.6 V	3,582 RPM	0.25 IPS (o-p)	0.22 IPS o-p	196 °	0.01 IPS o-p	45 °
Motor	A02 ips	11.0 V	3,582 RPM	0.21 IPS (o-p)	0.18 IPS o-p	158 °	0.00 IPS o-p	50 °
Pmp	H03 ips	10.9 V	3,582 RPM	0.32 IPS (o-p)	0.15 IPS o-p	337 °	0.01 IPS o-p	247 °
Pmp	V03 ips	10.9 V	3,582 RPM	0.24 IPS (o-p)	0.06 IPS o-p	258 °	0.05 IPS o-p	276 °
Pmp	A03 ips	10.8 V	3,582 RPM	0.27 IPS (o-p)	0.16 IPS o-p	7 °	0.01 IPS o-p	95 °
Pmp	H04 ips	11.1 V	3,582 RPM	0.22 IPS (o-p)	0.08 IPS o-p	352 °	0.01 IPS o-p	254 °
Pmp	V04 ips	10.9 V	3,582 RPM	0.19 IPS (o-p)	0.04 IPS o-p	294 °	0.03 IPS o-p	278 °
Pmp	A04 ips	10.8 V	3,582 RPM	0.31 IPS (o-p)	0.18 IPS o-p	3 °	0.01 IPS o-p	101 °

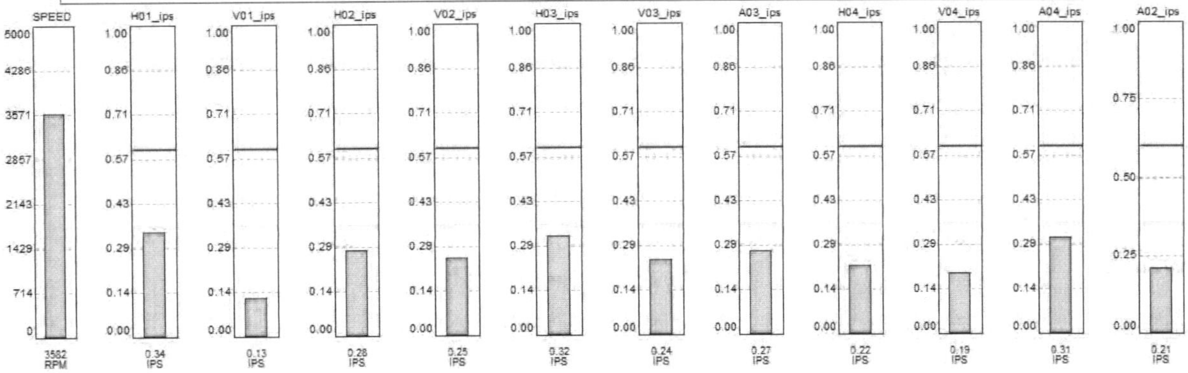

Figure 9

With the support jack, the vibration was below our alerts (Figure 9), but we were pre-loading the bearings. During one of the data gathering tests, the BUL, Plant Manager, top cheese in the refinery (those guys don't talk with us) starts asking me what I think. I told him that, based upon the data, the base is flexing. He was straight forward. If you find it, we will fix the base. The base was tested, and hollow spots were found. Right underneath the motor drive end was a hollow pocket. Figure10 shows hollow spots circled in yellow. When the pump was pulled for the seal change, we did a complete FRF of the motor and pump mountings, and measured elevation, co-planers, and level across the entirety of the base. The right pump support was low by 0.154 inches.

The Whiting Refinery is located on Lake Michigan and its sandy soil. So, the sounding conducted identified that a hollow spot existed (Figure 11). The foundation was completely removed, and a new concrete pump foundation was constructed. The original base plate was measured. The co-planer of the plate was within tolerance of API recommendations and reused. The baseplate was grouted. The pump and motor were installed and aligned. The piping strain was verified. The pump has been in service with no seal problems. Vibration levels have been within normal operational values (Figure 12).

Figure10

Sounding Hole

When the concrete slab was cut out, it was quite clear what happened. This is the corner that lowered allowing the pump base to sink causing twisting of the pump base. The rectangular box with the pipe is an old drain line, water was seeping and washing away the soil. Fig.8

Figure 11

Base Testing

Accelerometers used for testing pump centerline mounting pedestals, evidence of loss grout and air pockets with no grout.

Figure 10

Vibration trending since 1-8-23, no seal, coupling, or bearing failures since repair of foundation.

Figure 12

Case Study Two Conclusion

The misalignment problem driving the 1X vibration would have never gone away with aligning the shafts. Further testing was required to determine the much deeper issue literally. You would never see this base issue sitting in the office analyzing charts. Following the information drove the right deduction. Yes, it's elementary my dear Watson.

Case Study Three Axial Runout or 1X Vibration

The SRU (Sulfur Recovery Unit) has three five-stage segmented blowers, driven by 3000 HP two-pole motors, providing combustion air. They supply 1600k scfm/hr for the process. All three blower systems have separate lube oil supply skids and are fully equipped with proximity probes. The instrumentation includes an entire suite with axial trip protection. Additionally, they are monitored by an online software platform for analyzing trending, orbits, phase, etc.

As Equipment Monitoring Technicians, our job is monitoring and analyzing. The first noticeable change occurred in the axial position probes (Figure 13) in May of 2016. Once a change is identified the drilling into what could cause the change begins. Breaking down a logical process path to begin eliminating the possibilities with accurate data is foundation to analysis. Mr. Barry Ardel pressed upon me very early in my predictive maintenance career, "I'd rather have no data than bad data." I certainly became a student of his methodology; I still count over 10 peaks of a waveform to get a good average of the frequency.

What are possible defects that could cause the rotor to drift?

- High temperature of thrust pads (Thrust Temperatures were stable)
- Lubrication restrictions (Verification of orifice plate at lube oil supply, but still possible internal restrictions)
- Motor Rotor not on magnetic center (Blower running can't check until offline)
- Balance line plugged (Temperature stable from High pressure to low pressure side of machine)
- Probes loose (trending together, but possible exterior mounted on stingers, verified lock nuts)
- Runout (Blower running can't check until offline)
- Thrust collar moving (can't verify until rotor is disassembled)

There are three blowers. They are all the exact model numbers built at the same time. Two of the blowers are experiencing this axial drift issue. One is not. When the units were placed in service in 2011, the axial position measurements were at +5.8 and +6.8 mils. When the value reaches -5 mils in the position trends, the blower was shut down. Over the 3-year span of the trend, the total rotor float was increasing.

The first alert is 20 mils, a NOL (normal operating range). When a NOL setpoint is triggered, operations are required to act. If there are no handles to pull to get out of the alarm, the machine must come down at a controlled manner. REG (rotating equipment group) developed and executed a maintenance plan to replace the thrust pads. A horrible oil leak was also occurring on the NDE of the blower. This was addressed with an anti-wicking dual thermocouple design with epoxy impregnated, Teflon encapsulated armor jacketed wire. The repair was very effective, and the oil leak was eliminated. The higher pressure inside the pressurized bearing is greater than the absolute atmospheric pressure. Oil simply flows to the lower pressure, and the coated impregnated thermocouple seals off the oil from wicking through the wiring.

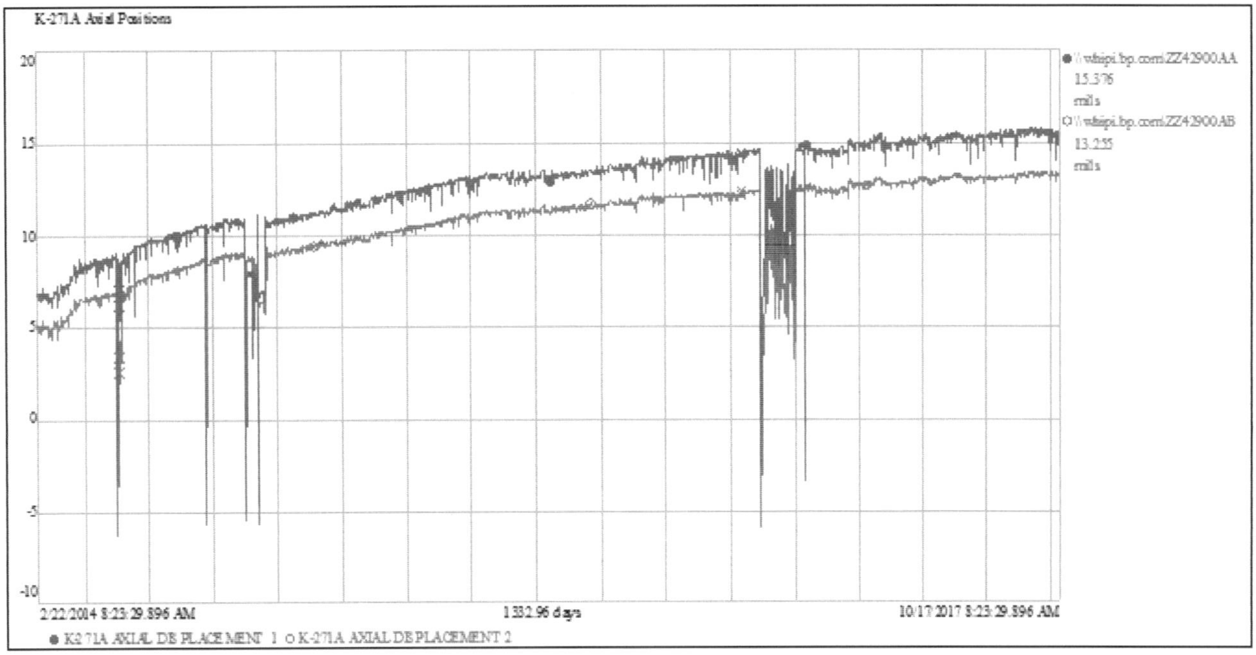

Figure 13

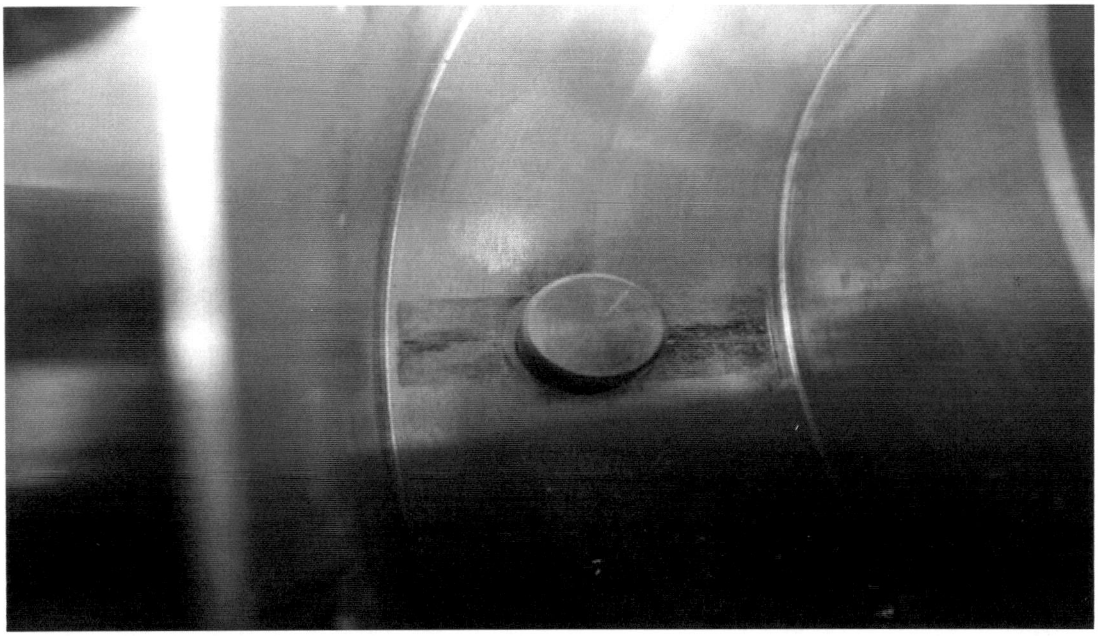

Figure 14

When removing the thrust package, we measured the entire thrust cage assembly and compared it to the axial drift. It matched exactly the loss of material within the cage, but the thrust pads measured to the exact dimensions of the new pads. Where was the material loss? The cage assembly had fretted (Figure 14) at the location of anti-rotation or alignment pin. As the rotor was moving, it would slowly wear out the cage over time. Fretting is the oxidation of material, polished off, oxidizing again, at the cycles of motion. Over time we see the drift of the loss of this material.

Figure 15

The thrust bearing components were measured with a height gauge, with the original cage and a new cage, and the loss of material in the old cage matched exactly to the axial drift measured with the axial probes (Figures 15 & 15a)

Figure 15a

The motor was uncoupled, and a motor solo run found the magnetic center was off. The DBSE (distance between shaft ends) was .015 more than the OEM recommendation. The motor was moved to the correct position. When the first cage was replaced, we identified the wearing component, and confidence was high that the problem would be resolved. However, when the unit was placed back into service, the axial position was moving away from the probes again. Our configuration for thrust position is that positive is away from the probe. The rotor was moving to the coupling end, the low pressures side of the blower. We still didn't know why it was moving and fretting the thrust cage. Data was collected from the monitoring system to evaluate the waveform, with the trigger. The high spots of the two sine waves are 180° apart. The trigger mark is changing position in the blue sine wave and increasing is negative amplitude 6 cycles and then returns to its original position. The rotor is moving back and forth causing the fretting, but what is causing it to move? The orange waveform remains constant. The trigger mark remains stable, but the amplitude changes at the negative peak (Figure 16). The rotor is also being restrained from going completely positive and its very sharp return from the bottom negative to moving back to forward. This data indicates a pre-load and an altered abnormal pattern. A runout should be clearly defined as a high and low spot without sharp peaks; something is causing the rotor to shuttle.

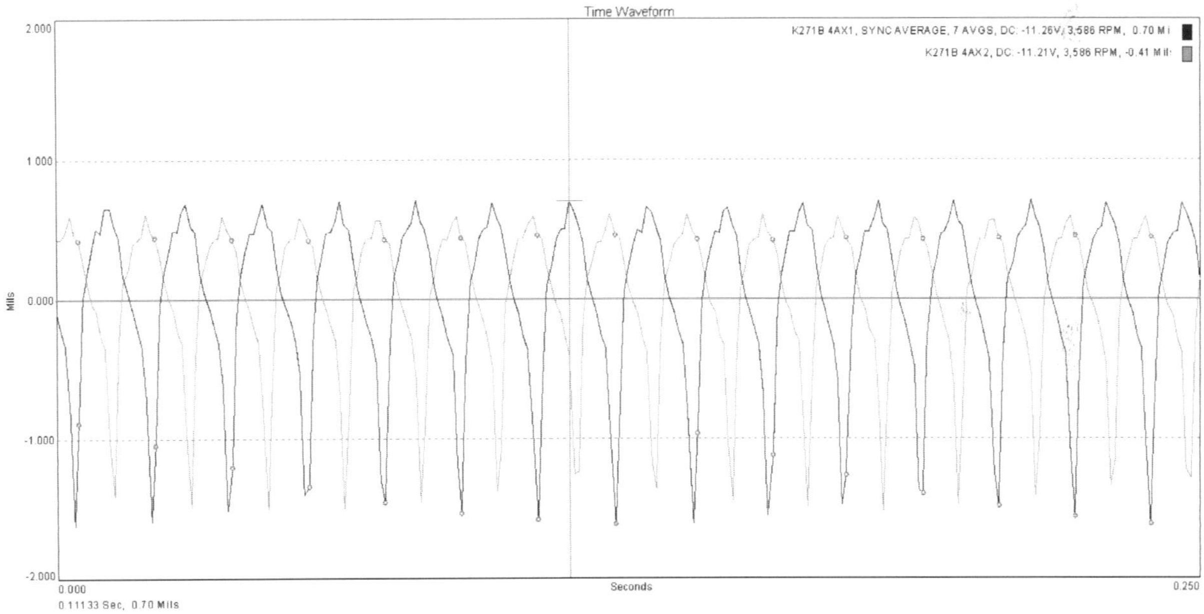

Figure 16

When one of the machines came out for a complete overhaul. The rotor was setup in a lathe on a centers at both ends. The head end of the lathe requires a dead center into the shaft and driven by a lathe dog in the chuck. You don't want to chuck the shaft. It needs to be turned on center. Both centers were indicated for runout before test, and indication of the end of the shaft revealed no runout. However, the thrust collar had 2 mils runout. When the collar keyway to the bore was measured the dimension was 2.354 inches. The key in the key seat and the shaft measured 2.340. The collar is heated for the interference fit over the shaft, but now the lack of clearance over the key was cocking the thrust collar causing a runout, resulting in the rotor shuttling back and forth wearing out the cage.

Thrust collar from first unit pulled for cycle ending key up. No evidence of scouring on either face of the thrust disc itself. Decision was to replace the collar with new component, even though disc was parallel. (Figure 17)

Figure 17

The thrust collar work on one blower was completed on 9/21/2018. A trend plot of thrust position is provided in Figure 18. The thrust collar keyway clearance over the key was confirmed to be 0.015 inches, and we are no longer seeing the axial drift. Axial position has remained steady for four years. We monitored closely and still do, but the rotor isn't moving. Another unit was taken out for key-up, and discovery found the same issue. The repair was completed, and the rotor position has been stable since. The trend plot in Figure 19 confirms that there is no axial drift.

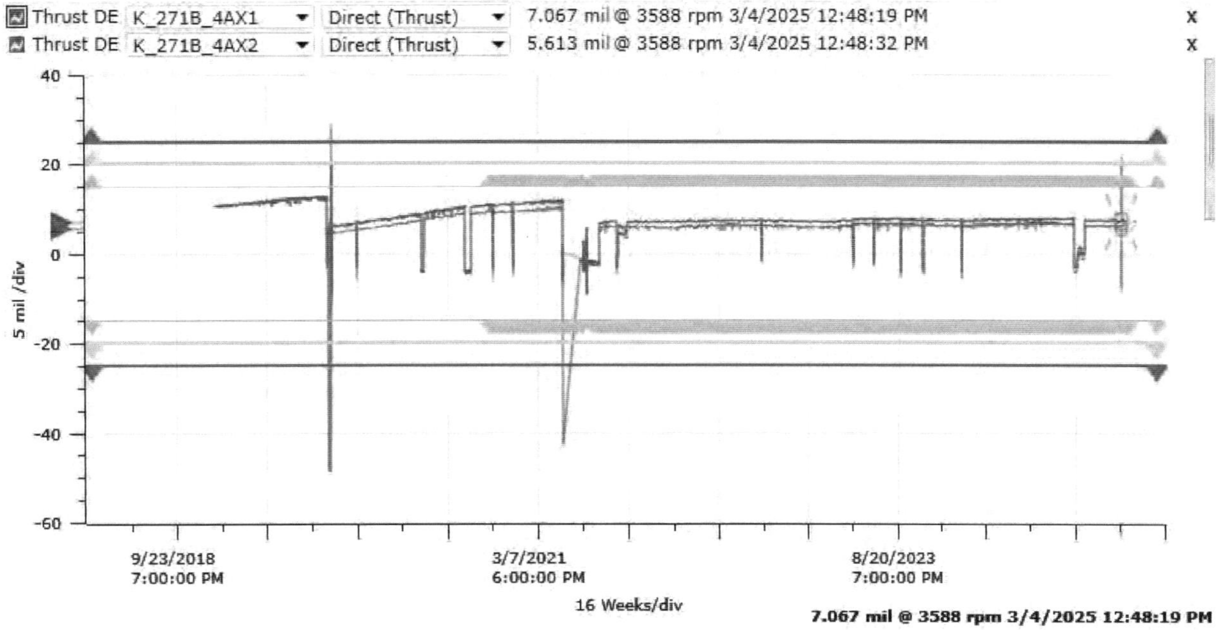

Figure 18

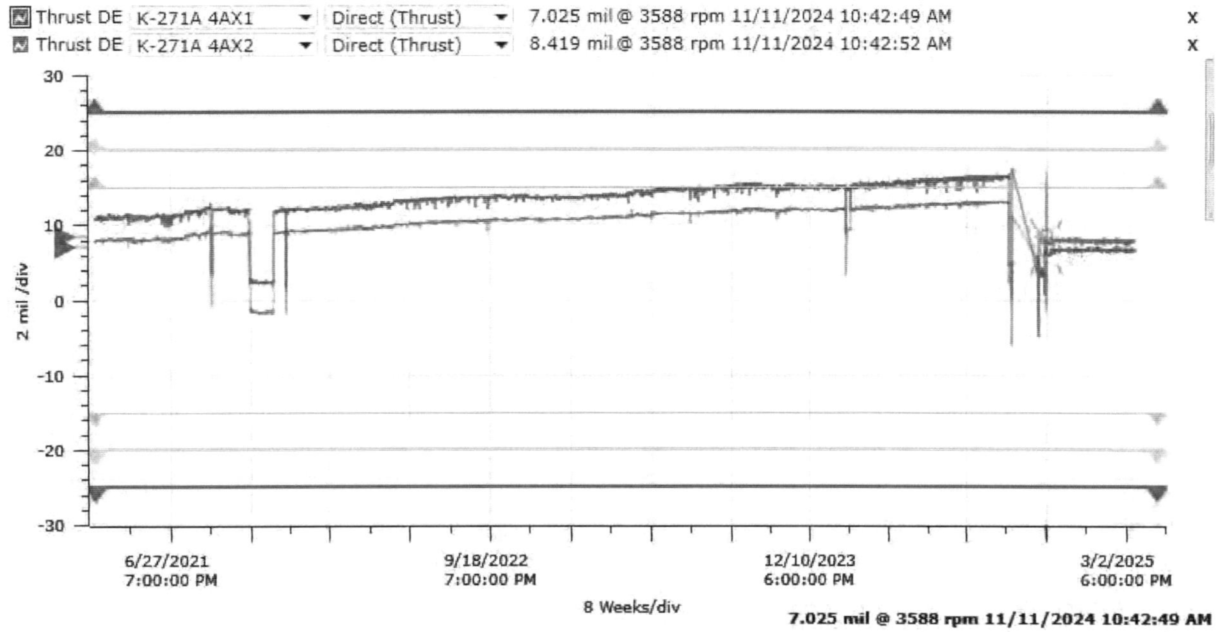

Figure 19

Case Study Three Conclusion

This case study showed that the wear on the thrust bearing was a symptom of an underlying problem, not the cause. The actual cause of the fretting and the resulting change in position was a manufacturing assembly error.

Lessons learned from this case history include the following:

- Keep digging
- Details are important.

Epilogue

I've always been a great proponent of multi-channel analyzers, virtual instruments. In my career as a vibration technician, I was introduced early to the use of a 4-channel analyzer. Even back then, I was somewhat criticized, "why do you haul all that extra stuff out"? I suppose it became a habit in some way, looking at applications, opportunities to sharpen skills, becoming creative in types of analysis. It has developed me into a more complete analyst. As my career is closing, I still will haul out an analyzer to the top of a cooling tower to power a spare sensor and collect as much data as possible with the trigger reference and pull in 7 channels of data. Is it work? Yes, but I can make a correct diagnosis, and that's what we are really after isn't it? I can confess with a clear conscience, I'm not the brightest, but I've always worked hard and made a multitude of mistakes. But the longer I've been around, I've seen attitude affects more outcomes than intelligence.

TRIM BALANCING AS A TEMPORARY SOLUTION TO RESONANCE BEHAVIOR

Author: Jerry Stieg

VIATC TRAINING CONFERENCE & EXPO

NEWPORT NEWS, VIRGINIA | AUGUST 6-8, 2025

VIATC | TRAINING CONFERENCE & EXPO

NEWPORT NEWS, VIRGINIA | AUGUST 6-8, 2025

TRIM BALANCING AS A TEMPORARY SOLUTION TO RESONANCE BEHAVIOR

Jerry Stieg
Lead Machinery Diagnostics Services Engineer – Bently Nevada, LLC.
Jerry.stieg@bakerhughes.com

Jerry Stieg is presently a Lead Machinery Diagnostics Services Engineer for Bently Nevada. He joined Bently Nevada in 2022. In prior job assignments, he worked for a power generation company where he was responsible for individual site condition-based maintenance programs. He has been in the vibration analysis, balancing, and equipment reliability fields since 2018.

Jerry earned a BS in mechanical engineering from Virginia Polytechnic Institute and State University. He is certified as a Category II Vibration Analyst through Technical Associates of Charlotte.

Abstract

Resonance behavior of ancillary components adjacent to a turbine generator may present an operability risk equivalent to unacceptable vibration of the machine proper. Support systems such as turning gear motors, fire detection systems, instrumentation conduit, excitation systems, and machinery shelters are seldom considered with regards to ensuring appropriate separation margin from running speed of the prime mover. In most cases, the transmission of vibration from the principal machine occurs through the support structure. Herein, a case was documented where resonance of a supporting structure common to the bearing housing, excitation system, and machinery shelter was posing an unacceptable risk to the operation of the machine. This risk was temporarily alleviated by trim balancing the generator rotor to a very small residual, in effect removing much of the forcing function to be transmitted. The result of the balancing effort was the ability to operate unhindered while options to obtain sufficient separation margin of the structure were explored. A secondary benefit was clear evidence that vibration of the static exciter was related to the imbalance force of the generator rotor.

Key Words: Ancillary, Balancing, Couple, Exciter, Generator, Resonance

Introduction

Excessive vibration due to fundamental structural resonance behavior is typically presented with two basic remediations in most literature. The options of adding stiffness to increase the natural frequency of the system or adding mass to decrease the natural frequency of the system have been successful remedial measures in many cases. Less commonly deployed remedial measures have included solutions such as system dampers, dynamic vibration absorbers (DVAs), or lastly reduction of the input force. For the use case of rotating equipment, often the easiest of the above solutions to deploy would be reduction of input force. Many machine trains have factory provided provisions for reducing the mass imbalance of the rotor, which often dominates the force input. However, in order to temporarily alleviate structural resonance related excessive vibration, rotors may require balancing well beneath the original equipment manufacturers (OEM) specification for "a well-balanced machine." Balancing a resonant system can be non-linear, however, the time and expense of multiple runs may still prove to favor balancing over more invasive engineering or maintenance techniques.

For the purposes of this paper, distinction must be made between the system resonance associated with the combination of the flexible rotor and bearing support structure versus structural resonance behavior of ancillary equipment. Rotordynamic analysis of the bearings and line shaft of a machine train would identify expected natural frequencies of the rotor and support system. The response at these natural frequencies causes increased sensitivity to imbalance at antinodes along the rotor, which would commonly be referred to by misnomer as the "critical speed." Often the critical speeds would be shop balanced and potentially trim balanced at field installation. In contrast, structural resonance behavior of ancillary equipment would rarely be evaluated as a complete system, as the components would typically be shop tested without the benefit of the oddities of the final support system. In the case study presented herein, the static exciter was shop tested on a solid concrete shop foundation, while field installed on a hollow metal floor. Furthermore, the static exciter was replacing an operating unit of similar footprint yet larger mass.

The author first encountered the concept of balancing atypical vibration responses in a commonly cited text, "Machinery Malfunction Diagnosis and Correction." In Case History 38, the author balanced a rotor in order to address the axial machine casing response. An excerpt reads: "Hence, when all else fails, go back to the basics, and try to eliminate the 1X driving force. This includes any eccentricity, bow, or mass unbalance in the rotor system." [1].

Herein, this methodology was applied to an air-cooled turbine generator with a newly upgraded static excitation system, illustrated in Figure 1. The new exciter installation was sufficiently different in stiffness and mass to tune the structure to a poorly damped 63 Hz horizontal exciter mode. In order to return the unit to service, the generator second mode was balanced utilizing a static-couple technique to a very small residual, resulting in acceptable vibration of the ancillary components.

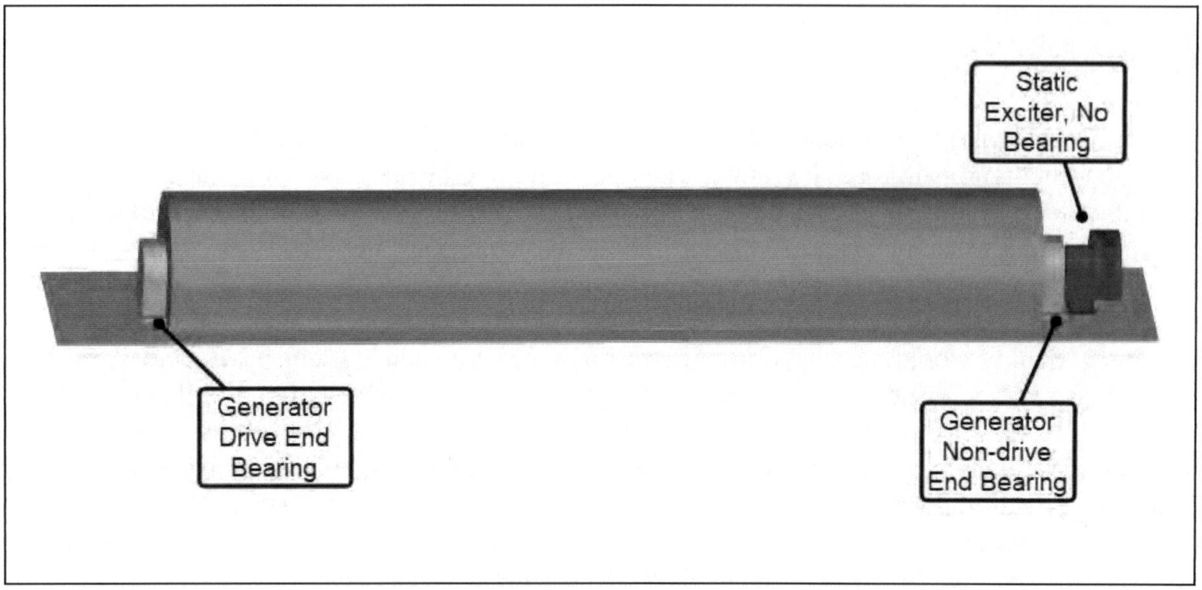

Figure 1: Machine train representation

Underlying Principles

In order to support both the approach and outcome of the subject, a brief overview of the underlying mathematical relationships was explored in the following section as simplified to a single degree of freedom (SDOF). When considering the response of the generator bearings, the equation of motion for a rotating imbalance most succinctly summarizes the variables driving the system response. Tongue summarizes a SDOF, single rotating mass imbalance with the following equation of motion [2]:

$$\left(m_1 + m\right)\ddot{x} + c\dot{x} + kx = me\omega^2 cos(\omega t)$$
$$where \#(1)$$
$$m\, represents\, mass\, of\, rotational\, imbalance\, \&\, m_1\, represents\, the\, "base\, mass"$$

Note the right side of the equation generally represents the force input to the SDOF system. As the mass of the imbalance approaches zero, so also must the response (x), regardless of the system rotational (ω) or natural frequency (ω_n). In the subject application, the generator rotor imbalance mass would be simplified as m, while the bearings and support structure would be simplified as m_1.

When considering the response of the static exciter, no rotating imbalance would exist for the exciter structure, due to the absence of an exciter bearing. In this case, the generator bearings carried most of the rotor forces, therefore the exciter structure received its excitation from the shared floor (neglecting the minor magnetic component). The bulk of the exciter response would be simplified as a seismically excited system with damping, which would be represented by the following equation of motion [2]:

$$\ddot{z} + 2\zeta\omega_n\dot{z} + \omega_n^2 z = \omega_n^2 y + 2\zeta\omega_n\dot{y}$$

where #(2)

z represents motion of the exciter & y represents motion of the floor

Intuitively, the motion of the shared floor between the static exciter and the generator bearings (y) would be related to the motion of the bearing structure (x), although the relationship may be complex. While SDOF equations do not effectively model the entire system response, the underlying principle of minimizing the forcing function from Eq. (1) by reducing imbalance mass in order to reduce the motion of both the generator bearing housing (x) as well as the shared floor (y), should produce the required outcome of less exciter motion (z). The discussion of coordinates was illustrated in Figure 2. Although illustrated in the vertical direction, the subject behavior was primarily horizontal.

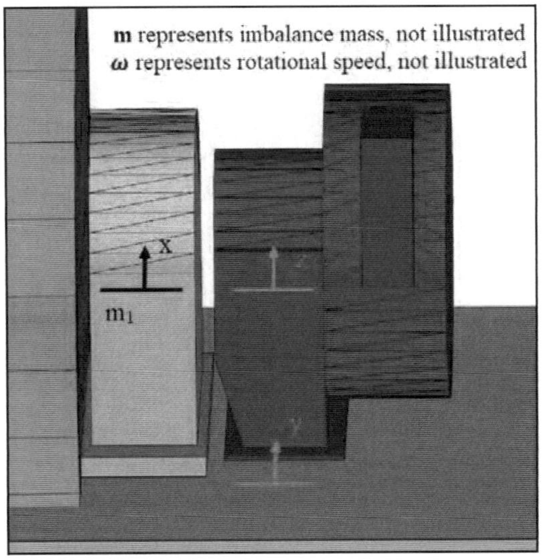

Figure 2: Illustration of terms used in Eq. (1) & Eq. (2)

There are several practical limitations to the approach of attempting to reduce the forcing function in order to reduce overall system response. As either or both structures approach the natural frequency (ω_n), the required force (or mass as illustrated in Eq. 1) to drive an unacceptable response becomes smaller. In some cases, the author has witnessed sensitivity to imbalance beyond the practical limitations of field balancing, where any accumulation of debris or minor structural change of the machine would once again produce an outsized response. Resonance behavior should be properly remediated, when possible, as machines operating with a sufficient margin to the natural frequency run in a more reliable manner. Repeated need to balance the same rotor has often been a sign of failure to address the underlying cause of the unacceptable response.

Initial Machine Response Evaluation

Following the replacement of the original static exciter with a similarly footprinted non-OEM model, a service request was made with a complaint of a malfunctioning new static exciter causing distressing vibration of the entire machine enclosure. The initial service request included a complaint of malfunctioning fire detectors due to the severity of enclosure vibration. Hardware on conduit straps was vibrating loose that had been properly secured for years. Furthermore, the response of the system was qualitatively detectible and uncomfortable to observe up close. Subsequently, discussions were held with the aftermarket static exciter manufacturer to understand how the unit might perturbate the enclosure and floor. The aftermarket manufacturer indicated that the magnetic (force) coupling from the stator to the rotor was weak, and that the most common fault frequency for the static exciter unit would be two times line frequency ($2XF_L$) rather than run speed frequency (1X). Upon evaluation, Figure 3 illustrated the elevated representative exciter stator response on the 3600 RPM machine train.

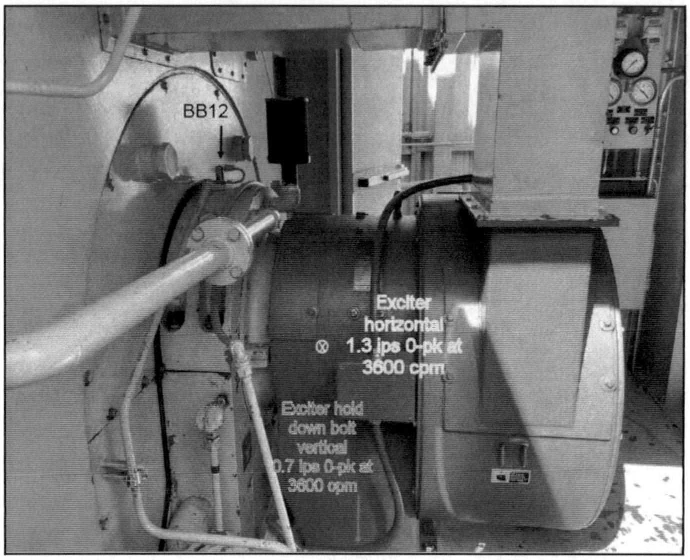

Figure 3: Photograph of exciter stator marked up with representative seismic amplitudes

Despite the concerning amplitudes measured on the static exciter stator, the generator drive end (DE) and non-drive end (NDE) bearing caps were found to be operating within OEM specifications, with no alarm (greater than 0.50 $ips_{0\text{-}pk}$) violation as measured by the permanently installed seismic moving coil protection devices at top dead center (TDC) of the bearing cap (Figure 4).

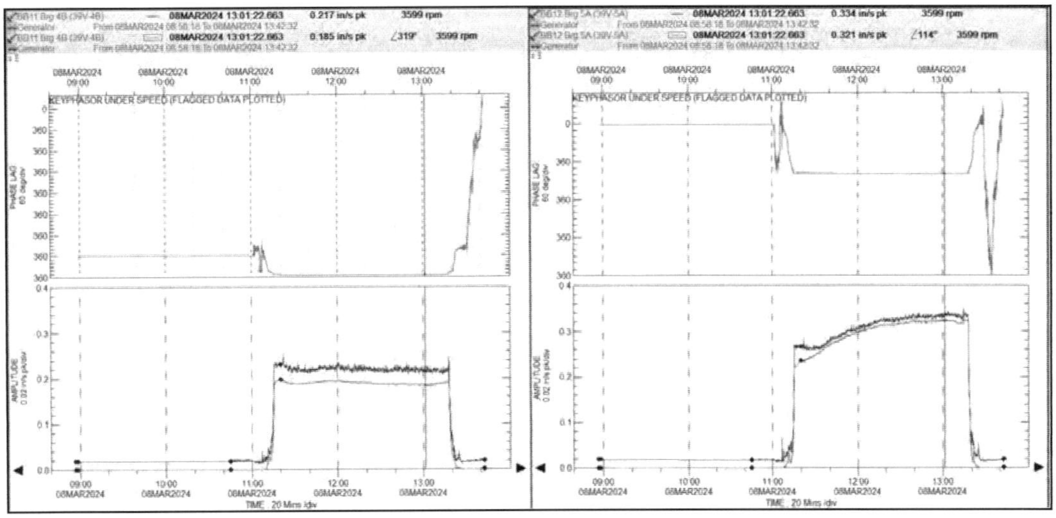

Figure 4: Trend plot of the generator bearing cap seismic response DE (left) & NDE (right)

Similarly, the permanently installed relative vibration probes did not indicate a concerning amplitude for the machine train type. Values were considered just outside of a "well-balanced machine" (less than 2 $mils_{pk\text{-}pk}$) with a dominant 1X frequency component (Figure 5).

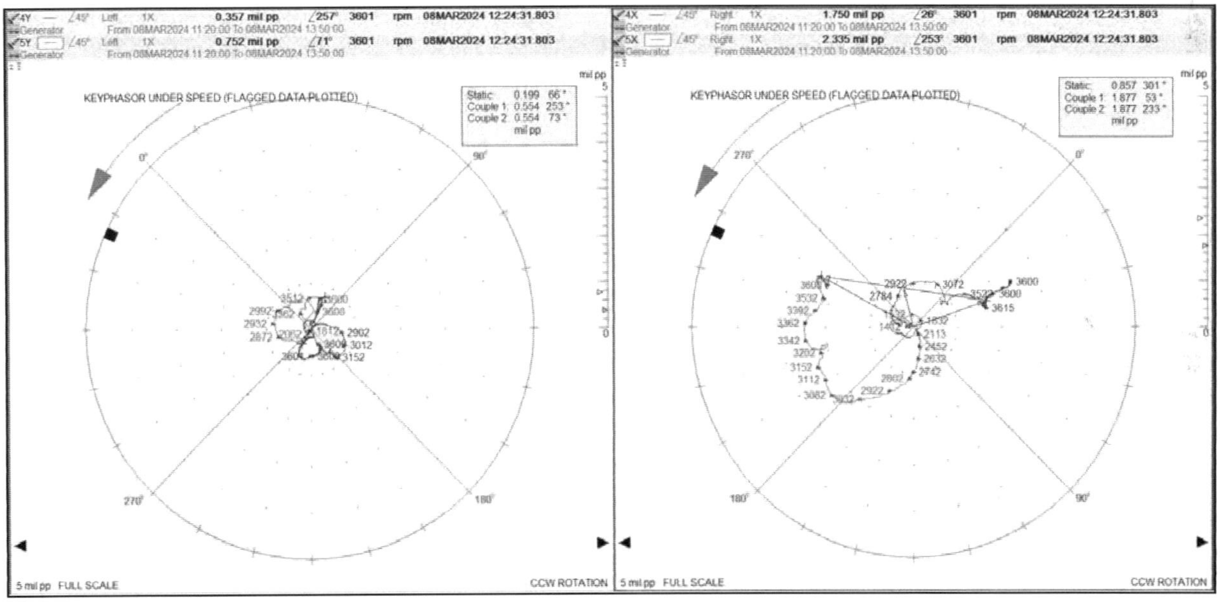

Figure 5: Polar plots of the generator 1X-filtered relative vibration response for the Y (left) & X (right) directions. DE was bearing 4, and NDE was bearing 5.

Due to the abnormally elevated and highly directional response of the AC exciter stator in the presence of relatively low generator vibration, a resonance condition was speculated. In order to attempt to identify the natural frequencies of the structure, a modal impact test was conducted utilizing a triaxial accelerometer and a modal impact hammer. Testing was performed with lube oil in service, which aids in prevention of the rotor coming off the oil wedge and adversely affecting (stiffening) results, but harms test coherence for the speed of the lube oil pumps. Additionally, the noise from lube oil pumps increases the noise floor of the measurement. Therefore, low amplitude response frequencies also have impacted coherence, due to an overall reduced signal to noise ratio. Figure 6 showed the mobility frequency response function (FRF) of the exciter as measured at two separate points at the top of the exciter in the horizontal direction. Figure 7 illustrated how the coherence reflected poor averaging at the run speed of the oil pumps.

Overall, the exciter stator was found to have a natural frequency corresponding with a horizontal mode at approximately 63 Hz, as illustrated in Figure 8.

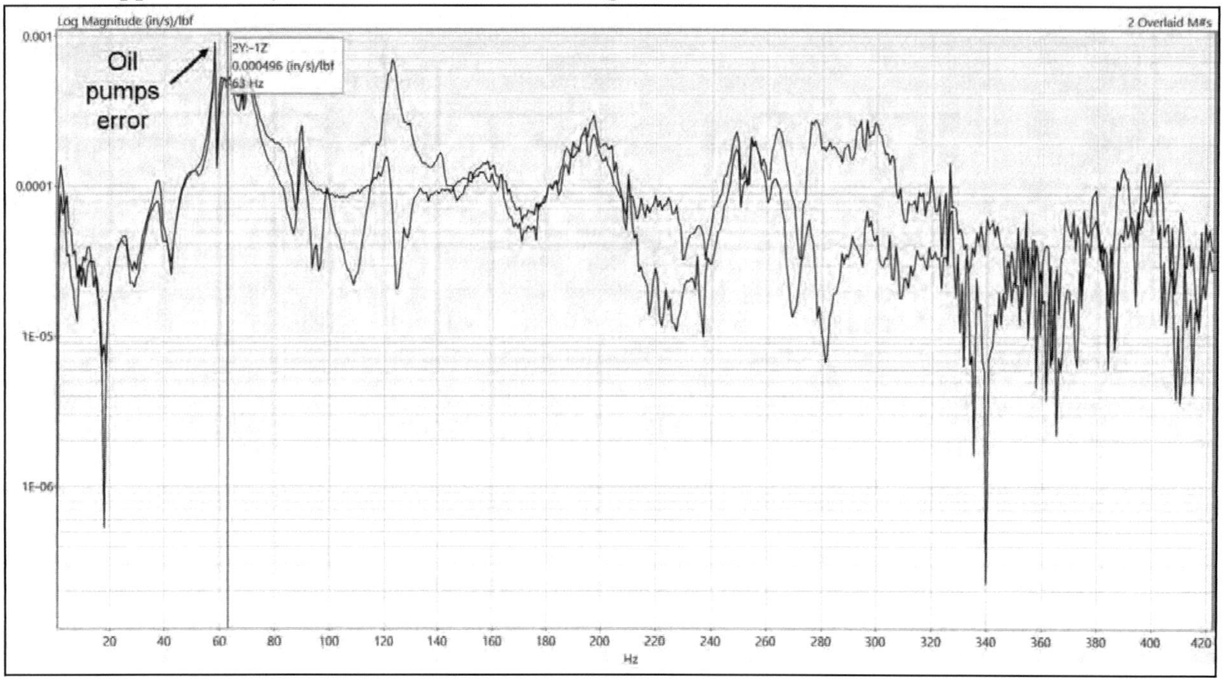

Figure 6: Mobility FRF of the exciter stator as measured at two independent points in the horizontal direction.

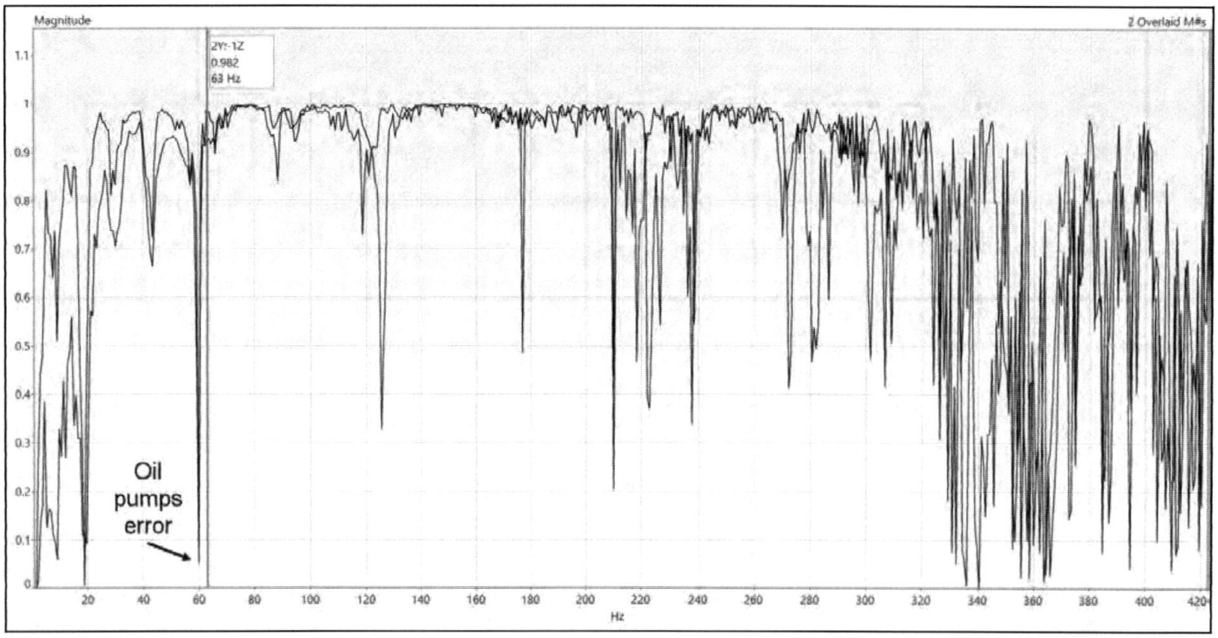

Figure 7: Coherence data of the two measured horizontal exciter stator modal impact tests across four averages.

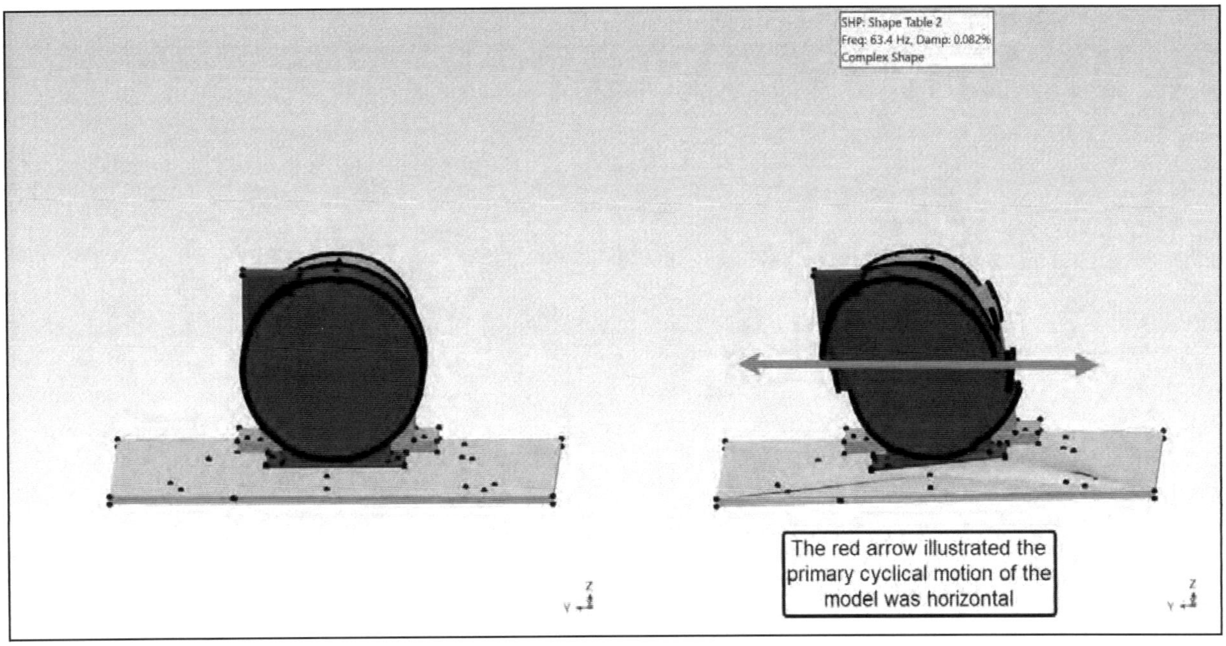

Figure 8: Undeflected (left) and deflected (right) models illustrating the modal response of the static exciter stator at 63.4 Hz.

Balancing Methodology

Several obstacles existed to a firm resolution of the observed resonance behavior, such as: a disbelief from the end user that the vibration was not caused directly by the static exciter, absence of an OEM representative capable of engineering a proper stiffener or appropriately adding mass, and a need to expediently return the unit to service. Therefore, field balancing the generator rotor was proposed to much resistance, as no maintenance had been performed on the generator rotor during the outage.

Due to the seismic nature of the response, the author elected to arrange temporary seismic transducers in plane with the permanent relative vibration transducers for shaft absolute (SA) vibration. Utilization of SA vibration was speculated by the author to have the highest likelihood of eliminating as much force from imbalance of the rotor possible, which had nearly equal seismic and relative contributions. In subsequent data presentations, data will be split into individual seismic and relative components for more intuitive observation and emphasis will be given to the dominant response direction, “X” (45 degrees right from TDC).

Following the decision to balance the unit, static-couple balancing was identified as the method likely to require the least number of runs to achieve significant reduction of response, based on the following criteria [3]:

- Symmetric rotor
- Rotor response was highly influenced by a single mode, see Figure 9

- Field accessible balance planes were symmetric, i.e. generator fan hubs on each end of the rotor of identical diameter and distance from the bearings

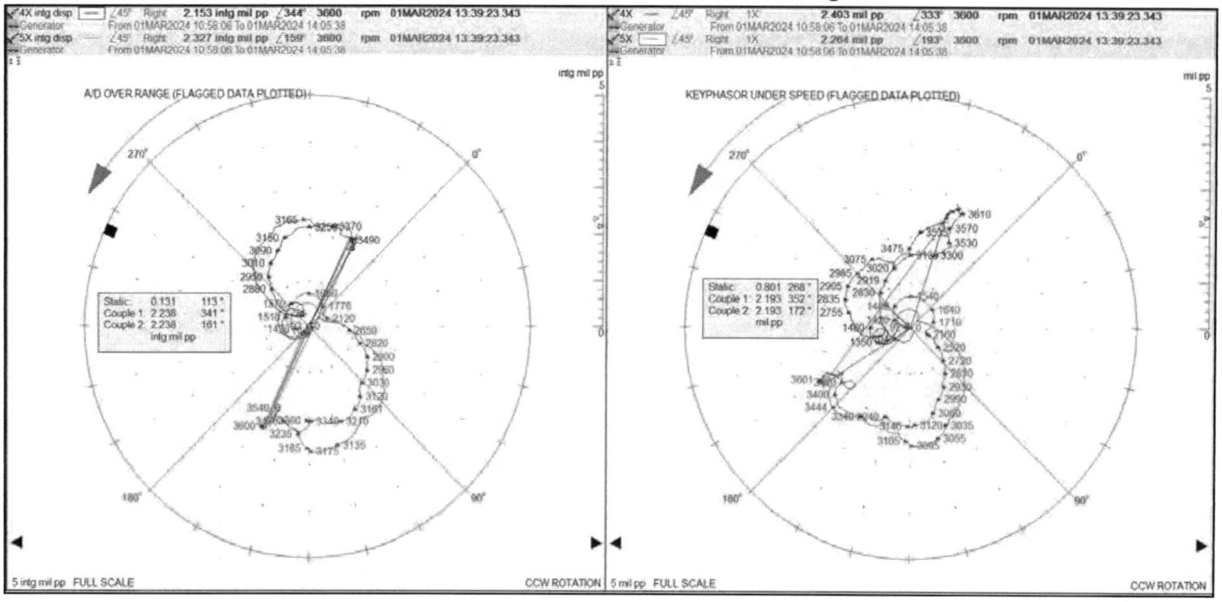

Figure 9: Seismic (left) and relative (right) polar plots illustrating the R0 (before balancing) generator bearing amplitudes. DE was bearing 4, and NDE was bearing 5.

A calibration weight location was selected based on influence derived from a peer machine, as detailed in Figure 10. In cases where the rotor system has been noted to be operating nearby to its natural frequency, each machine train may have a wider statistical distribution of influence than typical to machine trains operating with adequate separation margin. In this case, placing a calibration weight based on polar plot analysis would have created an improved first run. However, machine specific influence was derived for an improved correction weight, and the minimum goal of deriving influence while avoiding worsening of the response was achieved, as illustrated in Figure 11.

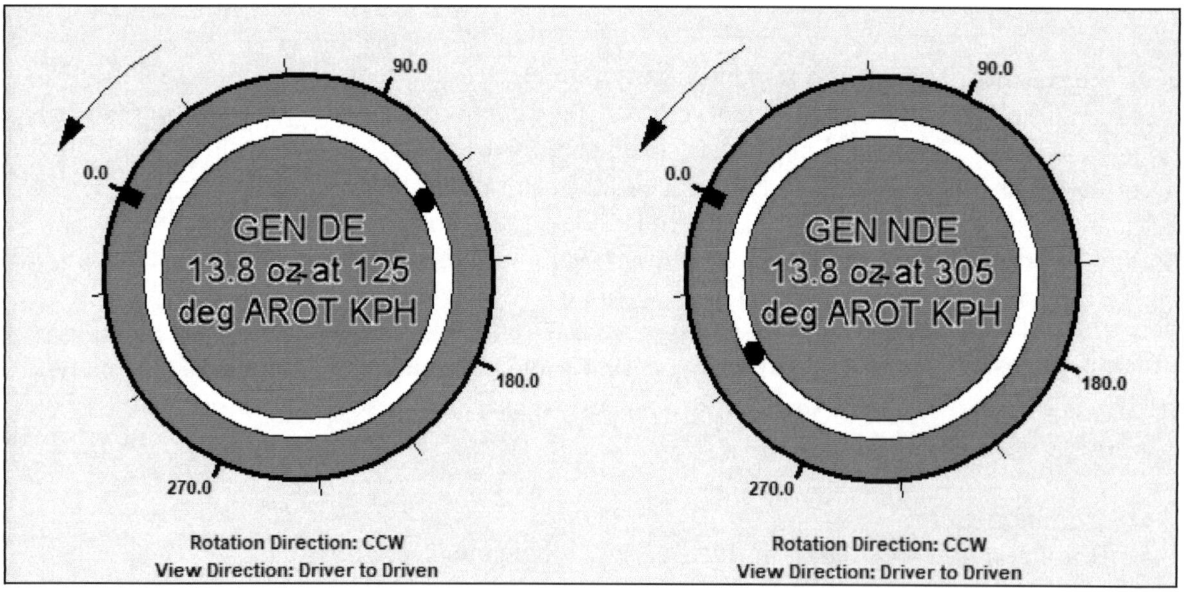

Figure 10: Generator DE (left) and Generator NDE (right) calibration weight placement illustrations. Weight placement was identified as against rotation from the Keyphasor® notch (AROT KPH).

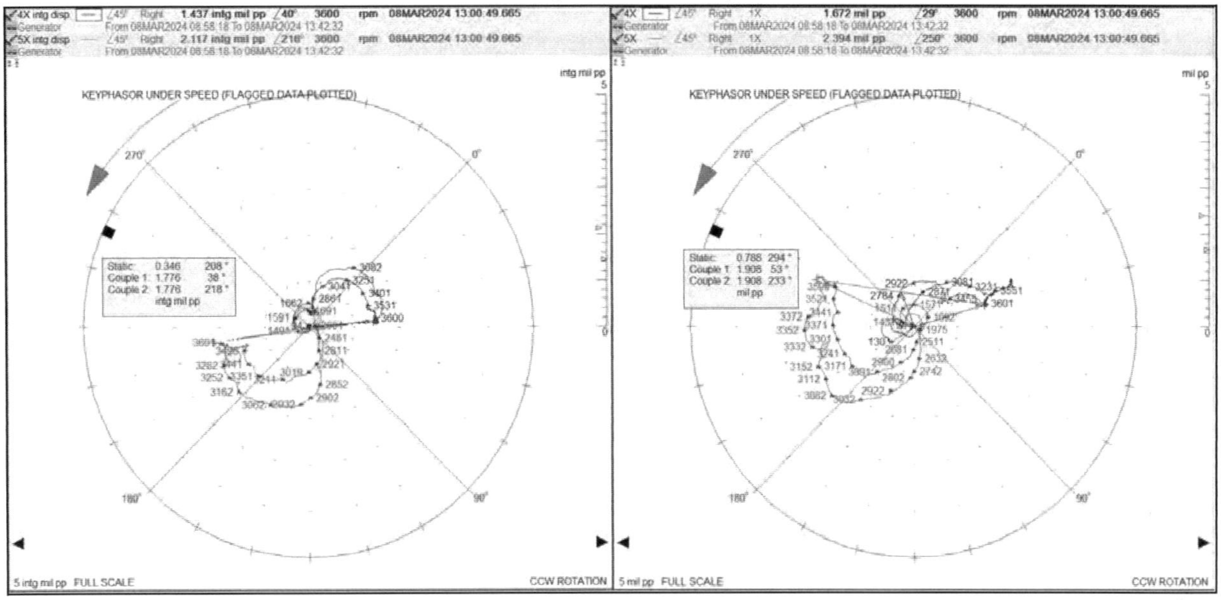

Figure 11: Seismic (left) and relative (right) polar plots illustrating the R1 (calibration weight) generator bearing amplitudes. DE was bearing 4, and NDE was bearing 5.

Utilizing the relative and seismic components, SA vibration can be estimated by adding the discrete filtered vectoral components as detailed in Eq. 3. This relationship only exists if the transducers are located in the same plane. As indicated by the polar plots in Figure 11, both the X direction seismic and relative transducers were mounted at 45 degrees right of TDC.

$$\vec{SA} = \vec{Rel} + \vec{Seis} \ \#(3)$$

If following the static-couple field balance procedure detailed by Ciesla and Titone [3], the following steps could be taken to calculate the correction weight. Note in the subscript, couple was referred as CP. Additionally, slow roll runout was found to be negligible in the balance calculation (less than 0.25 $mils_{pk\text{-}pk}$) and therefore remained unaccounted. Influence was calculated utilizing the DE AROT X-Probe weight location. As convention for a couple pair weight placement, the rotor was viewed from driver to driven and weight placement was simplified to the DE weight, with the assumption that the NDE weight placement would be moved the same angular rotation from its calibration position. The balance calculations were completed assuming the calibration weight would be removed, therefore the null vector opposed the R0 original vector. Furthermore, static-couple cross effects were observed to be negligible.

$$\vec{O}_{SA,CP} = \vec{Seis}_{CP1,R0} + \vec{Rel}_{CP1,R0} = 4.4\ mils_{pk-pk\ SA} \angle 346° \ \#(4)$$

$$\vec{O + C}_{SA,CP} = \vec{Seis}_{CP1,R1} + \vec{Rel}_{CP1,R1} = 3.7\ mils_{pk-pk\ SA} \angle 46° \ \#(5)$$

$$\vec{O + C}_{SA,CP} - \vec{O}_{SA,CP} = \vec{C}_{SA,CP} = 4.1\, mils_{pk-pk\, SA} \angle\, 115° \#(6)$$

$$W_{cal,CP,DE} = 13.8\, oz \angle\, 125°\, AROT\ KPH\ or \#(7)$$

$$W_{cal,CP,DE,X} = 13.8\, oz \angle\, 15°\, AROT\ X\ Probe \#(8)$$

$$\vec{H}_{SA,CP} = \frac{\vec{C}_{SA,CP}}{W_{cal,CP,DE,X}} = 0.3 \frac{mils}{oz} \angle 100° \#(9)$$

$$\vec{N}_{SA,CP} = - \vec{O}_{SA,CP} = 4.4\, mils_{pk-pk\, SA} \angle\, 166° \#(10)$$

$$W_{corr,CP,DE,X} = \frac{\vec{N}_{SA,CP}}{\vec{H}_{SA,CP}} = 14.8\, oz \angle\, 67°\, AROT\ X\ Probe\ or \#(11)$$

$$W_{corr,CP,DE} = \frac{\vec{N}}{\vec{H}} = 14.8\, oz \angle\, 177°\, AROT\ KPH \#(12)$$

After further reflecting on the influence from the permanently installed seismic transducers installed at TDC of the bearing cap, the correction weight to be installed was adjusted slightly. Figure 12 illustrated the actual correction weight installed, after removal of the calibration weights. The measured influence in Eq. 9 indicates that the generator rotor was operating only slightly above its second mode.

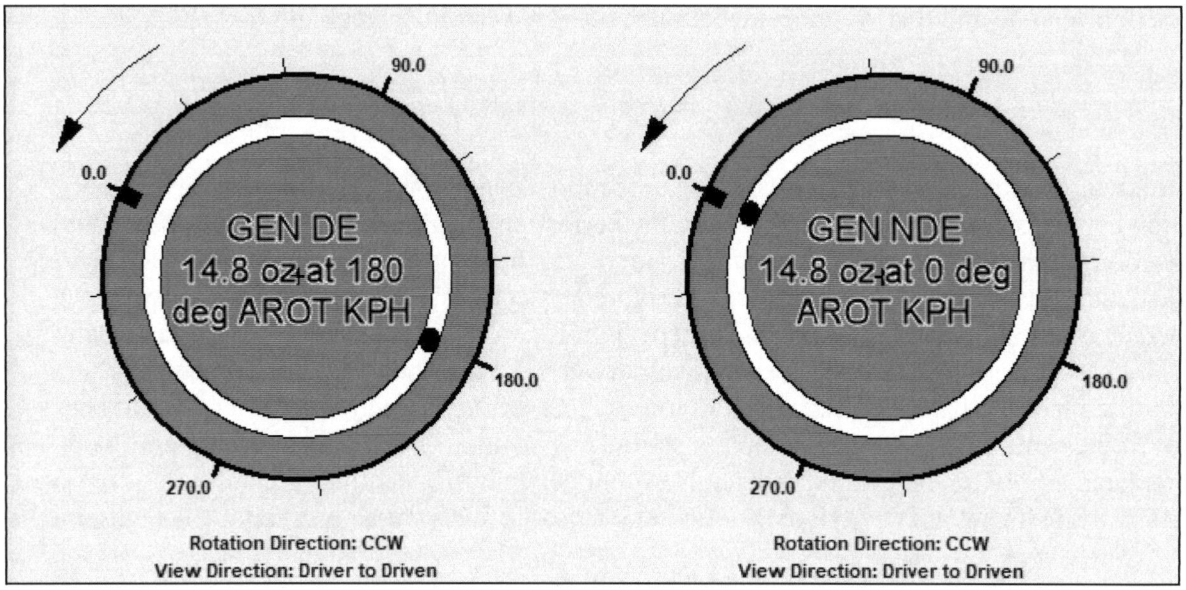

Figure 12: Generator DE (left) and Generator NDE (right) correction weight placement illustrations. Weight placement was identified as against rotation from the Keyphasor® notch (AROT KPH).

The resulting amplitudes from application of the correction weight were detailed in Figure 13. A linear and predicted result was achieved. Not all machinery behaves linearly or repeatably, however those two base assumptions were met in this case.

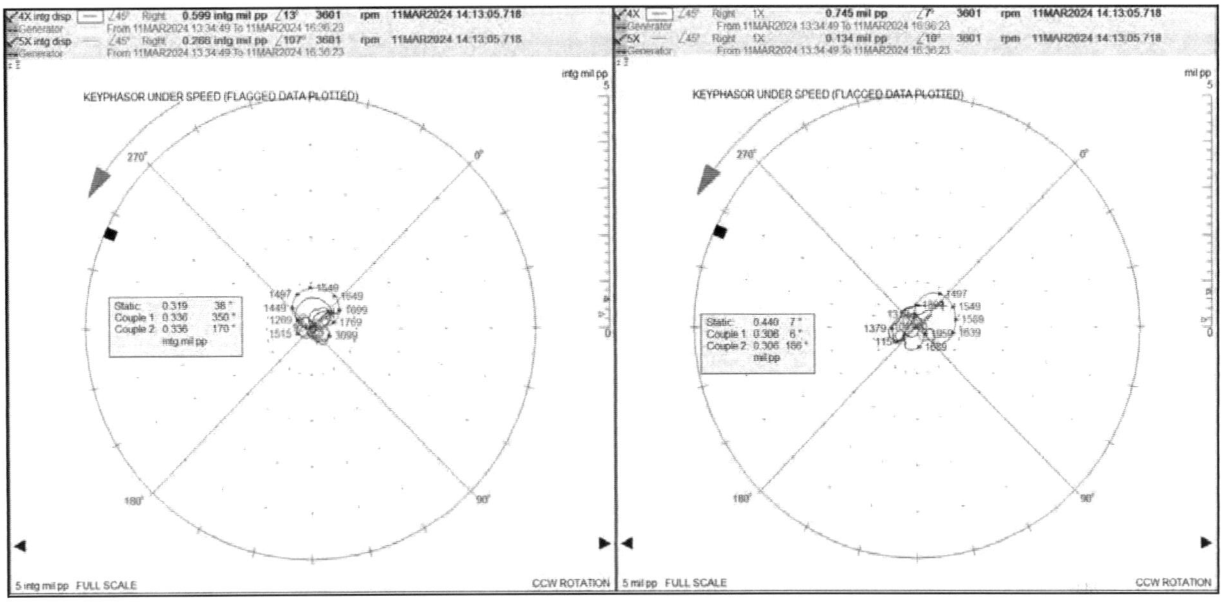

Figure 13: Seismic (left) and relative (right) polar plots illustrating the R2 (correction weight) generator bearing amplitudes. DE was bearing 4, and NDE was bearing 5.

Exciter Evaluation Following Balancing

Recall the purpose of balancing the generator was not to reduce the generator rotor response, but rather to quell the static exciter stator and surrounding housing. During the balancing effort, seismic data was taken on the static exciter stator, as detailed in Figure 14 & 15.

As illustrated, an 89% reduction in seismic amplitude of the exciter stator was realized through balancing activities of the generator. Before and after exciter amplitudes were taken between R1 and R2 for the most precise single run to run comparison. Qualitatively, the response of the surrounding enclosure and contents was no longer presenting an obvious risk to the various attachments.

Figure 14: R1 (left) and R2 (right) exciter stator seismic amplitudes.

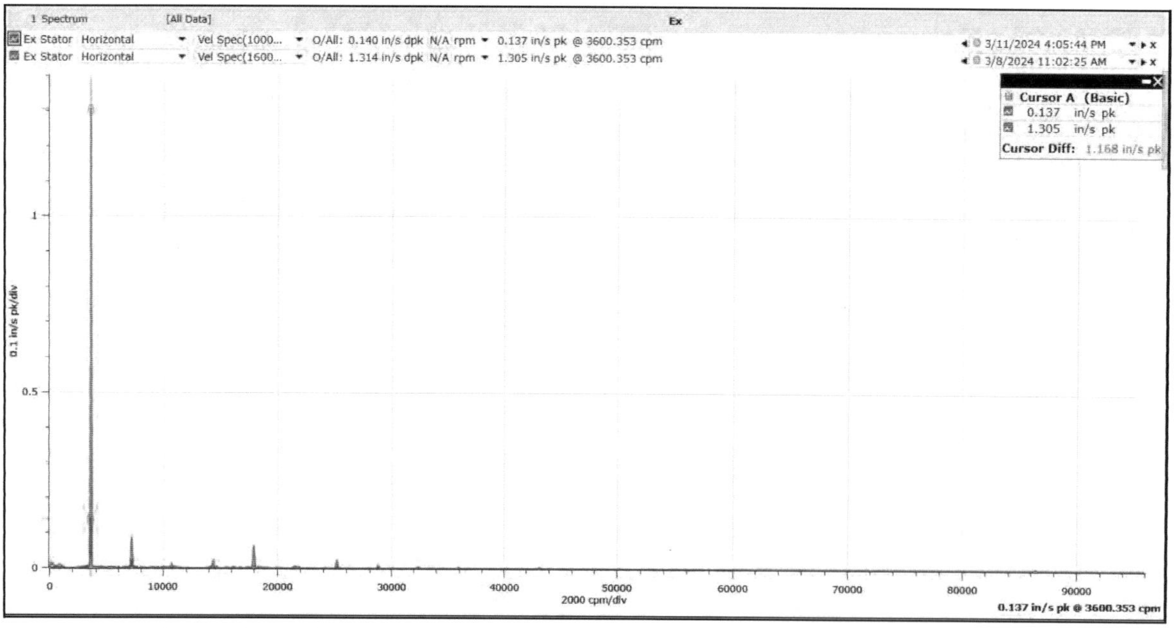

Figure 15: R1 (red) and R2 (blue) exciter stator seismic amplitudes.

Conclusions

Utilizing existing balance planes on the generator in order to reduce the vibration response of the exciter had several immediate benefits. It was obvious to the end user at the conclusion of the effort that the source (or forcing function) of the vibration was not related to malfunction of the static exciter. The risks associated with elevated enclosure vibration were mitigated, including the cessation of false fire detector alarms and reduction of component loosening risks associated with elevated seismic vibration. The unit was allowed to return to service with no further immediate maintenance. However, the author still recommended achieving adequate separation margin of the exciter stator and associated structure. Consider the response to minor upsets of the balance state of the generator, such as generator thermal sensitivity, increasing levels of rotor bow, or a generator major overhaul would likely be recurrence of the elevated structural vibration. Without proper follow up and documentation, an unsuspecting new maintenance team may encounter again elevated vibration of the exciter stator with no obvious cause. The transmissibility of the generator vibration to the exciter structure was not altered by balancing efforts, thus resulting in a temporary remediation.

Acknowledgements

Much appreciation is given to William Trevino, whose decades of field experience and counsel guided this troubleshooting effort to a swift resolution.

Nomenclature:

1X = vibration filtered to the operating speed/frequency
$2XF_L$ = two times line frequency, 120 Hz for US
C = vibration caused by calibration weight or weight pair
AROT = against rotation, as a reference for angular weight placement
CP = couple vibration component
CP1 = couple vibration component end 1
CPM = cycles per minute
DE = drive end
DVA = dynamic vibration absorber
FRF = frequency response function
H = influence vector
H_{cp} = couple influence vector due to a couple balance weight pair
KPH = Keyphasor®
m = generator rotor imbalance mass
m_1 = representative mass of generator bearing structure
N = null vector

N_{cp} = couple null vector
NDE = non-drive end
O = original vibration
O + C = vectorial combination of original and calibration response vibration
OEM = original equipment manufacturer
R0, R1, R2 = Original run, first run after balancing, second run after balancing
Rel = shorthand for relative vibration
SA = shaft absolute
SDOF = single degree of freedom
Seis = shorthand for seismic vibration
TDC = top dead center
ω = rotational frequency
ω_n = natural frequency
$W_{cal,CP,DE}$ = couple calibration weight (pair) represented by the drive end placement with reference to the KPH notch angular location.
$W_{corr,CP,DE,X}$ = couple correction weight (pair) represented by the drive end placement with reference to the X probe angular location.
x , y , z = generator bearing displacement, foundation/floor displacement in the exciter structure, displacement of the static exciter stator

References

[1] R. C. Eisenmann Sr. and R. C. Eisenmann Jr., *"Machinery Malfunction Diagnosis and Correction: Vibration Analysis and Troubleshooting for Process Industries*", Reliabilityweb, 2015.
[2] B. H. Tongue, "*Principles of Vibration: Second Edition",* Oxford University Press, 2016
[3] M. J. Titone and S. P. Ciesla, "Static/Couple Balancing Using Commonly Available Modern Tools," in *Vibration Institute Annual Training Conference*, Covington, KY, 2024.

THE CONSIDERATION OF ECCENTRICITY RATIO AND ATTITUDE ANGLE IN TURBOMACHINERY MISALIGNMENT

Author: Robert A. Shannon, PE

VIATC TRAINING CONFERENCE & EXPO
NEWPORT NEWS, VIRGINIA | AUGUST 6-8, 2025

VIATC TRAINING CONFERENCE & EXPO

NEWPORT NEWS, VIRGINIA | AUGUST 6-8, 2025

THE CONSIDERATION OF ECCENTRICITY RATIO AND ATTITUDE ANGLE IN TURBOMACHINERY MISALIGNMENT

Robert A. Shannon, PE
Manager- Customer/Corp Training – Mechanical Dynamics & Analysis, LLC
rshannon@mdaturbines.com

Robert A. Shannon, P.E., has nearly forty years of extensive engineering, analysis, technical management, and customer interfacing experience as a mechanical design, performance, diagnostic, and test engineer in the power generation and heavy industrial sectors. Prior to joining Mechanical Dynamics & Analysis (MD&A) in 2016, he held several mechanical engineering team leader, design, and field service/diagnostic management positions as well as professional engineering consultant of Shannon Diagnostic Engineering, PLLC, which he founded in 2002. Robert earned a Bachelor of Science degree in Mechanical Engineering from the State University of New York at Buffalo and is a registered Professional Engineer in New York State. His current responsibilities include corporate administration and development of MD&A's technical training programs. He is lead instructor of MD&A's Turbine-Generator Vibration Analysis training course, and the author or co-author of several technical articles on turbomachinery diagnostics, vibration, and design.

Abstract

This paper is a case history which investigates the relationship of shaft eccentricity, bearing design, and bearing attitude angles with turbomachinery misalignment. Bearing clearances, eccentricity ratios, shaft centerline excursions, and shaft orbit plots are evaluated to determine alignment characteristics and determine corrective action for a large turbine generator set.

Key Words: alignment characteristics, axial movements, bearing attitude angles, bearing clearances, bearing design, coupling misalignment, diagnostics, eccentricity ratios, shaft centerline excursions, shaft orbit plots, turbine misalignment

Introduction

The characteristics, effects, and diagnostics of turbomachinery misalignment are widely acknowledged and understood. It is well known that misalignment can result in a change in vibration, excessive bearing wear, and reduced equipment operability. The key characteristics of turbine-generator misalignment include elevated axial and radial vibrations, abnormal shaft centerline excursions, elevated bearing metal temperatures, and undesirable variations in available bearing clearances due to high shaft centerline eccentricity. Prolonged operation of turbomachinery in this condition can accelerate bearing wear, damage rotating components (shafts, rotors) as well as stationary components (diaphragms and seals). When diagnosing misalignment, analysis of bearing eccentricity ratios and attitude angles can augment typical vibration diagnostic data formats.

Misalignment Characteristics and Failure Modes

Misalignment repercussions range from slight to severe. Excessive misalignment conditions may manifest as light, intermittent rubbing between rotating and stationary parts, rotor bowing, or in extreme cases, part extrication and catastrophic failure. See Figures 1 through 4.

Figure 1. Bearing Misalignment-Edge Loading Damage

Misalignment of the bearing to the shaft may cause edge loading as in Figure 1. A cocked, pitched, or nonconcentric bearing will often produce edge loading. Edge load will result in much higher local bearing temperatures, local pressures and much thinner oil film thicknesses. High bearing temperatures may not always be detected on turbomachinery supervisory systems, depending on the relative location of RTDs and the edge loading (progression of edge loading may go undetected without notification). When any of these factors exceed acceptable limits, damage will occur normally as a breakdown in the local oil film and result in metal-to-metal contact. The uneven contact is clearly seen by the frosted appearance on the bearing surface as shown.

Shaft full annular rubs due to misalignment may occur when the rotor is not properly aligned with the bearing, causing continuous contact around the entire internal bearing circumference. See Figure 2. Full contact misalignment will lead to increased friction, heat generation, and excessive babbitt wear. As rubbing contact is made, the relatively softer bearing liner or babbitt material will melt and smear due to high friction and leave deposits and surface scoring of the bearing and shaft.

Figure 2. Bearing Misalignment- Full Annular Rub Damage

Over time, the misalignment exacerbates the damage, potentially causing deformation of the shaft and seal (see Figure 3), compromising the seal integrity, and reducing the machine's efficiency and lifespan.

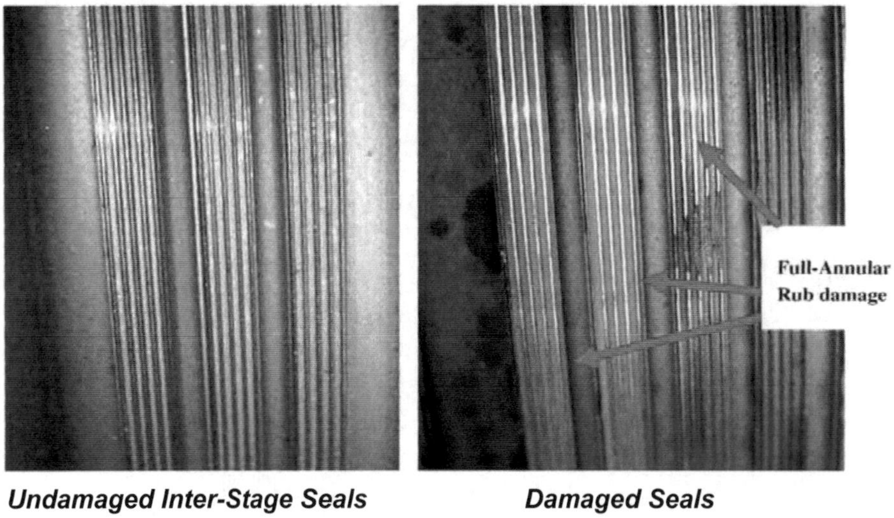

Figure 3. Seal Rubs Due to Misalignment

Catastrophic shaft failure due to coupling misalignment is a very serious issue that can lead to significant component damage and operational downtime. A severely misaligned coupling can create uneven stress distribution at the mating halves and along the shaft, causing excessive vibration, increased friction, and heat. Operation in this condition with continuous high levels of stress and strain can lead to shaft cracks, material deformation, and ultimately, catastrophic failure of the shaft, coupling, and bearing components (see Figure 4).

Figure 4. Misalignment- Catastrophic Damage

The consequences of such a failure are severe, as it can result in the complete breakdown of machinery, leading to costly repairs and refurbishments. Additionally, the sudden failure can pose significant safety risks to personnel and damage to surrounding equipment.

Misalignment Conditions and Causes

Improper bearing elevation or offset or angularity, coupling misalignment, and other conditions introduced at assembly, re-assembly, or during abnormal operating conditions, such as excessive or constrained thermal growth or temperature stratifications, can affect the alignment (Figures 5 through 8). An example of vertical bearing positions is shown in Figure 5(a), when the machine train is viewed from the side (turbine rotor to the generator rotor). The turbine bearing is shown vertically high, and the generator bearing is vertically low. In Figure 5(b) with the machine train viewed from the top, the turbine bearing is shown offset horizontally to the left, and the generator bearing horizontally toward the right. Abnormal mechanical and thermal offsets can be present in vertical or horizontal planes, or a combination of both.

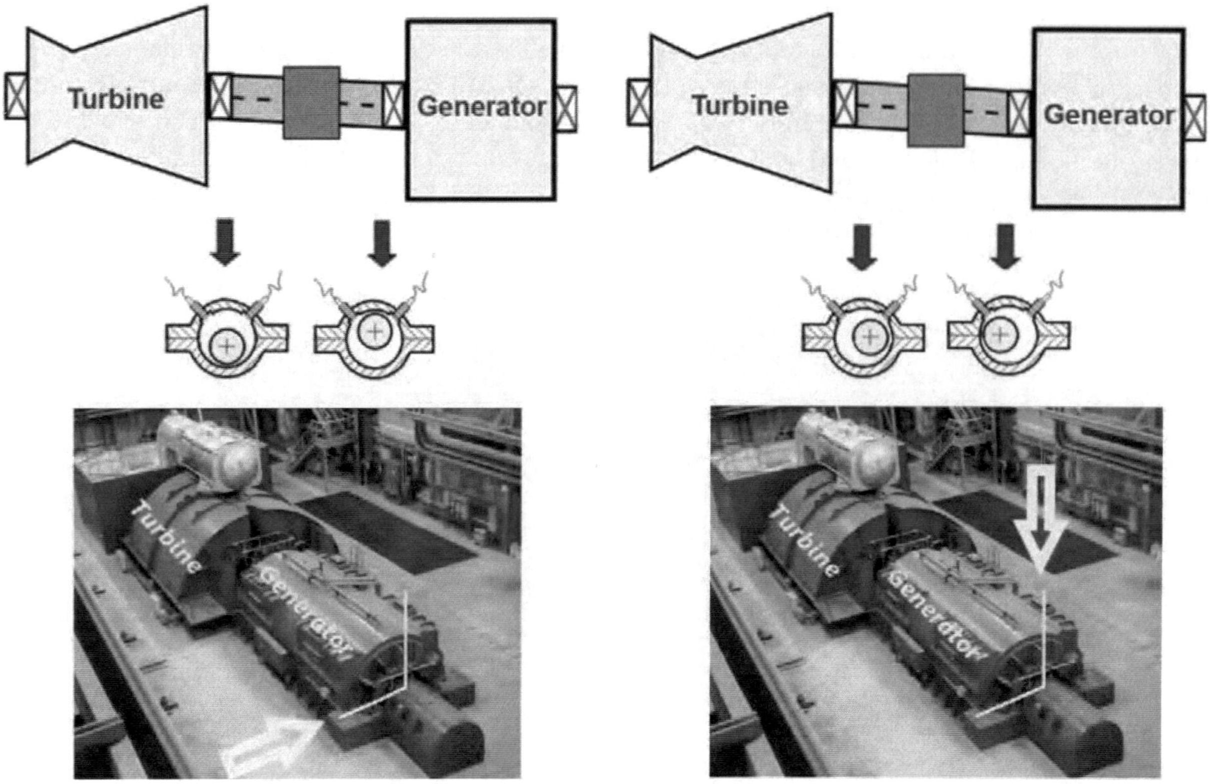

(a) Side View- Vertical Bearing Position *(b) Top View – Horizontal Bearing Position*

Figure 5. Misalignment Due to Vertical and Horizontal Bearing Offset

Less common but just as detrimental to alignment state is a weakened, compromised, or inadequate support structure for the coupled machine train. Severe cracks (Figure 6) and fissures in the cement, substrate or grout base, as well as deteriorated or damaged anchor bolts and fasteners at the bearing pedestals and foundation bases reduces the structural integrity of the rotor support system and creates uneven load distribution affecting the alignment.

Figure 6. Concrete Support/ Foundation Cracks

Localized foundation settlement, where one section of a foundation or support bed no longer provides adequate support can cause the supported equipment to sink or recess, and change the relationship of the rotor centerlines and respective bearing loadings.
Long term creep of the concrete or aggregate material typically found in turbine pedestals can cause changes to bearing elevations. Heavily loaded concrete will expand over time due to thermal cycling. Creep is most prevalent when foundations operate at elevated temperatures, such as near an HP/IP turbine section (typically 100 to 125°F).

Coupling misalignment occurs when the centerlines of connected shafts are not perfectly aligned (Figure 7). The three main types are angular misalignment, where shafts intersect at an angle; parallel misalignment, where the shafts are parallel, but the centerlines are offset; and combined misalignment, which is includes both angular and parallel offsets. Intermediate connecting spool pieces (Figure 8) that are occasionally employed between bearing and coupling spans can also introduce misalignment forces and excessive vibration if the mating flanges are not flat, square, or have fastener mismatch.

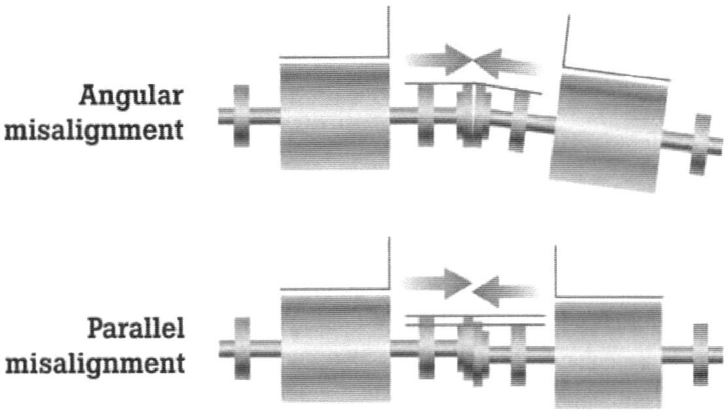

Figure 7. Coupling Misalignment

Finally, human error can cause misalignment issues when elevations and offsets are misinterpreted or not set in accordance with OEM specifications. Close adherence to published alignment specifications is mandatory to ensure proper bearing-to-bearing centerline positions.

Figure 8. Spool Piece Assembly Showing Match Marks

Misalignment Diagnostic Methodologies

Diagnostic methodologies for detecting turbine generator misalignment involve several techniques and interpretation of many different diagnostic data formats. One such data format is shaft orbits. The shaft orbit is the dynamic trace of the geometric shaft center vibration at the corresponding radial measurement plane. Shaft orbits are characterized by orbit orientation and geometric shape and help to visualize the movement of the average shaft centerline position within the bearing clearance, providing insights into the alignment condition.

When a shaft is properly aligned, its orbit at steady state operating modes typically appears as a slightly elliptical shape for horizontally oriented machinery. However, excessive misalignment can cause the orbit to become irregular, flattened, or elongated due to heavy preload, or other thermal or mechanical distortion. Extreme misalignment forces during steady state operation can produce a shaft orbit in a figure-eight shape with corresponding increased amplitude at 2X (twice running speed) vibration frequency, as well as additional harmonics.

Shaft orbits can be heavily influenced by bearing design. A tilting pad bearing that has symmetrical stiffness often has a circular orbit while other pad orientations might cause the orbit to be elliptical and oriented in a horizontal position. For common elliptical and cylindrical bearing designs, the ideal orientation of an elliptical orbit for a clockwise rotating shaft has the ellipse major axis pointing near the 10 o'clock position, and for counterclockwise rotation pointing near the 2 o'clock position. Large deviations from conventional orbit shapes and orientations yield very good insights into alignment condition. Figure 9 exhibits common shaft orbit shapes observed for gravity-loaded, horizontally supported machine trains from the lightly preloaded to heavily preloaded conditions, with counterclockwise shaft rotation.

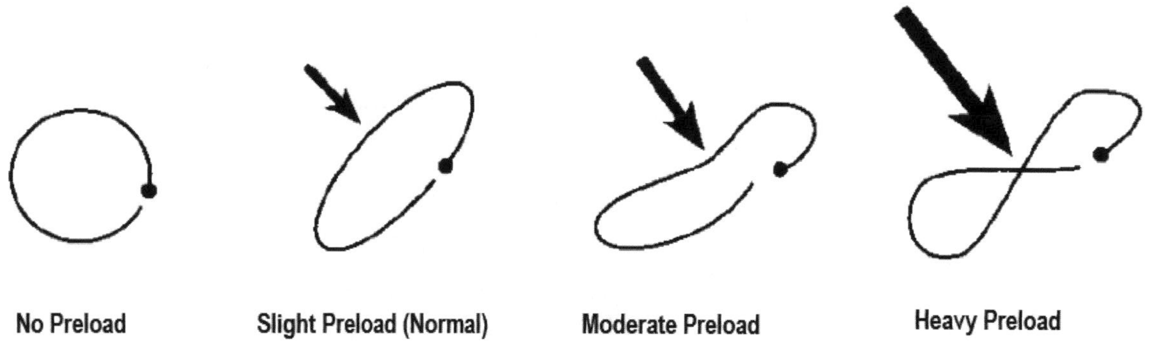

Figure 9. Shaft Orbit Traces vs. Increasing Preload – Cylindrical and Elliptical Bearings

Flattened and elongated shaft orbits can be symptomatic of a shaft position which is either too low, too high, or to a far left or right side of a bearing. Any time a rotor has restricted movement in an improper 'quadrant' of a bearing, this can reveal excessive preload in the form of a flattened elliptical orbit shape. The shaft is confined to a smaller area within the available bearing clearance accompanied with inadequate oil support stiffness. This predicates the shape and orientation of the orbit.

In a multi-bearing setup, such as a large turbine generator arrangement, preload characteristics and shaft orbit shapes can vary significantly depending on the alignment, bearing design, internal clearances, oil support stiffness, rotation direction, and operating conditions. See Figure 10. Journal bearings support the turbine rotor and generator rotor weight, maintaining radial position and rotor train alignment. Thrust bearings counteract axial thrust forces, mainly from steam pushing the rotor, maintaining axial position and proper axial clearances between the rotating and stationary components. Proper radial preload ensures that the bearings can handle the dynamic loads and thermal expansions that occur during operation, reducing the risk of excessive wear, misalignment, and potential failure.

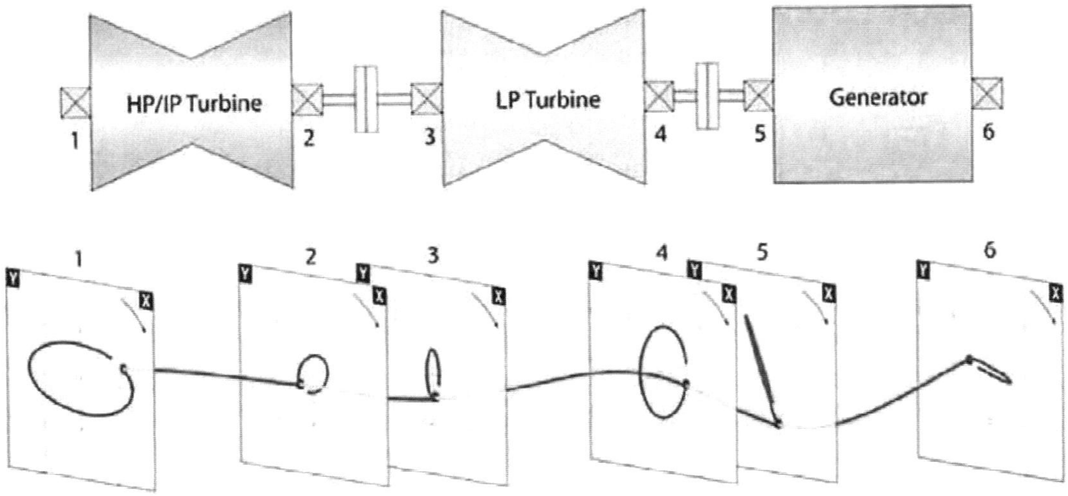

Figure 10. Shaft Orbits for Multi-Bearing Machine Train

Bearing design will also affect the preload and dynamic loading characteristics and influence the alignment condition, particularly if different bearing designs are employed across the machine train. Oil pressure profiles in cylindrical bore bearings typically show a single peak at the location of maximum load, with pressure decreasing symmetrically around the shaft. See Figure 11(a). This profile is influenced by shaft speed and position, with higher speeds increasing pressure due to greater hydrodynamic forces. Elliptical bore bearings differ in that two distinct pressure peaks are evident due to their non-circular geometry. See Figure 11(b). These peaks result in smaller cavitation zones and better thermal stability. Shaft position significantly affects pressure distribution, with variations in load and speed altering the pressure and temperature profiles.

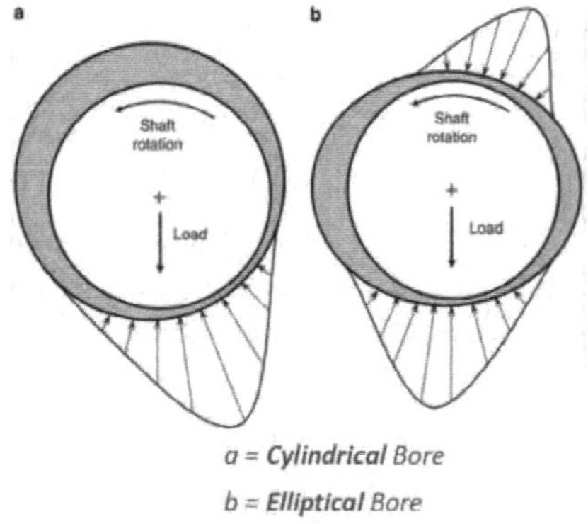

Figure 11. Pressure Profiles for Cylindrical and Elliptical Bore Bearings

Tilt-pad journal bearing configurations are much more mechanically complex and thus have multiple loading arrangements and pressure profiles dependent upon their design, e.g., load-on-pad (LOP) or load-between-pads (LBP) and the combinations of pad fixture variations such as spherical pivots, rocker pivots, and flexure pivots. Figure 12(a) shows a design with each pad with a simple point or line contact. In the non-operational mode, the pads are positioned concentrically with the journal. The journal is of radius R, and each tilt pad surface is positioned further radially from R by a clearance C (machined clearance). Each pad is of active length L. The active length of the pad refers to the portion of the pad's arc length that is in contact with the shaft and carries the load. The active pad length typically spans 55-60° for 5-pad bearings and 70-75° for 4-pad bearings. Each pad has a circumferential arc β, while the pivot is positioned at an angle α relative to the pad leading edge. Figure 12(b) shows a typical load configuration for a five-pad bearing operating with load-on-pad (LOP). The loaded pad at the bottom-dead-center position develops the greatest pressures and thus supports the largest percentage of the applied load W. The neighboring pads will share a larger percentage of the load as operating conditions change and speed is increased.

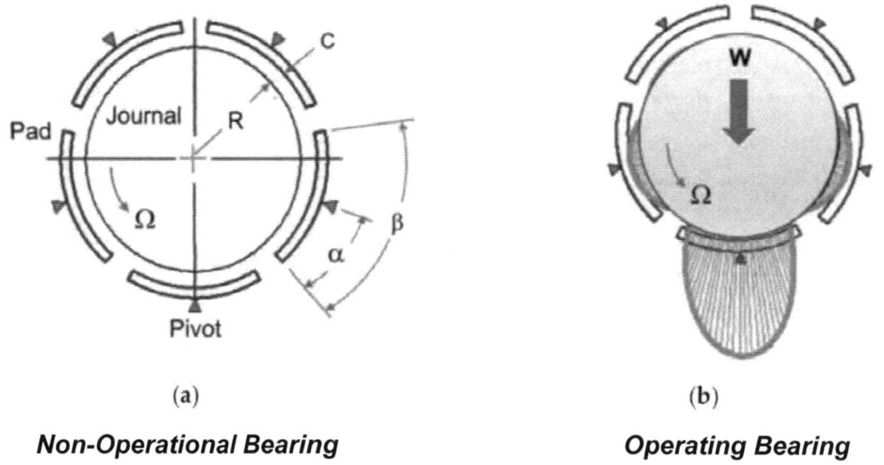

Figure 12. Geometry and Loading of a 5-Pad, LOP Tilting-Pad Bearing

In addition to the shaft orbit plot, another diagnostic data format commonly used to assess alignment condition is the shaft centerline plot. This plot is constructed from tracking the direct current (DC) gap information from the proximity probe signals. As the distance between the stationary probe tips and the rotating shaft changes with respect to speed or time, a two-dimensional graphical representation of the shaft movement relative to the internal bearing clearance is produced. A generalized assumption is that when the rotor is at rest, that it is resting at the bottom of the bearing, and that the relative thermal growth of the journal diameter to the casing diameter is negligible, such that the cold bearing clearance circle can meaningfully be superimposed on the shaft centerline plot. As the rotor starts to turn, the oil film is developed between the journal and the bearing. The oil film will lift the rotor off the bottom of the bearing, and as speed increases, continues to lift it up further. Figure 13 shows an example rotor (shaft) position trend within the bearing clearance that is tracked with increased speed. Note the progressive upward direction of the shaft centerline position as speed is increased from low (168 rpm) to high (3600 rpm) speeds. Like the shaft orbit plot, the shaft centerline excursion is highly dependent on the dynamic stiffness of the bearing, bearing design, internal clearances, oil support stiffness, interaction between adjacent bearings, rotation direction, and operating conditions.

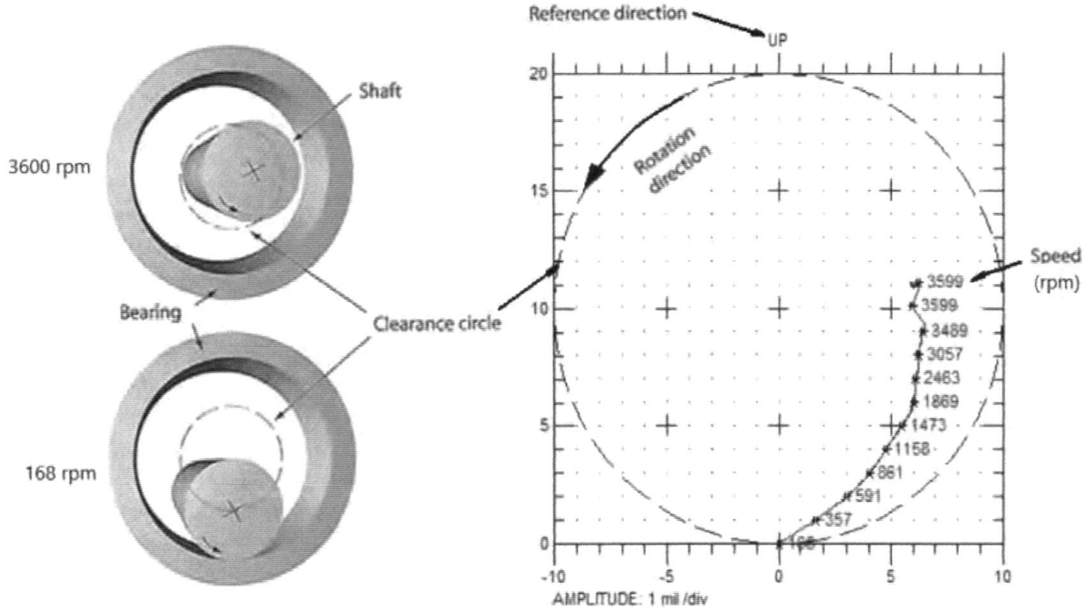

Figure 13. Shaft Centerline Plot and Rotor Position Movement – Cylindrical Bearing

Configuring the clearance values for shaft centerline plots for tilting pad bearings can be complicated by the potential movement of the shaft outside of the clearance boundary during zero and low speed conditions. Referencing the lift check factors will verify this. In addition, probe mounting configurations can sometimes impact the shaft centerline plot. Long probe 'stingers' used in some probe assemblies that are mounted away from the bearings can produce erratic or invalid shaft centerline plots. Probes mounted directly to the bearing sides will provide more useful and reliable information. Error sources will be reduced by the comprehensive analysis of supporting data such as shaft orbits, seismic vibration, and bearing metal temperatures. Valid baseline data acquired from a machine train with good mechanical integrity is highly recommended.

Shaft centerline plots are chiefly used to determine the severity of misalignment defined by two other concepts relative to rotor position and bearing clearance: the eccentricity ratio and attitude angle.

Shaft eccentricity is the offset distance from the bearing center to the average shaft center. The eccentricity ratio is a dimensionless number that indicates eccentricity relative to the bearing radial clearance. It is calculated by dividing the eccentricity by the radial clearance of the bearing. Values for the eccentricity ratio range from 0, where the shaft is centered in the bearing, to 1, where the shaft is touching the bearing.

The dynamic stiffness characteristics of different bearing designs are widely varied, as are the eccentricity ratios. Normal eccentricity ratio is dependent not only on bearing design but operating conditions, most notably load and speed, as well as lubricant viscosity. That said, a range of normal eccentricity ratios for various bearing designs are available as useful rules of thumb in analysis. A typical eccentricity ratio for a properly loaded, cylindrically bored bearing generally ranges from 0.6 to 0.8, and for an elliptically bored bearing 0.4 to 0.6. The typical eccentricity ratios for tilt-pad bearing designs can vary depending on their specific design and operating conditions. The hydrodynamic support in the tilt pad arrangement is proportioned such that the shaft centerline excursion tends to rise more linearly upward along the vertical centerline axis of the bearing clearance, producing small attitude angles and very low eccentricity ratios. For centrally pivoted pads, the operating eccentricity ratio is around 0.45, while for tilt-pad bearings with offset pivot it is approximately 0.3. Lower values of eccentricity ratio mean higher operating minimum film thicknesses and, thus, higher load capacity.

The attitude angle is the angle formed between the shaft's centerline and the line connecting the bearing center to the shaft center. This angle helps determine the direction of the shaft's loading and its relationship with the internal bearing surfaces. Normal bearing attitude angles typically fall between 20 degrees and 50 degrees for cylindrically and elliptically bored bearings (Figure 14). Tilt-pad bearings often exhibit bearing attitude angles that can vary widely depending on the design (load between pads, load on pads, central or offset pivots). For a LOP arrangement, normal attitude angles fall between -10° and 10° due to the near-BDC location of peak oil wedge pressure and minimum oil film thickness, as illustrated in Figure 12.

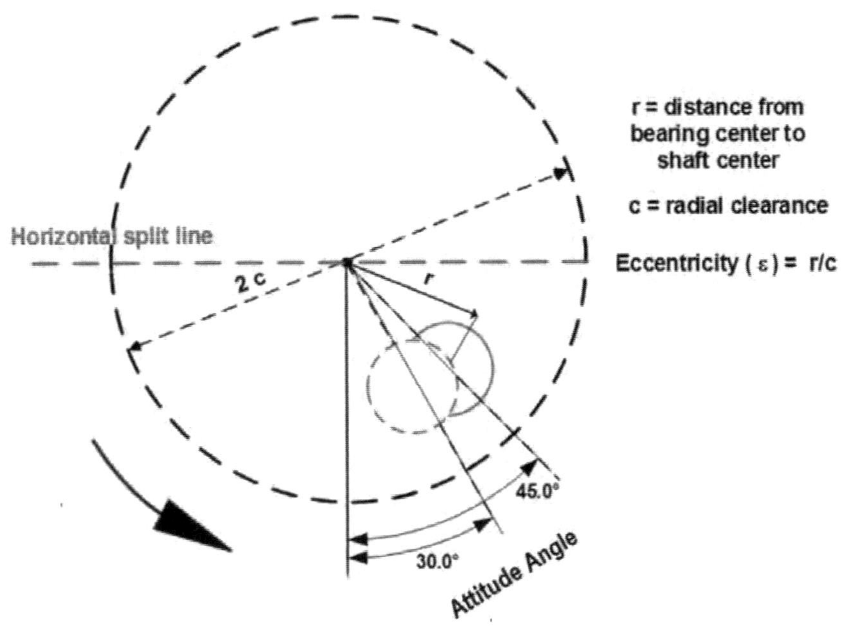

(CCW Shaft Rotation Shown)

Figure 14. Eccentricity Ratio and Attitude Angle- Cylindrical/Elliptical Bored Bearings

High eccentricity ratios and abnormal attitude angles can lead to elevated axial vibration as well or cause the shaft to operate in the upper half or wrong load quadrant of the bearing, leading to potential issues with lubrication and increased friction. A higher eccentricity ratio and uncharacteristic attitude angle signifies a more severe grade of misalignment that is established by an abnormal load distribution across the bearings (bearings either too lightly or too heavily loaded). See Figure 15.

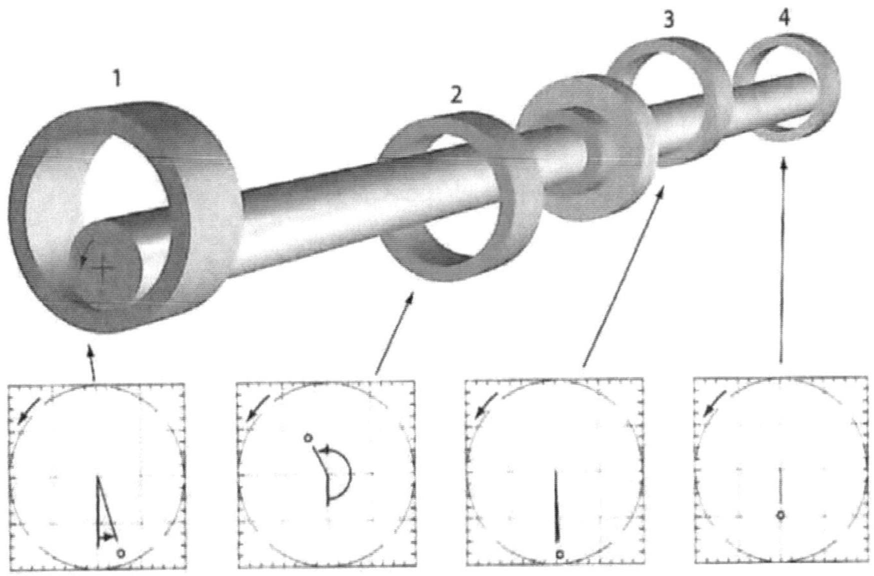

Figure 15. Alignment Condition Along a Machine Train

It is worthy to note that the change in radial shaft position can occur with or without changes in the observed vibration responses. To understand the total response of the rotor system, shaft orbit plot analysis is very often used in conjunction with shaft centerline review. In severe cases, the high eccentricity ratio and improper attitude angles can result in shaft orbits that are enlarged or vibration precessions that are reversed, potentially leading to rubs and other mechanical failures.

Case History – 225 MW Steam Turbine Generator

For this review, a General Electric® Type D-11 steam turbine generator arrangement (Figure 16) was analyzed for alignment condition defined by eccentricity ratio and attitude angles. This machinery was installed and operating in a midwestern US combined cycle power plant. The machine train features an opposed-flow high pressure/intermediate pressure (HP/IP) section, a double-flow low pressure (DFLP) section, and a generator with 225 MW power output rating.

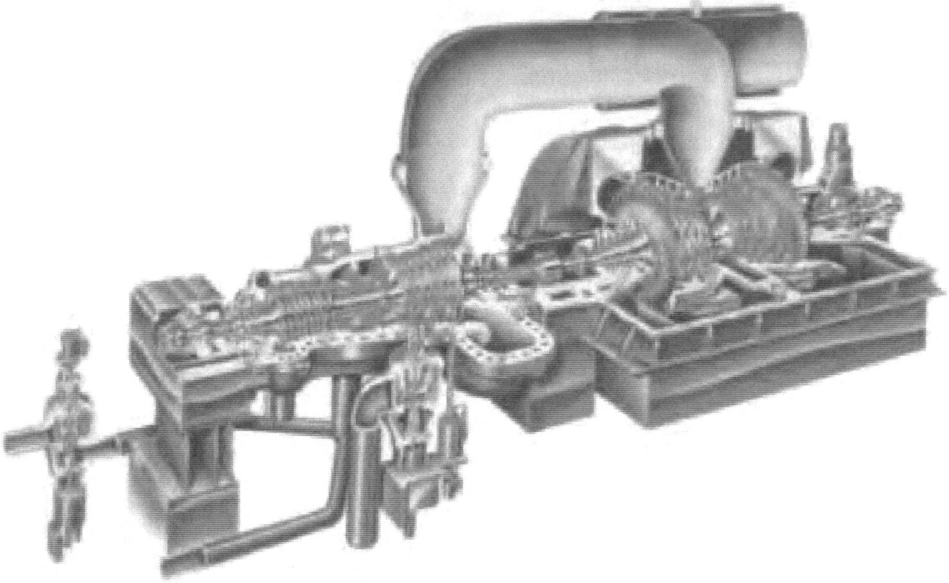

Figure 16. General Electric® Structured D-11 Steam Turbine

The vibration transducer suite (Figure 17) consisted of orthogonal pairs of X-Y proximity probes (scale factor = 200 mV/mil) installed at each of the upper halves of six radial bearings supporting the turbine and generator rotors. Bearings T1 and T2 are tilt-pad design and support the HP/IP rotor, bearings T3 and T4 are tilt-pad design and support the DFLP rotor, and bearings T5 and T6 are elliptically bored design and support the generator rotor. The exciter rotor is unsupported on the outboard side. X-Y proximity probes are oriented 45 degrees from top-dead-center, and a once-per-turn speed reference probe (Keyphasor®) is installed at top-dead-center. Shaft rotation direction is counterclockwise when viewed from the turbine (driver) to generator (driven).

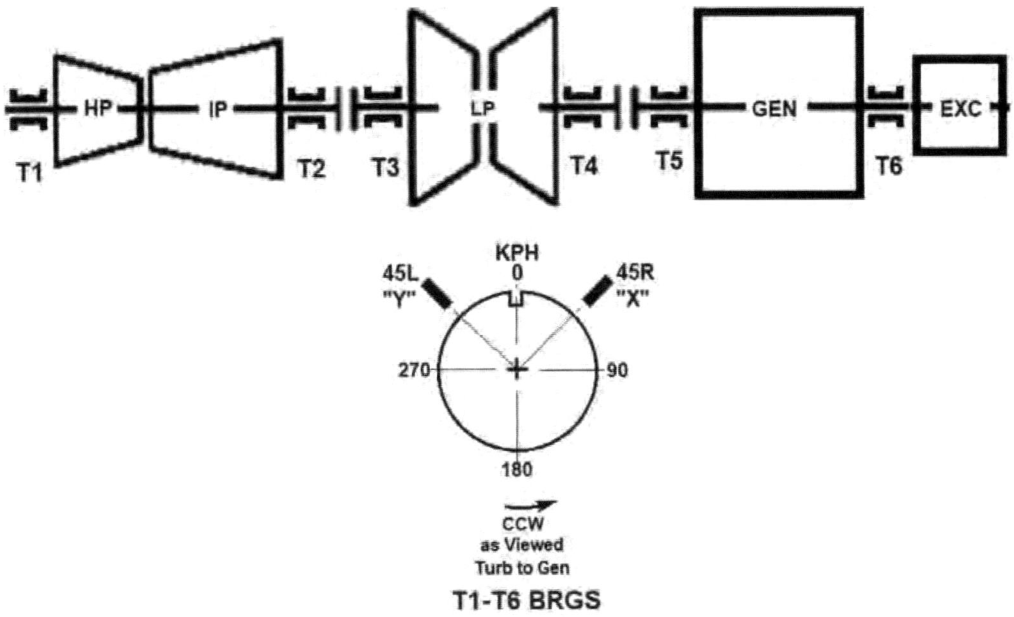

Figure 17. Machine Train Configuration and Transducer Suite

The shaft centerline plot and attitude angle observed of the initial startup to 225 MW for tilt-pad bearing T1 is shown in Figure 18. An exploded view of the endpoint shaft centerline position with respect to 255 MW defines a final attitude angle of approximately 5 degrees and with the distance from the bearing center to the shaft center, r, at 0.0005" (0.5 mils). Similarly, the remaining shaft centerline excursions within the bearing clearances for each bearing are shown in Figure 19. The plot starting point is normalized at the bottom of the plot (bottom of the bearing) for each case and indicated as the steam turbine generator was spun from 126 rpm to 3600 rpm, generator synchronization, and load sweep to 225 MW. Attitude angles at the tilt-pad bearings are observed for a wide range, between 5 to 122 degrees (T1 – T4 bearings), whereas the attitude angles for the elliptical generator bearings are 35 to 42 degrees (T5 & T6 bearings). This variance is dependent upon several factors but is mostly typical of the loading characteristics and restraints between these bearing mechanical designs.

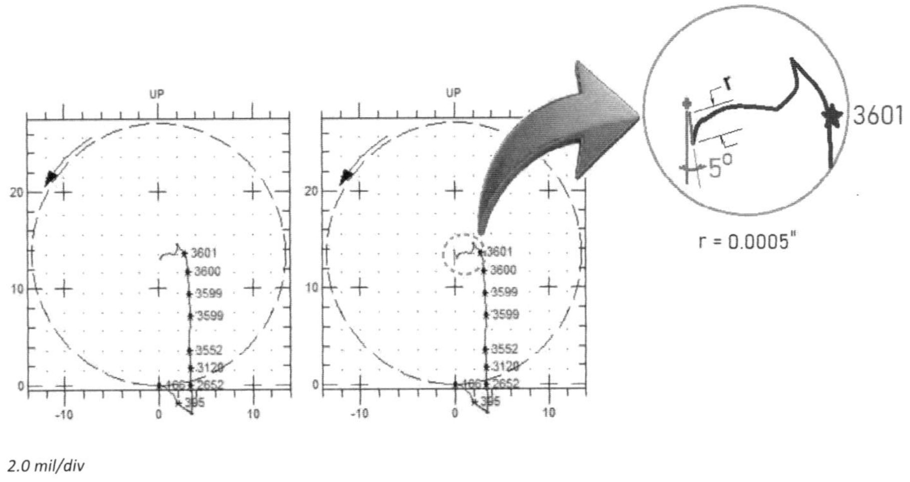

Figure 18. Bearing T1 (HP/IP) Shaft Centerline Plot and Attitude Angle,

Start Up from 126 rpm to 3600 rpm/ 225 MW

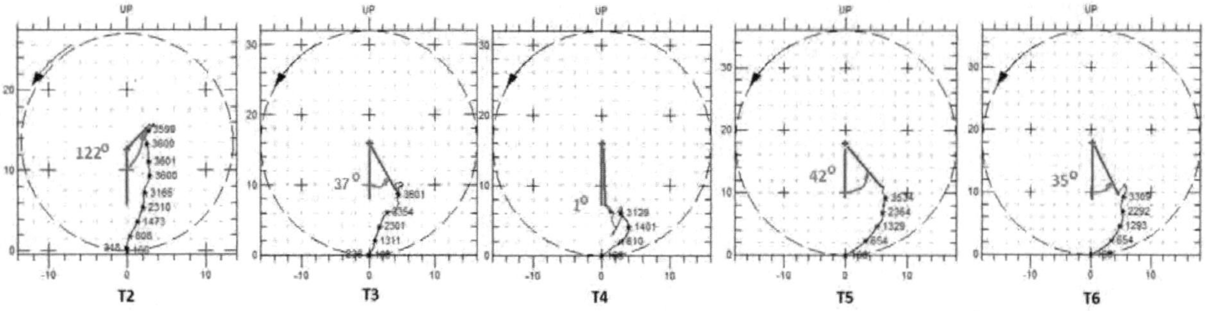

Figure 19. Shaft Centerline Plots and Attitude Angles, Bearings T2 to T6, Start Up from 126 rpm to 3600 rpm/ 225 MW

The vertical shaft elevations to near full load and operating temperatures are seen measured in Figure 20 in mils (thousandths of an inch) from slow roll (126 rpm) to the maximum endpoint at 225 MW.

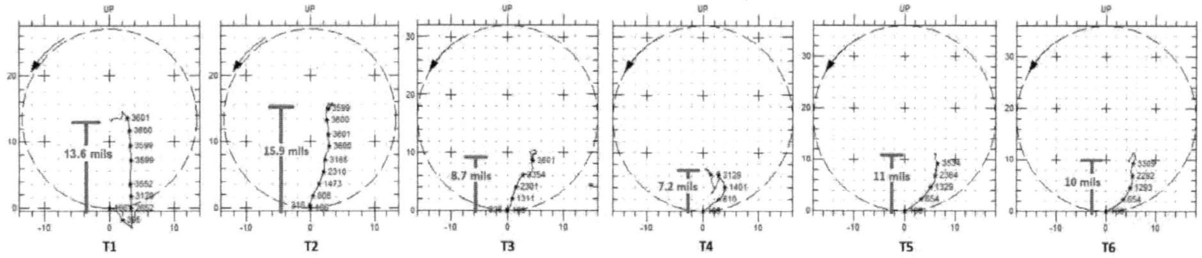

Figure 20. Shaft Vertical Movement/ Elevation, Bearings T1 to T6, Start Up from 126 rpm to 3600 rpm/ 225 MW

From the plots in Figure 20 it can be observed that the vertical lift at each bearing differs based upon factors such as rotor weight, bearing design and thermal expansion characteristics. The lighter weight of the HP/IP rotor and thermal conditions yield a resultant movement along the central axis of the bearing configuration with approximately 50% more displacement than its elliptical bored counterparts at the generator. Figure 21 shows the maximum vertical displacement "map" of the shaft axis to the known bearing clearances and centerlines at each bearing.

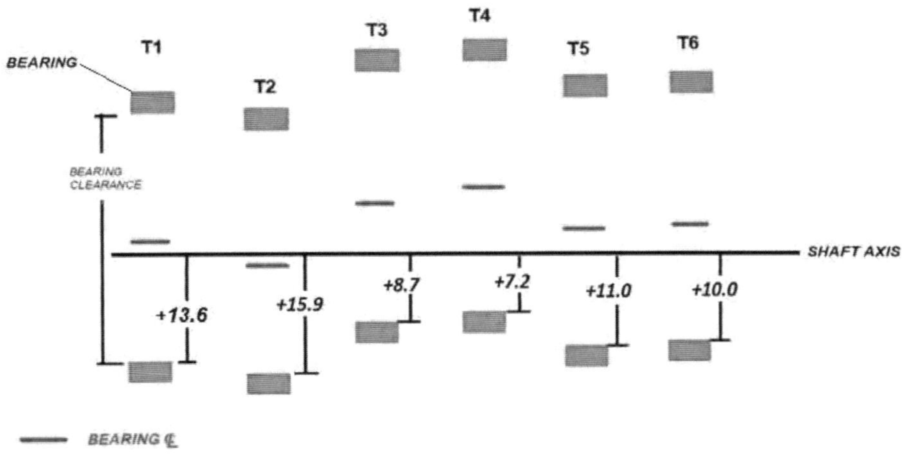

Figure 21. Alignment of Bearings to Shaft, Vertical Plane (Side View)

Similarly, the horizontal movements/offset at each radial bearing plane can be measured and mapped out to show the maximum horizontal displacement of the shaft axis relative to the bearing clearances and centerlines as seen in Figures 22 and 23. The hydrodynamic side-loading characteristics of the elliptical bearings at the generator move the shaft more predominantly into the lower right quadrant of the bearing clearance due to oil film pressure development, shaft rotation direction, and comparatively heavier generator rotor weight.

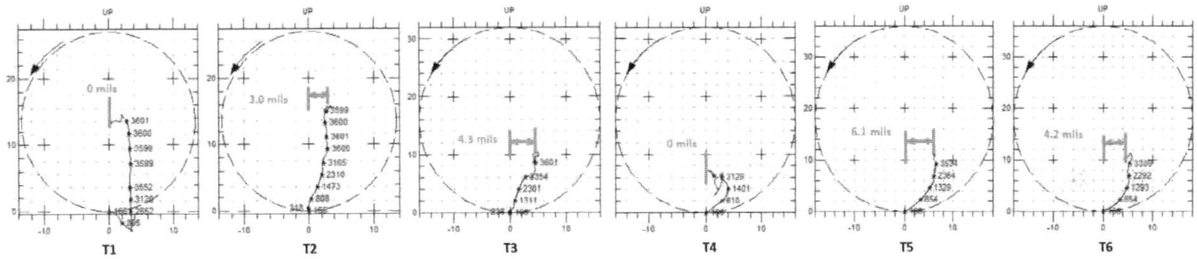

Figure 22. Shaft Horizontal Movement/ Offset

Both the vertical and horizontal movements of the DFLP rotor (bearings T3 and T4) are observed to be at a composite intermediate position between the lighter, hotter, HP/IP rotor and the heavier generator rotor.

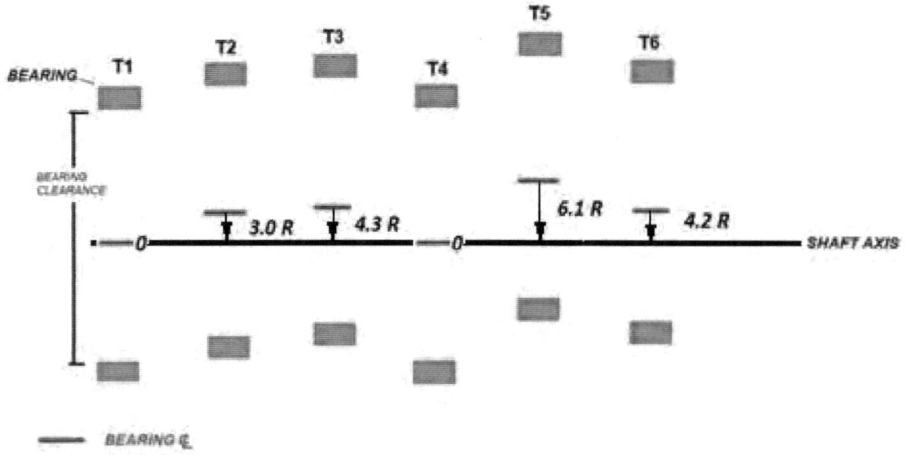

Figure 23. Alignment of Bearings to Shaft, Horizontal Plane (Top View)

Combining the information from Figures 18 through 23 yields the calculated eccentricity ratios, and attitude angles tabulated below:

TABLE 1. Eccentricity Ratios and Attitude Angles

Bearing Number	*Design*	*Radial Clearance [in]*	*r [in]*	*Eccentricity Ratio* ξ	*Attitude Angle* α
T1	*Tilt-Pad*	*0.014*	*0.0005*	*0.036*	*5°*
T2	*Tilt-Pad*	*0.014*	*0.0043*	*0.307*	*122°*
T3	*Tilt-Pad*	*0.016*	*0.0078*	*0.488*	*37°*
T4	*Tilt-Pad*	*0.016*	*0.0083*	*0.519*	*1°*
T5	*Elliptical*	*0.018*	*0.0097*	*0.539*	*42°*
T6	*Elliptical*	*0.018*	*0.0085*	*0.472*	*35°*

The eccentricity ratios and attitude angles observed are typical except for T1 and T2; these values indicate that a potential misalignment condition exists at the HP/IP rotor; this can be further validated by reviewing the vertical and horizontal displacements shown in the shaft centerline trends, bearing metal temperatures, as well as the shaft orbit plots (Figure 24). Shaft orbit plots reveal somewhat abnormal shaft orbit shapes and orientation particularly for the HP/IP rotor. The orbit orientations are nearly horizontal flattened ellipses, indicative of abnormal loading characteristics present.

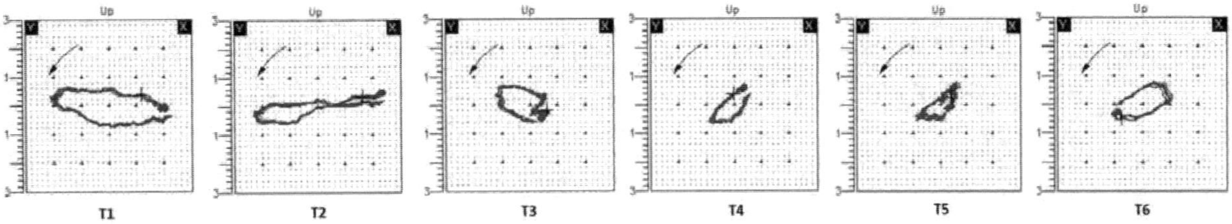

Figure 24. Shaft Orbit Plots, 3600 rpm/ 225 MW. ***[Full Scale = 6 mils]***

The horizontal offsets and thus horizontal alignment characteristics are considered typical and acceptable for this machine train configuration. Recommendations for improved alignment may consider a minor vertical adjustment at the T1 and T2 bearings by 0.0015 to 0.002" to reduce the attitude angle at the T2 bearing and provide a more favorable eccentricity ratio at the T1 bearing. It should be noted that caution must be exercised when making any alignment adjustments at one bearing plane as that may cause adverse preload at an adjacent bearing plane. A thorough understanding of thermal growth, mechanical characteristics, operating parameters, and recommended OEM alignment specifications is necessary for establishing acceptable machine train alignment.

Summary

Addressing misalignment in turbomachinery is crucial for maintaining optimal performance, avoiding damage, and extending the lifespan of the equipment. Eccentricity ratios and bearing attitude angles are very useful tools in assessing mechanical alignment. These parameters are governed by many important characteristics such as rotor weight, thermal expansion, operating conditions, and bearing and coupling mechanical design. Advanced diagnostic analysis tools such as shaft centerline excursions and shaft orbit shape and orientation should be included to perform a comprehensive alignment assessment of any machine train. Finally, regular maintenance and precise alignment of bearings, shafts, and couplings during installation and reassembly activity are essential to minimize the risks associated with abnormal eccentricity ratios and attitude angles.

References

[1] Colgrave, D. et al, *"Structured Steam Turbines for the Combined-Cycle Market"*, GE Power Systems GER-4201 (05/01).

[2] Bently, D., Hatch, C., Grissom, B., *"Fundamentals of Rotating Machinery Diagnostics"*, ASME Press, ISBN 9780971408104, 2002.

[3] Jordan, M., *"What are Orbit Plots, Anyway?"*, Bently Nevada Orbit, Vol.14, No. 4, December 1993.

[4] San Andrés, L., *"Static and Dynamic Forced Performance of Tilting Pad Bearings: Analysis Including Pivot Stiffness"*, Texas A&M Technical Note, 2010.

[5] Eisenmann Sr., R.C., Eisenmann Jr., R.C., *"Machinery Malfunction Diagnostics and Correction",* ISBN 0132409461, 1998.

[6] Shannon, R.A, *"Turbine-Generator Vibration Analysis",* Author Course Notes, Mechanical Dynamics & Analysis LLC, 2024.

[7] Bachschmid, N. et al, *"Diagnostic Significance of Orbit Shape Analysis and its Application to Improve Machine Fault Detection"*, Journal of the ABCM, Vol. XXVI, No.2, 2004.

UNDERSTANDING BEARING WEAR MECHANICS

Author: David Sheffler, CMRP, EIT

VIATC TRAINING CONFERENCE & EXPO
NEWPORT NEWS, VIRGINIA | AUGUST 6-8, 2025

VIATC TRAINING CONFERENCE & EXPO
NEWPORT NEWS, VIRGINIA | AUGUST 6-8, 2025

Understanding Bearing Wear Mechanics

David Sheffler, CMRP, EIT Nuclear Facilities Engineer,
Los Alamos National Laboratory

David Sheffler is a facilities engineer leading the predictive maintenance program for the nation's premier nuclear research facility at Los Alamos National Laboratory, implementing PdM technologies including vibration monitoring and ultrasonic compressed gas leak detection. David has over 10 years of maintenance and reliability engineering experience in automotive, paper, steel, and nuclear industries. He is a Certified Maintenance and Reliability Professional as well as a Vibration Analyst Category III.

Abstract

Common causes of bearing failure are fairly well known and guides to the analysis of bearing damage are easily available. However, something that these guides do not typically address is the fundamental mechanics behind the material failures that result in bearing wear. An understanding of these fundamentals can help with the prevention of failures in existing bearings and the selection of an appropriate bearing for a given application. This paper aims to articulate in relatively simple terms the mechanics behind bearing wear.

Key Words: Bearing wear mechanics, fatigue, abrasion, plastic deformation, fracturing, lubrication, stress, strain

Material Failure Modes

There are many material conditions that would constitute "failure" of the material; however, for the purposes of this discussion we will focus on one single condition with three main ways to attain that condition: the bulk of the material of a bearing is no longer in its intended shape within manufacturer specifications. The three primary ways that materials can reach this condition are through permanent (plastic) deformation, fracturing leading to a loss of material, and abrasion leading to a loss of material.

This writing will focus on ductile metallic materials, because most bearing components are primarily made of metal alloys such as steel and bronze. In ductile materials there are two material properties primarily involved with these failure modes, being a material's yield strength (Sy) and its ultimate tensile strength (Sut). Yield strength is the maximum stress that a material can withstand before it cannot return to its original shape and experiences permanent (plastic)
deformation. A material's ultimate tensile strength is higher than its yield strength and is the maximum stress that it can withstand before it fractures and essentially rips apart. These strength properties are demonstrated on the stress-strain curve example below.

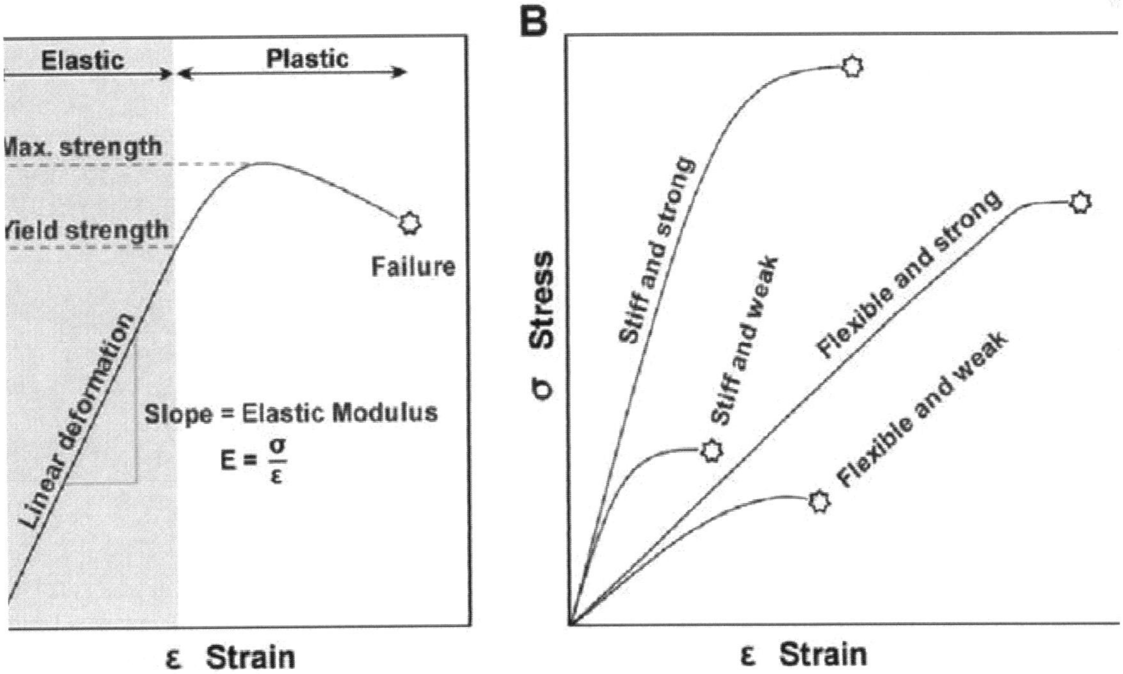

Figure 1: Stress-Strain Curve Example [1]

Stress is simply defined as an amount of force per unit area, such as pounds per square inch (PSI) or newtons per square meter (Pascal, Pa). Since stress is a force distributed over an area, it is directly related to the size of the part being placed under force. Two parts of different size, but otherwise identical, experiencing the same amount of force, will experience different stresses.

The larger part will experience less stress than the smaller part because there is more cross-sectional area to distribute the force across in the larger part.

Strain is the amount that a part deforms when acted on by a stress, expressed as a percent change. For example, if a wire is stretched its strain will be the change in length divided by the original length. Part B of Figure 1 has good examples of the relation between stress and strain. Flexible items, like rubber bands, will stretch more and experience more strain with the same stress compared to a stiff item like a piece of steel bar stock.

Plastic Deformation

All ductile materials deform when placed under stress. The amount of deformation, or strain, is directly related to how much stress the material is experiencing. Below the yield strength of a material, a part experiences elastic deformation and can return to its original shape when the stress is removed. When stressed above the yield strength, a part will experience plastic deformation and permanently change shape.

A good example of plastic deformation is the pre-fracture "necking" stage of tensile tests, which occurs as a ductile material passes the ultimate tensile strength and is about to fracture, as shown in Figure 2 below.

Figure 2: Necking of a Ductile Material Prior to Fracture During a Tensile Test [2]

Plastic deformation does not only work when an object is placed under tension and stretched. It can also be observed when an object is placed under compression and crushed. For example, if a ball is pressed into a solid surface with enough force, a divot will be left in the solid surface when the ball is removed. Incidentally, the size of this resulting divot from a specified ball and force is how a material's Brinell hardness is measured, and the divot is also a bearing wear type known as brinelling. For example, a large bearing is dropped from several feet up prior to installation. Installation continues and goes smoothly other than the initial drop, but the machine does not run smoothly. Deconstruction of the bearing would likely find a few dents in the bearing races spaced apart by the distance between rolling elements, such as in Figure 3.

Figure 3: Example of Brinelling in a Bearing [3]

Fracturing

When enough force is applied to a material to stress it beyond the material's ultimate tensile strength, the material will fracture, essentially ripping apart. Fractures have a starting point and will grow from there, sometimes joining with other fractures if more than one has formed. If a fracture grows large enough, a part can break in two, constituting a catastrophic failure. The below diagram in Figure 4 is an example of the initiation and propagation of a fracture in a ductile material tensile test.

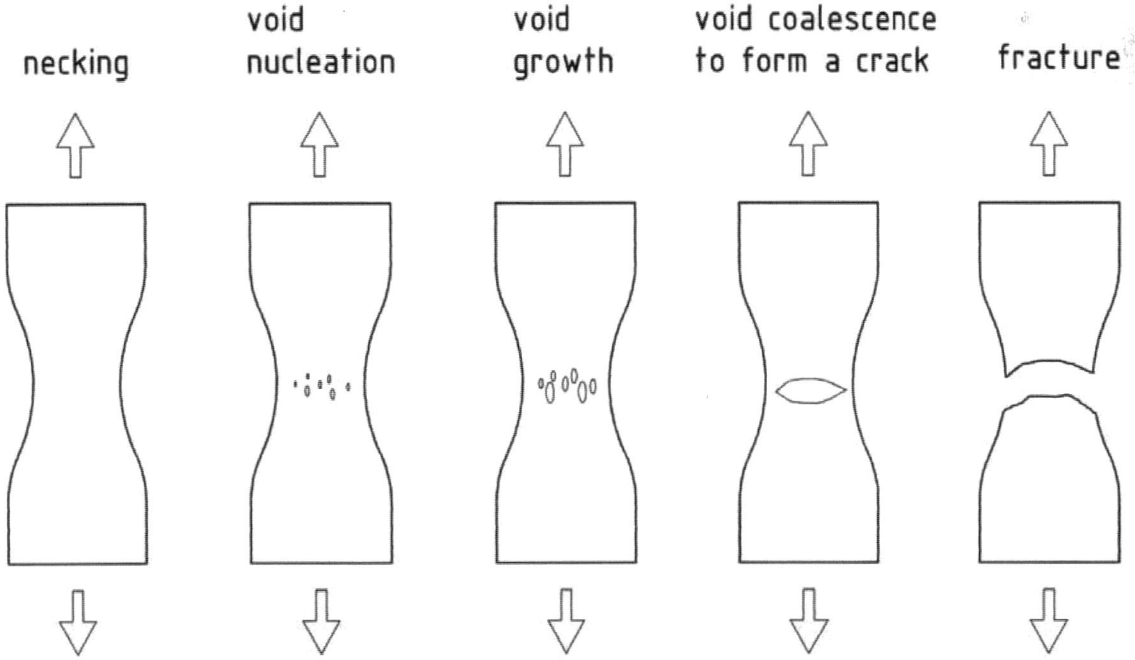

Figure 4: Fracture Growth [4]

In a bearing, fracturing can result in cracks that may go through a part completely or result in surface spalling so that a raceway or rolling element is no longer smooth. The raceway in Figure 5 is exhibiting both through-cracks and surface spalling.

Figure 5: Through-Cracks and Surface Spalling [5]

Abrasion

Abrasion occurs when two surfaces rub against each other and wear material away. Debris is often the rubbing agent, but the surfaces themselves can also be the rubbing agent if there is either no lubrication layer or the lubrication layer breaks down. A high degree of surface roughness with a broken down lubrication layer leads to an increase in abrasion. The removal of material through abrasion could be viewed as a repeated small scale version of fracturing at the surface of a material. This is because in order for material to be rubbed or scratched away, fractures must form and either return to the surface or coalesce with other fractures resulting in particles separating from the original block of material. Abrasion is a wear mechanism that builds on itself, as the generation of particles through abrasion can also abrade the material surface. As shown in Figure 6, abrasion from contact between two surfaces is known as two body abrasion, while that stemming from debris is known as three body abrasion.

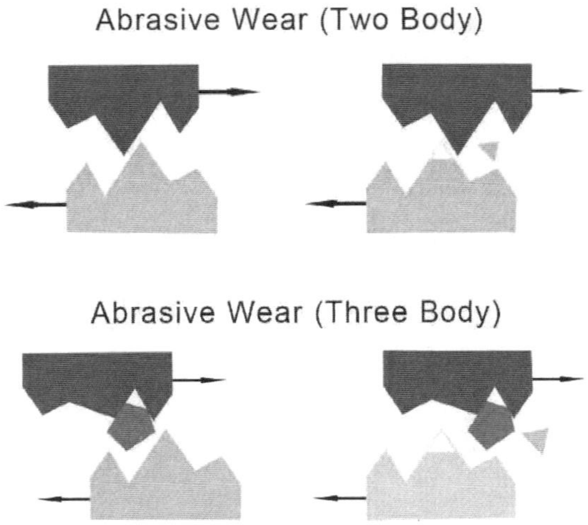

Figure 6: Abrasion Diagram Example [6]

A real world example of what surface abrasion looks like up close is given in Figure 7, where a bearing raceway has undergone surface peeling due to abrasive wear from insufficient lubrication.

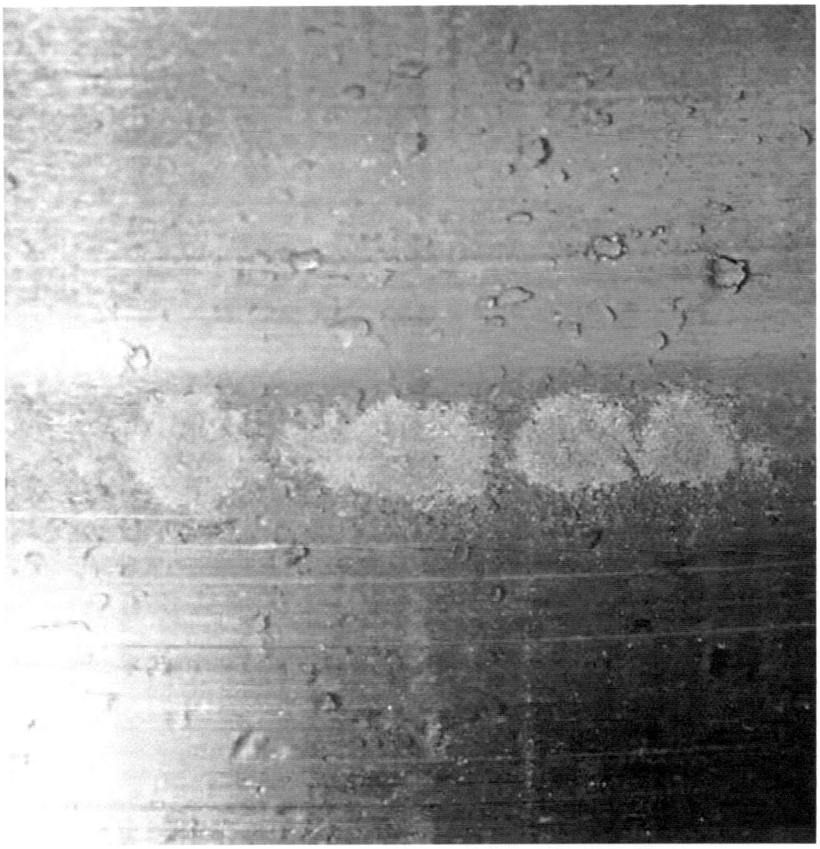

Figure 7: Surface Peeling From Abrasive Wear [7]

Forces and Stresses

Although a rolling element bearing is composed of many parts and multiple rollers are under load at a given time, at the most basic level the bearing can be viewed as a single rolling element resting on a single raceway. This is not to ignore the existence of the other rollers or raceway, but a way to simplify analysis of the forces involved in bearing operation down to the interaction between two parts. Most simply, a ball bearing can be viewed as the interaction of a sphere in a cup, while a cylindrical roller bearing can be viewed as the interaction of two cylinders, or a cylinder in a trough depending on which raceway is being looked at. When under load, the rolling element has a force, F, acting on it, with a corresponding reactive force from the raceway. Examples of what this may look like can be seen in Figure 8.

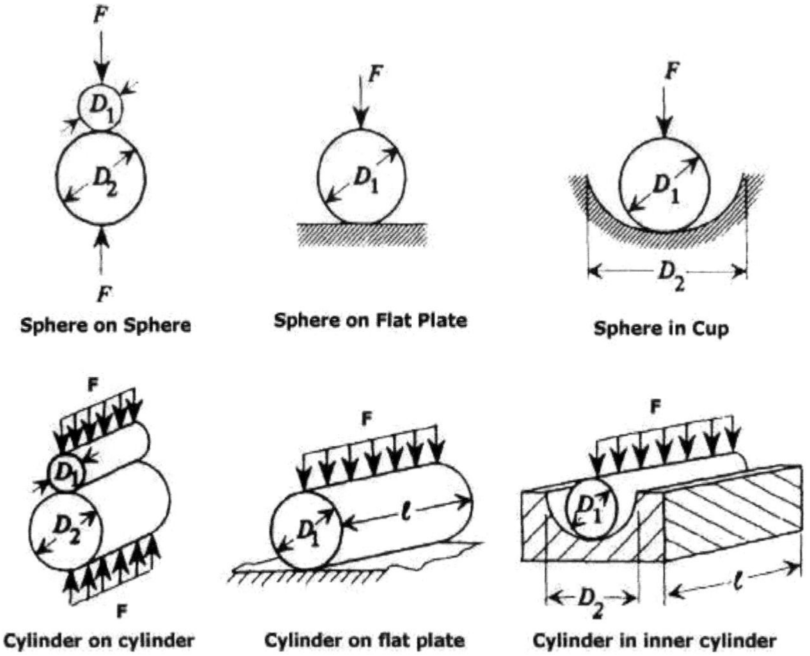

Figure 8: Examples of Spherical and Cylindrical Contact [8]

Since the rolling element is round in shape, with no flat surfaces, when it comes in contact with another object only a small spot is the actual contact area. Therefore the force F from above is transferred to the raceway through a very small area. This means that even if F is a small force, comparatively massive stresses are experienced at the contact spot. Most bearings are made of elastic materials, like steel, which are not truly rigid but slightly deform under stresses and return to their original shape when relieved of stress. Therefore, at the contact spot between the rolling element and raceway, the contact area changes size depending on the magnitude of the force F, as seen in Figure 9.

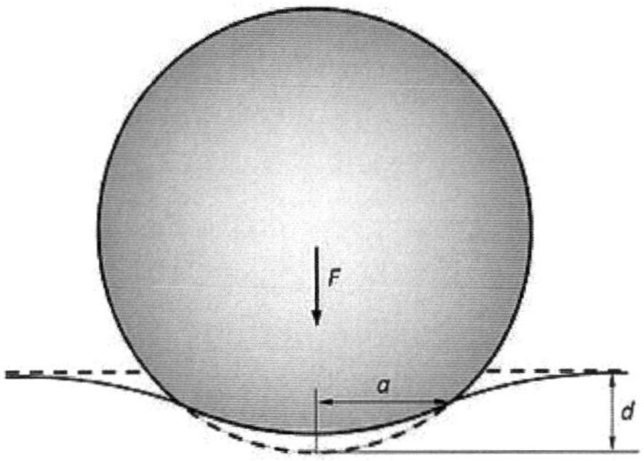

Figure 9: Contact of an Elastic Sphere with a Flat Elastic Surface [9]

Stresses in the contact area between the rolling element and raceway are known as contact stresses, or Hertzian stresses. Stress in the contact area is not uniform, but maximum at the center of the contact area and tapering off to zero at the edges as shown below in Figure 10. If the materials of the raceway and the rolling element and the applied force is known, then the size of the contact area and all stresses can be calculated. There are many possible cases for Hertzian contact stresses depending on geometry. Examples and equations can be found in Roark's Formulas for Stress and Strain Chapter 14, Table 14.1 In fact, the complex geometry of an actual ball bearing is better described by Case 4 within Table 14.1.

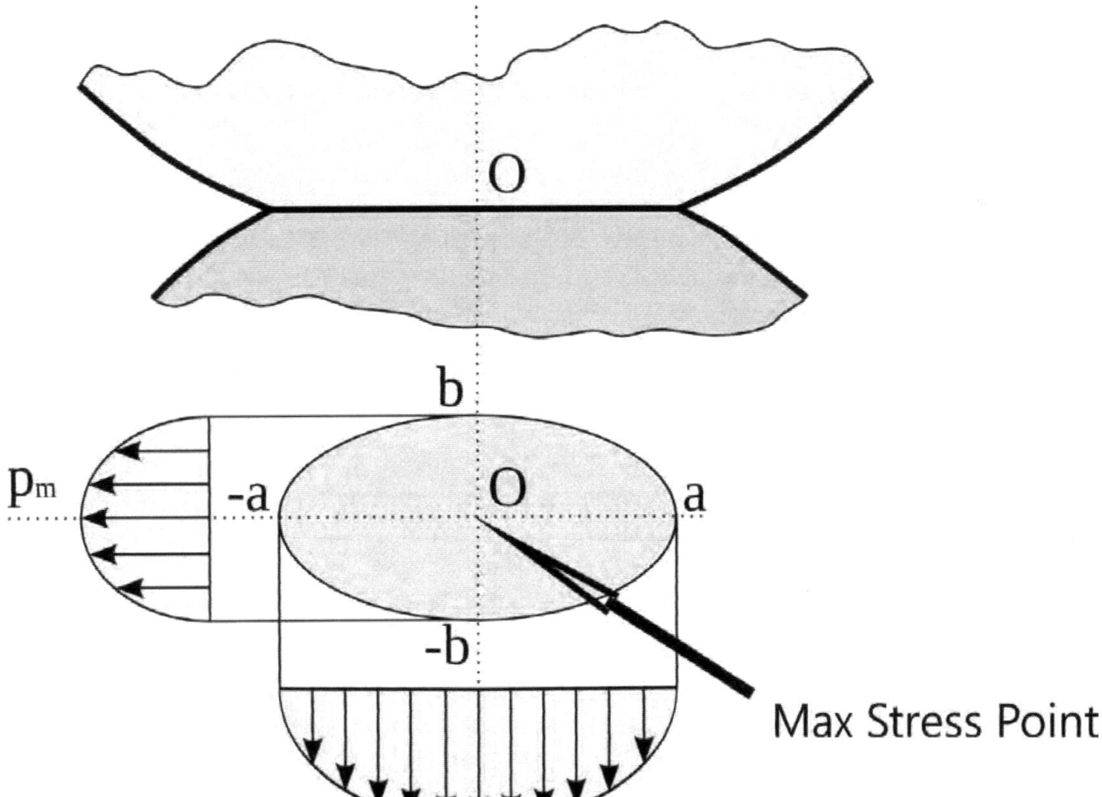

Figure 10: Hertzian Contact Stresses Between Two Spheres [9]

Bearings are designed and rated for certain load cases, which should involve forces resulting in stresses well below the material failure stresses we discussed earlier. If, however, the maximum contact stress experienced inside a bearing is higher than the yield strength of either the raceway or the rolling element, there will be permanent deformation. If that stress is greater than the ultimate tensile strength of either the raceway or the rolling element, cracks may form which could result in material breaking away from the surface. With enough build-up of permanent deformation and/or material removal, the bearing will no longer be within manufacturer specifications, and it will be considered a failed bearing.

Cyclic Loading

The stresses discussed above are for a static or very slow moving bearing and also applied to shock loads (i.e. the bearing is dropped on the ground). Most bearings, whether in storage, transport, or operation, will experience some level of repetitive, cyclic loading. In storage, such loading can come from nearby truck or train traffic vibrating the shelf the bearing sits on. In transport, a bearing can experience cyclic loading from the vibration of the transport truck driving down the road. Bearings in operation typically experience cyclic loading as a result of the shaft they are on turning and the consequential motion of rolling elements. When cyclic loading is experienced over long periods of time, materials undergo a phenomenon known as High Cycle Fatigue, where it takes smaller and smaller stresses repeated over time to cause a material to fracture than under simple static stress. This is commonly seen when bending a paperclip back and forth until it finally breaks. If a bearing is subjected to a very high cyclic load, it will not last as long as one that is subjected to a very low cyclic load. However, the bearing experiencing a very low cyclic load will still eventually wear out and fail because it has experienced high cycle fatigue.

The lifespan of a material in high cycle fatigue depends on the number of cycles and not the amount of time. So if two components of the same material are experiencing the same cyclic load, but one is cycling at twice the speed as the other, one will last twice as long as the other. Cyclic loading for many materials has been studied many times, and formulas for determining the lifespan of parts of different materials are available in Shigley's Mechanical Engineering Design. Often there are charts available for the allowable cyclic stress on a given material vs the total number of cycles, as shown in Figure 11. The chart in this figure is for SAE 4130 steel, and has what is called a "fatigue limit" at the right end of the N (cycles) axis. The existence of a fatigue limit is common in ferrous alloys such as steels, and represents a stress level below which a part made of this material can theoretically last indefinitely. This does not mean that bearings made of alloys that have a fatigue limit and are only ever operated below that stress level will indeed last forever, which will be discussed later in this paper.

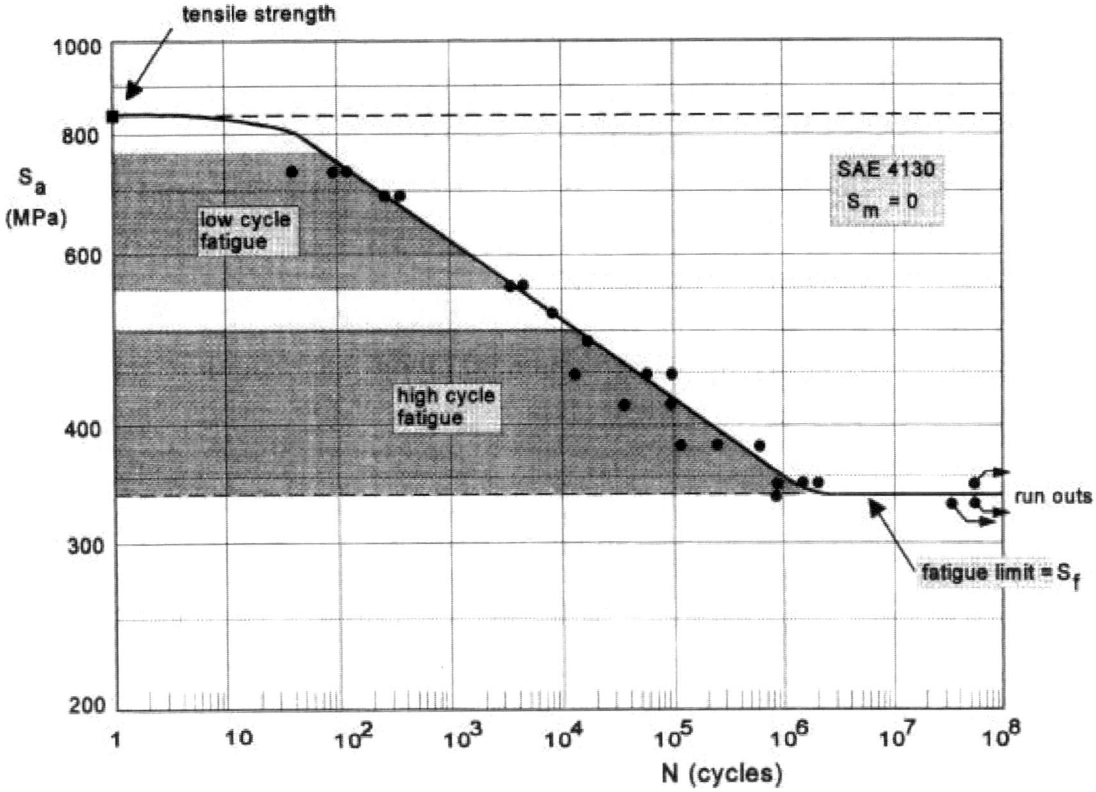

Figure 11: Example of Strength vs Number of Cycles Experienced Chart [10]

Charts such as the one in Figure 11 can be used to estimate the expected lifespan of a part. Once the number of cycles exceeds the expected lifespan or the repeating stress is above the line in the chart at the expected number of cycles, fractures will begin to form and propagate. This is very similar to if the part were subject to a stress greater than its ultimate tensile strength. The propagation of cracks to eventual spalling is illustrated in Figure 12.

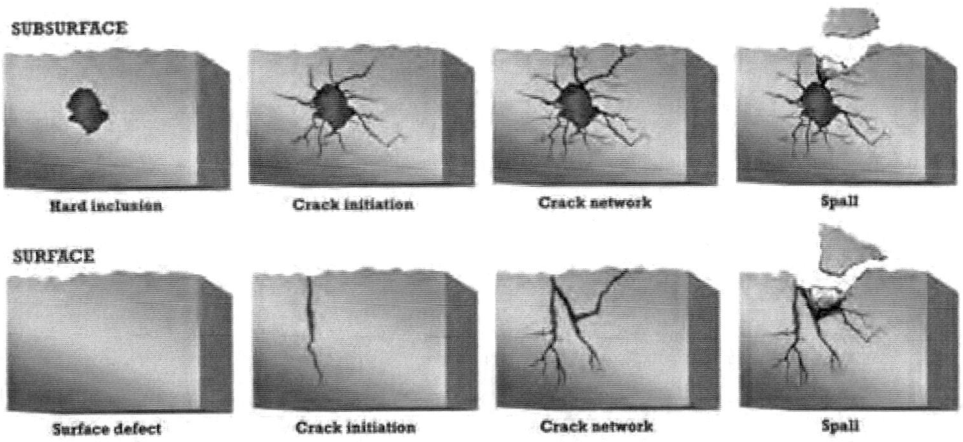

Figure 12: Propagation of a Fatigue Fracture to Eventual Spalling [11]

The eventual result of fatigue wear over a long period of time can be seen below in Figure 13. In this case the cause was excessive axial loading [12].

Figure 13: Spalling Due to Fatigue From Excessive Axial Loading [12]

Sometimes the cracks forming because of fatigue can progress until they penetrate through the entire depth of a part. When this occurs, the new surface formed by the crack will leave a unique pattern that can reveal the crack origin and the type of repetitive stress the part was experiencing. An example of such a through-crack can be seen in Figure 14.

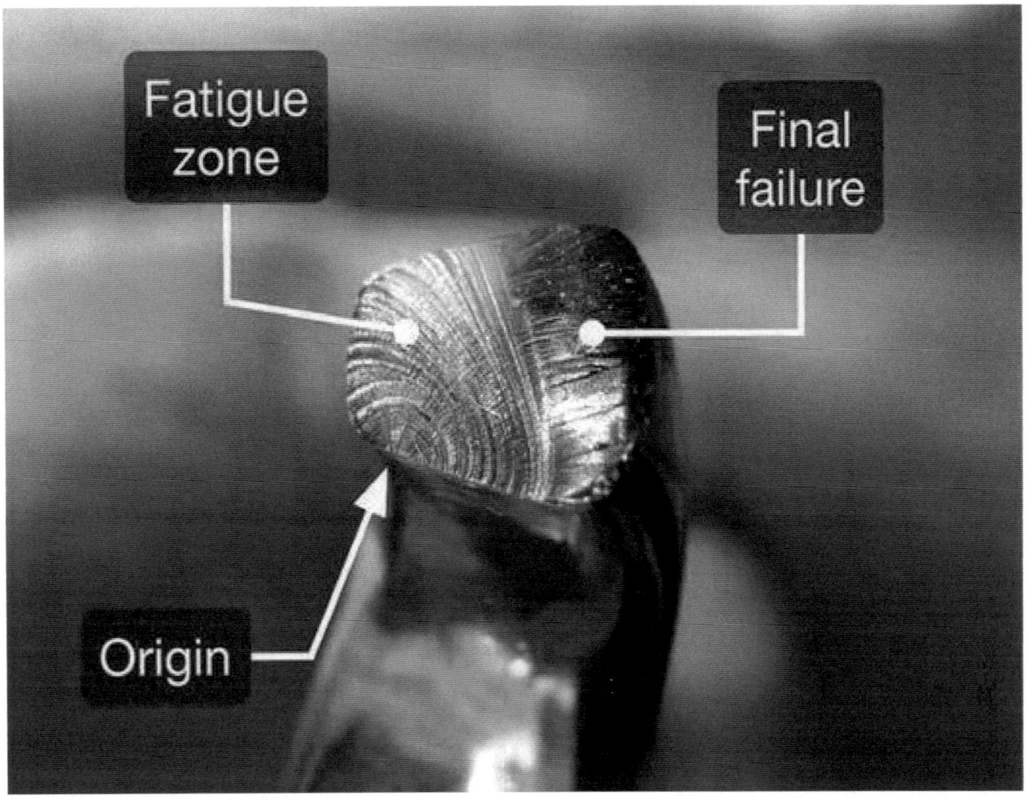

Figure 14: Pattern Left by a Fatigue Fracture Progressing Through a Part [13]

Cyclic loading without rotation, such as a bearing in storage in a high traffic warehouse, over a long enough period of time, is the cause of the “false brinelling” type of bearing wear. The bearing was never subjected to enough force to cause single-event plastic deformation as in true brinelling as discussed earlier. Instead, the rolling elements vibrate on the raceway surface with a small amount of force enough that the raceway material slowly creeps into forming slight dimples on its surface similar to Brinell dents. An example image of false brinelling is seen in Figure 15.

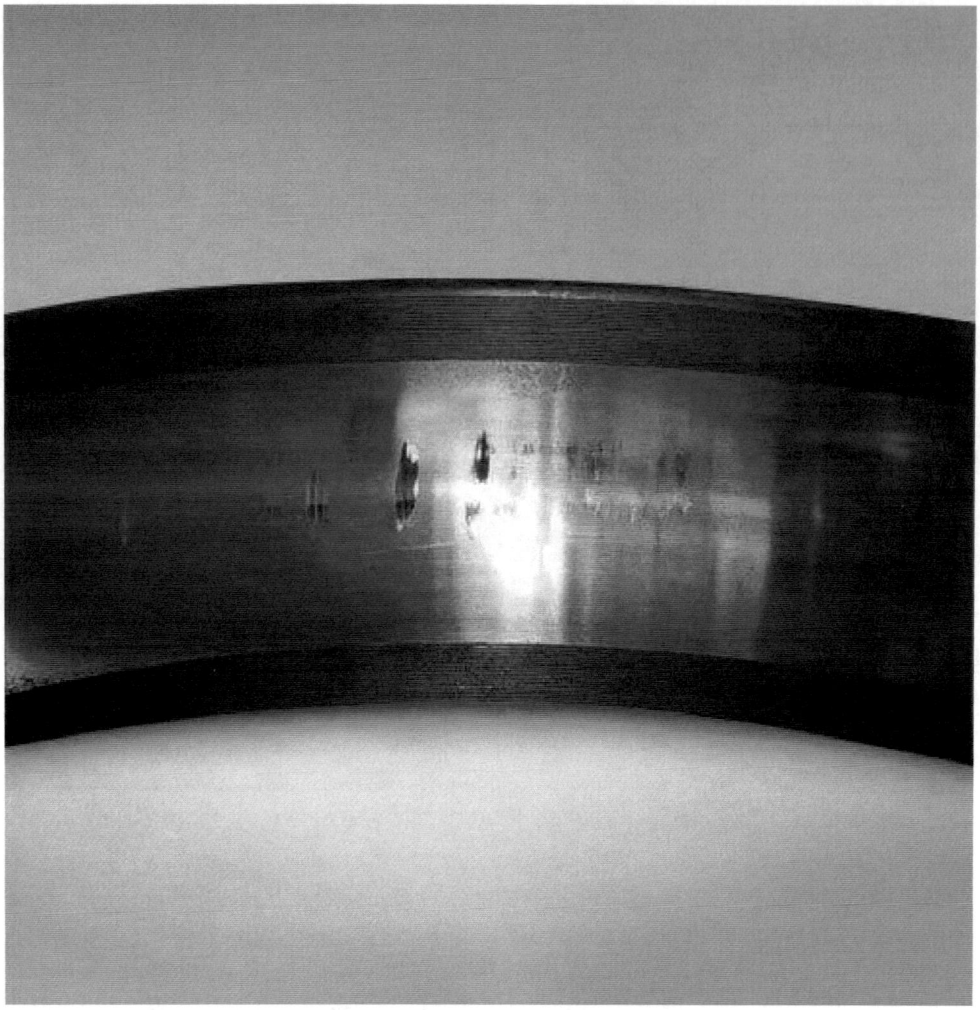

Figure 15: False Brinelling [14]

Real World Stresses for Rotating Bearings

So far we have discussed types of material failure modes, contact mechanics, and part lifespan for cyclic loads. These are useful for understanding the fundamental mechanics experienced by a part, but there is yet another dimension that adds complexity to the stresses experienced by a bearing in use. That additional complexity comes from the fact that rolling elements *roll*, and are also likely to experience some degree of *sliding* during operation. Typically, there is a small amount of clearance available in the unloaded area of a bearing, so rolling elements that are not in the load zone are unlikely to rotate much as there is little to no force acting on them to make them rotate. This condition does not exist in bearings with a significant pre-load, such that all rolling elements are under some load at all times. Ideally, as soon as a rolling element comes into nip as it enters the load zone, it would begin rolling and never experience sliding. However, this is not the case. A combination of imperfect forces and the lubrication fluid generally results in a short period of time where a rolling element experiences sliding as it enters the load zone of a bearing. As the rolling element rolls, a different portion of the rolling element is in contact with the race at any given point in time, and it may take several revolutions for the same area of the rolling element to be in contact during load. Under ideal conditions, a rolling element will rotate a set number of times before the same area is in contact with the raceway again, however, since there is likely to be some degree of sliding experienced by the rolling elements, it could be any number of rotations before the same area of a rolling element is once again in contact during load.

During sliding, the surface of a rolling element is experiencing a shear stress in addition to stresses as a result of normal loading as discussed previously. This results in a higher total stress that can be unpredictable depending on factors such as the amount of sliding that may be experienced, the lubrication layer integrity, debris infiltration into the bearing, internal clearance changes due to thermal growth [15], etc. So in the real world, a bearing will experience a complex and rarely fully predictable set of stresses resulting from a combination of normal load stresses and shear stresses from a variety of sources. Because of this complexity and non-predictability, bearing manufacturers such as SKF often have helpful bearing selection tools. These selection tools are based on statistics from manufacturer testing, for instance: "90 percent of bearings of this specification under this much load will last this long at this speed." This 90 percent survival rating is referred to as an L10 rating.

Some Ways to Mitigate Bearing Failure Due to Material Stresses

Previous discussion in this paper has pertained to the material stresses that eventually cause bearing failures, but what are some ways that these failures can be delayed or prevented? In this section several mitigating measures from part procurement to design and operational changes will be discussed.

At the very beginning of the life of a machine part, procurement, its failure or success can be determined. If a bearing other than the one specified by the designers is purchased and installed, it may result in premature failure. An example from experience – a flying shear drive gearbox at a steel mill was being rebuilt after successfully operating for over 10 years. The bearing specified by the manufacturer had a bronze cage. However, a bearing of the same size but with a stamped steel cage was purchased because it was less expensive. After careful installation, the new steel cage bearing was in service for less than 6 months before it failed and locked up the gearbox leading to a complete production stop. A root cause analysis determined that the bearing cage had failed. It was overall weaker than the bronze cage from the original specifications. Ensuring the correct part is acquired extends to material receipt. Receipt inspection to ensure that a newly delivered part matches the order manifest and the design bill of materials is key to preventing incorrect or even counterfeit parts from being installed.

Proper storage of parts prior to installation can also prevent premature failure. Storing bearings in sealed containers, oil films, or environmentally controlled spaces prevents surface rusting and dust intrusion thereby preventing abrasion on operation. To prevent false brinelling in high traffic warehouses, store individual bearings lying flat on the shelf and rotate the shafts of motors, gearboxes, pumps, etc. periodically.

Careful and proper installation of bearings will prevent overly pre-loading bearings that results in a significantly shortened lifespan. For instance, if the machine specifications call for an interference fit of the bearing onto a shaft, it is advisable to either heat the bearing in an oil bath or dip the shaft in liquid nitrogen so that thermal expansion/contraction allows for the bearing to be slipped on. It is not advisable to beat the bearing onto the shaft with a hammer – this delivers repeated shock loads to the bearing, possibly resulting in brinelling, and may possibly misalign the bearing on the shaft resulting in a possible axial overload. For bearings with adjustable internal clearances, research by NASA has shown that clearance setting effects the overall lifespan of a bearing due to spreading the load across multiple rolling elements and limiting or increasing the maximum load on an individual rolling element at any given time [15]. The graph in Figure 13 shows bearing life vs internal clearance after accounting for factors such as thermal expansion.

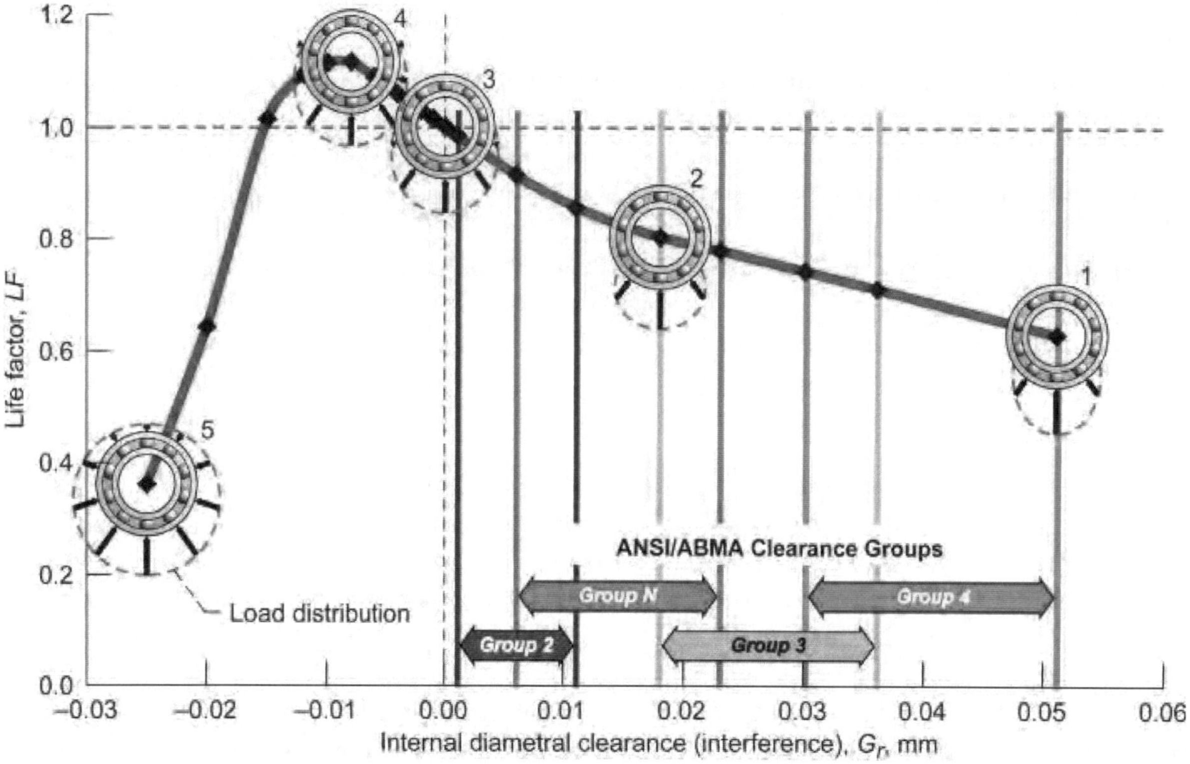

Figure 16: Effect of Internal Diametral Clearance on Life Factor for 210-size, Deep-groove Ball Bearings Radially Loaded to 1720 MPa Inner-Race Maximum Stress at the Zero-clearance Condition [15]

During operation, correctly maintaining the lubrication of a bearing can dramatically extend the bearing life. The correct amount of oil or grease of the correct type at the correct interval will help maintain the lubrication layer between parts, preventing metal to metal contact and abrasive wear. Lubrication needs vary depending on factors such as machine type, operating speed, temperature, etc.

Finally, if needed, there are a number of design and operation choices that can help reduce stresses in bearings and extend the lifespan. First, increasing the number of rolling elements in a bearing spreads the load through more contact spots, reducing the maximum stress experienced by a single rolling element. This can be achieved by selecting a bearing with more rolling elements in a single row or changing to a multi-row bearing design. Second, changing the shape of the rolling elements from a ball to a cylindrical roller changes the shape of the contact spot from a circle to a rectangle, and changes the maximum stress point from a singular point to a line, which reduces the maximum stress experienced by the rolling element. A trade-off to be considered in changing to cylindrical rollers from ball rollers is that radial cylindrical roller bearings have a lower axial load capacity. Third and finally, lifespan for cyclic loading can be extended by reducing the rotating speed of a bearing. This change may not change the total number of cycles that a bearing can be expected to survive, but it would extend the *time* that the bearing could be expected to survive. This option may or may not be possible depending on process needs. For instance, the rotating speed of a belt driven fan shaft may need to remain constant to deliver a required amount of air, but the driving motor can have its rotation slowed by changing sheave diameters on both the motor and the fan.

Acknowledgements

Ken Starry of IVC Technologies and Scott Dyar of Stratton Lumber, mentors who taught me much of my knowledge about bearings and vibration analysis.

References

[1] Steinbrecher, T. & Leubner, G. *The biomechanics of seed germination.* Journal of Experimental Botany. December, 2016.

[2] "*Necking & the Ultimate Tensile Strength.*" Plastometrex, February 18, 2021, https://www.plastometrex.com/blogs/necking-the-ultimate-tensile-strength

[3] Langnau, L., "*Brinelling and Why Bearings Fail – How Bearings Fail Part 6 of 6*", Linear Motion Tips, Apri 28, 2017, https://www.linearmotiontips.com/how-bearings-fail-a-closer-look- at-brinelling/

[4] Kossakowsky, P., "*Influence of Initial Porosity on Strength Properties of S235JR Steel at Low Stress Triaxiality*", Archives of Civil Engineering. September 2012.

[5] "*Damage by Type – Cracks*", NSK Troubleshooting Tools, https://www.nsk.com/eu-en/tools- resources/troubleshooting/damage-by-type/cracks/

[6] William, "*An Overview of Bearing Wear*", Lily Bearing, https://www.nsk.com/eu-en/tools- resources/troubleshooting/damage-by-type/cracks/, September 19, 2023.

[7] "*Damage by Type – Peeling*", NSK Troubleshooting Tools, https://www.nsk.com/eu- en/tools-resources/troubleshooting/damage-by-type/peeling/

[8] Budynas, R.G. & Nisbett, J.K., "*Shigley's Mechanical Engineering Design*", McGraw-Hill Education, 2014.

[9] "*Contact Mechanics*", Wikipedia, Wikimedia Foundation, October 2024, https://en.wikipedia.org/wiki/Contact_mechanics

[10] Broer, A. "*Fatigue Life Prediction of Carbon Fibre-Reinforced Epoxy Laminates Using a Single S-N Curve*", Master of Science Thesis, Delft University of Technology, May 23, 2018.

[11] Zaretsky, E., "*How to Determine Bearing System Life*", Machinery Lubrication, December 2012

[12] "*Damage by Type – Flaking*", NSK Troubleshooting Tools, https://www.nsk.com/eu- en/tools-resources/troubleshooting/damage-by-type/flaking/

[13] "*Fatigue Physics*", Fatigue Life LLC, 2020, https://fatigue-life.com/fatigue-physics/

[14] "*Damage by Type – False Brinelling*", NSK Troubleshooting Tools, https://www.nsk.com/eu-en/tools-resources/troubleshooting/damage-by-type/false-brinelling/

[15] Oswald, F., et al., "*Effect of Internal Clearance on Load Distribution and Life of Radially Loaded Ball and Roller Bearings*", Proceedings of the 65th Annual Meeting and Exhibition sponsored by the Society of Tribologists and Lubrication Engineers (STLE), Las Vegas, Nevada, May 16-20, 2010.

Bibliography

1. Budynas, Richard G., Ali M. Sadegh, et al. *Roark's Formulas for Stress and Strain*. McGraw- Hill, 2020.

2. Gulati, Ramesh. *Maintenance and Reliability Best Practices*. Industrial Press, 2013.

3. "SKF Product Select." *SKF Product Select*, productselect.skf.com/#/bearing-selection-start. Accessed Jan. 2025.

WHEN BAD VIBRATIONS HAPPEN TO GOOD MACHINES
REQUIRED DILIGENCE DURING DETAIL ENGINEERING

Author: Mantosh Bhattacharya

VIATC TRAINING CONFERENCE & EXPO
NEWPORT NEWS, VIRGINIA | AUGUST 6-8, 2025

WHEN BAD VIBRATIONS HAPPEN TO GOOD MACHINES - REQUIRED DILIGENCE DURING DETAIL ENGINEERING

Mantosh Bhattacharya
bhattacharyamantosh@yahoo.co.in

Mantosh Bhattacharya is designated Subject Matter Expert on Turbomachinery in UAE . He is a chartered engineer CEng (MIMechE-UK), Intl. PE (UK), a member of ASME, Vibration Institute and ISCM – (BINDT UK). He is member of advisory committee of ATPS [TEES – Texas A&M]. He is a member of HI 9.6.4, 9.6.5 and 14.5.1 standards. He is a member of API SOME standard committee. He holds 4 U.S. patents related to structural dynamics. He has published and presented several technical papers in journals and at conferences.

Abstract

A well designed and well-balanced rotating machine can show up with high vibrations when properly installed at site. Despite accepted factory testing of equipment in compliance to standards and specifications, high vibrations were experienced in bearing caps of equipment when installed on contractual support structure. Another complex issue faced was high vibration on pulsating line support structure. Various methods known to industry such as standard bump testing / modal impact testing, operational deflection study [ODS] and motion magnification video method were used to detect the root cause of the issue. Post identification, a large amount of time and effort were consumed to mitigate the issues. To enable the unit to operate at site as installed, a few deviations had to be taken from project specifications. These had caused considerable schedule impact to the projects. This paper presents when and how due diligence is required to review the predicted machinery performance during test at site and how the test methods set up should be selected to diagnose and mitigate the problems before tow out of units from test beds of OEM works.

Key Words: Resonance, FEA, ODS, VFD, Motion magnification, Modal testing, Monolithic connection.

Introduction

There are some failure cases where machinery unit distresses could have been avoided with due diligence (excluding a detailed design audit) during the detailed engineering stage. Designed as "fit for purpose" and well-balanced rotating machines can show up with high vibrations when properly installed at site. Despite a factory testing of rotating equipment unit, high vibration can be experienced on machines when installed on contractual support structure. Due to reasons attributed to limitation of test bed, the tested unit can show high vibration at site. In a similar way even with proper pulsation study, the pulsating pipe support structure and associated support and utility lines can show high vibration. First, identification itself to find the root cause of the problems takes time, external support, and escalation of cost. Post identification, a large amount of time and effort are consumed to mitigate the issues, adding difficulties to mobilizing a team due to remote locations. These issues cause considerable schedule impact to projects in all aspects. This write-up describes a few such cases where issues could have been mitigated before towing out of unit at site.

Case 1 -Electric Motor Vibration

Two pole electric motor driven 750 KW screw compressor units were designed to be installed on a metallic modular skid as illustrated below as models where motor, compressor, and lubrication skids are shown as lumped masses in Figure 1.

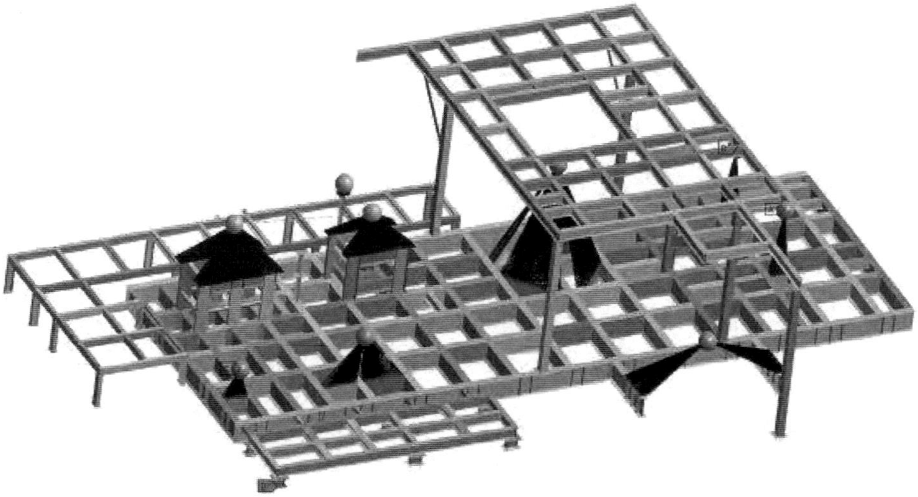

Figure 1 – Solid Model of Complete skid of unit

A finite element study was conducted for modular structure including the installed motor, compressor, and other auxiliary units. The purpose of the finite element analysis is to evaluate mode shapes and natural frequencies in the compressor skid structure. A separation margin of +/-15% was employed in this analysis for all modes to facilitate the mechanical design of the Compressor Skid, ensuring a distinction between the inherent frequency of the motor support structure and the working speed. If the interference points due to structural resonance are projected to lie within the separation margin (+/-15%) of the operating speed (2980 rpm), a

harmonic response study must be conducted to verify that the possible resonance does not produce detrimental effects.
In the finite element study, the maximum displacement response occurred at a frequency of 50 Hz, which was close to the compressor motor's working frequency of 49.61 Hz. Consequently, a displacement profile of the modular skid was plotted at 50 Hz.

The displacement profile indicated no detrimental or elevated vibrations on the motor support structure.

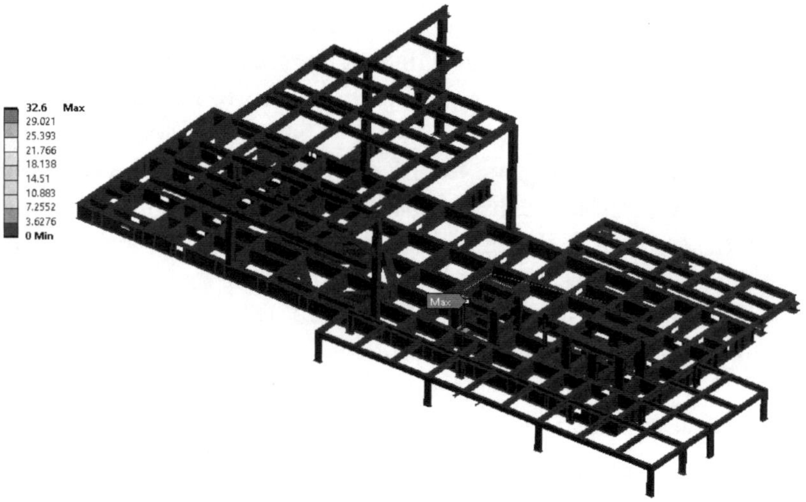

Figure 2- Harmonic response of the unit at 50 Hz

A solo motor run test for the screw compressor motor was conducted at the site installation with the whole skid. The findings during the test were as follows.

Shaft displacement measurements were within the specified limits according to the OEM settings.

The vibrations of the bearing housing exceeded the commissioning limitations established by ISO-20816-3. Repetitive vibration fluctuations were detected in shaft vibration trends occurring at approximately 7-minute intervals. The same was noted about the direct value and the 1x and 2x vectors.

2x line frequency LF [100Hz] was recorded with significant amplitude, exhibiting dominance in cyclic fluctuations. The 2xLF vector decreased to a minimum immediately upon the motor being shut off, indicating a line frequency issue. Intermittent axial hunting of the rotor was noticed, with the rotor operating distant from the magnetic center designated by the OEM. Motor Bearing cap vibration exceeded ISO 20186-3 acceptance criteria. However, when testing of the identical motor was conducted at the manufacturer's facility on a solid base, no such abnormalities were observed.

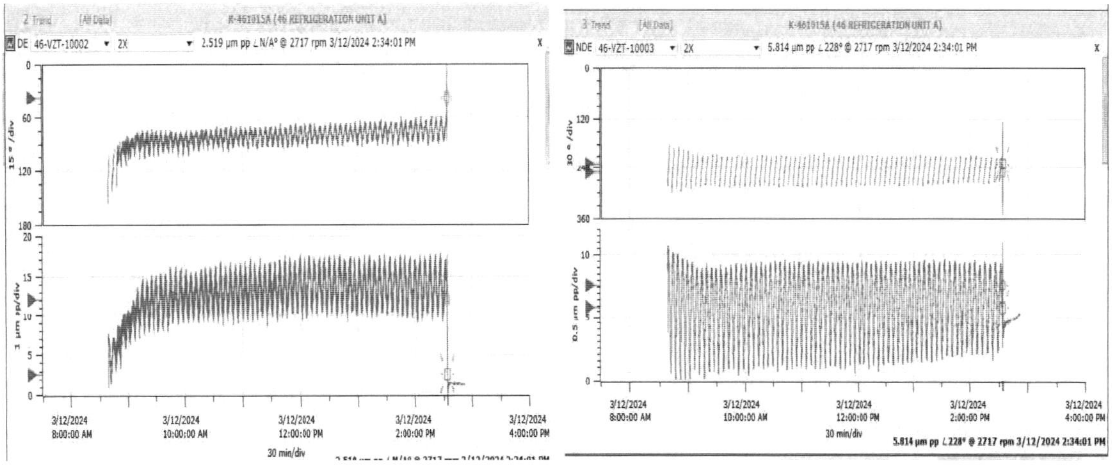

Figure 3 -2LF trend of Electric Motor during solo run with fluctuating phase .

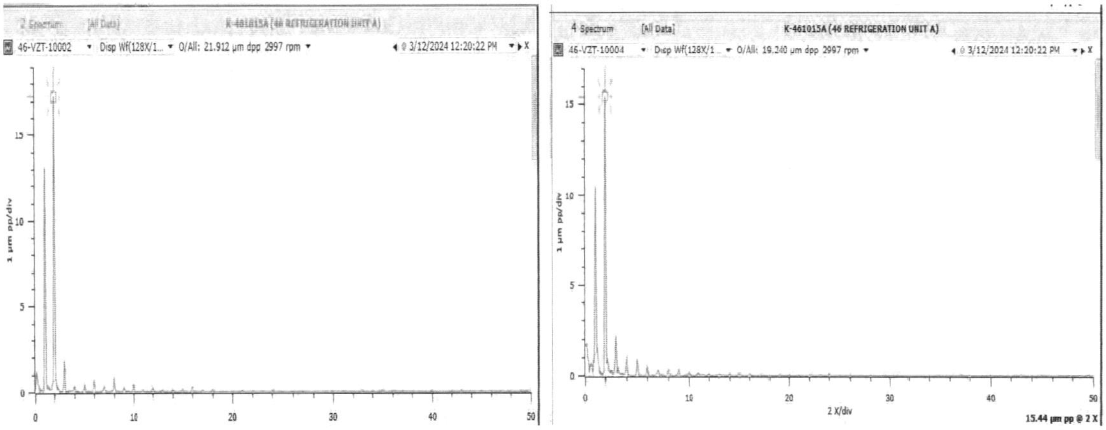

Figure 4- High 2LF peaks dominating the overall vibrations on motor bearing caps

Intuitively, we had suspected the presence of soft foot or welding imperfections in the motor support structure. This possibility was eliminated following a re-examination of alignment and an inspection of welds. Nevertheless, upon testing the other unit with a solo motor run, analogous vibration patterns were detected. To further examine the issue and its underlying cause, we implemented motion magnification video techniques. Motion Magnification, referred to as video amplification, is the technique of enhancing minute, imperceptible movements in a video sequence to a level detectable by the human eye. This technique has proven essential for visualizing vibrational motion in machinery and related static components such as pipelines and foundations, offering a thorough visual comprehension of the motion of these components.

The analyzed video files distinctly revealed the inadequate support structure of electric motors, which was not identified in the finite element analysis investigation. The FE report was subsequently revised using this information, and the Frequency Response and Phase Change Curve for Velocity on the Electric Motor was plotted. This plot clearly demonstrates that the vibration amplitude of the motor at the bearing caps exceeds the acceptability criteria stipulated in ISO 20186-3 near the operational frequency.

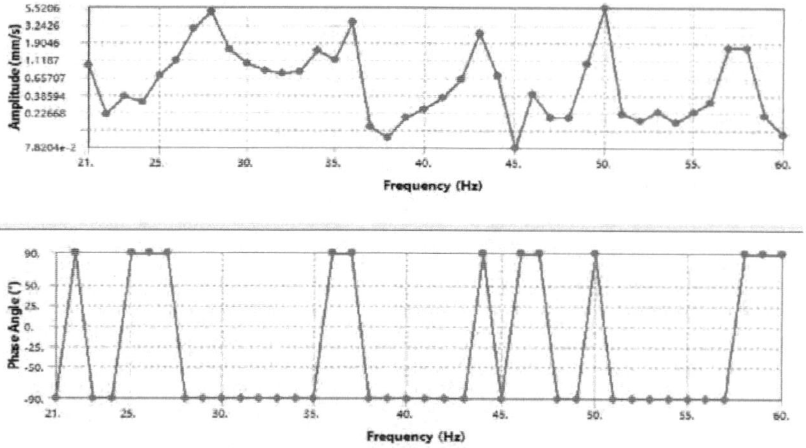

Figure 5- Predicted Velocity at motor bearing cap before modification

Following the reinforcement of the motor's support structure by controlled welding techniques, the response at the operating frequency decreased significantly, falling well below the acceptable requirements established in the analysis. Identical solutions were employed on other units to address the problem.

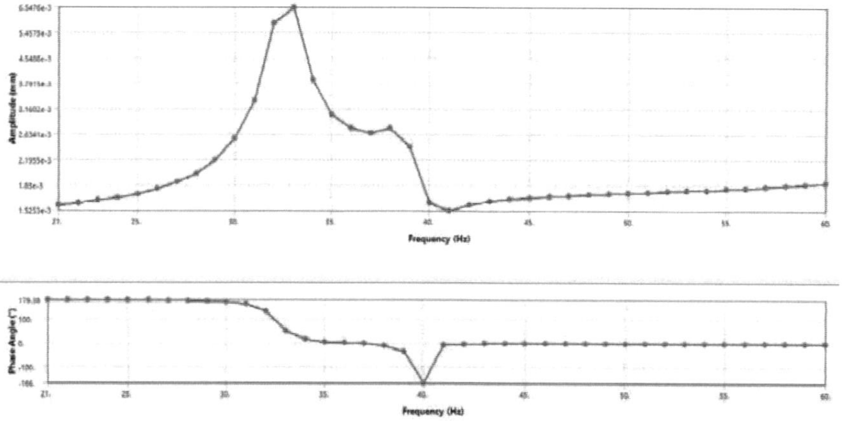

Figure 6 - Predicted Velocity at motor bearing cap after simulation of modification

Remarks

The Finite Element study identified the initial thirty-four frequencies of the structure, ranging from 21 Hz to 61 Hz. The measured natural frequency was determined to be in close proximity to the unit's operational frequency range; therefore, harmonic analysis was conducted to evaluate the frequency response. The frequency response noted in the harmonic analysis was within permissible limits, thereby confirming the structure's safety by finite element and vibration analysis.

How this issue could have been avoided

While building and analyzing the FE model and outputs, the motor support structures were overlooked in terms of required stiffness. The individual support stands were stand alone and were not braced. The target acceptance criteria on motor bearing cap as per ISO 20186-3 was overlooked as it was considered that the motor has hydrodynamic sleeve bearings.

Case 2 – Heavy duty plunger pump unit vibration

Problem statement -Trips caused by high vibrations were reported in VSD Motor / Gear box driven in all three Five Plunger pumps at site during start up and operating regime (70-80% of full RPM). The pump maximum operating frequency was 16 Hz.

Actions - Several attempts were made to undertake all possible corrective actions, Such as checking levels and correcting the positions of shims with no favorable results

Figure 7 – inadequate shimming

The vibration response during a third-party vibration survey run-up distinctly indicated that the motor experienced resonance at 70-80% of full RPM. Resonance was determined by the phase angle of the vibration shifting by 90 degrees at resonance and thereafter by 180 degrees after passing through the resonance.

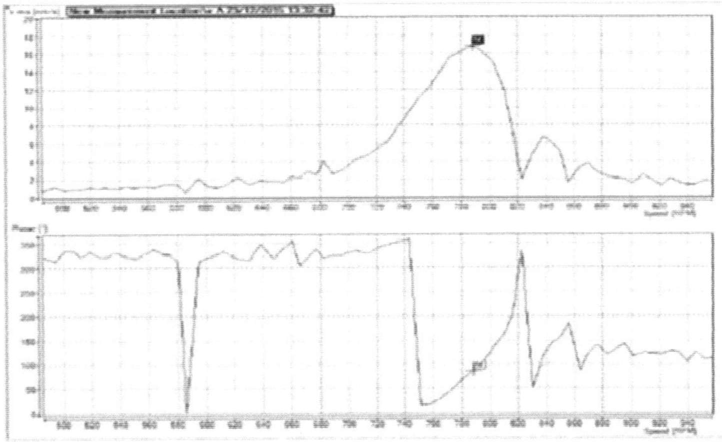

Figure 8- change of phase with high Vibration amplitude in RMS

The third party conducted Bump tests on all three pumps and performed an ODS survey on two of them as the third one was in operation. Upon review, it was confirmed that the deflection modes of all three pumps were analogous.

The entire issue was initially ascribed to deficiencies in installation, including poor shimming. The installation challenges were solely attributed to the springing motion of the base frame's periphery. Deflections recorded near the edge of the base plate were rather minor in relation to those measured at the locations of the motor and gearbox installations on the base plate.

Analysis of motor pitching and gearbox rocking at 13 Hz and 30 Hz indicated that the cross beams integrated into the baseplate structure beneath the motor and gearbox exhibited excessive flexibility. Nevertheless, the FEA analysis of the skid frames initially determined that there were no issues about resonance.

The following two solutions were evaluated:

1. To enhance the rigidity of these beams, an auxiliary support structure must be implemented to reinforce each cross beam by securing them directly to the concrete floor.
2. Reinforcing the entire baseplate by filling its interior cavity with cement grout and high-pressure injecting epoxy grout to occupy any leftover spaces. This option would provide a more reliable means of stiffening the entire bedplate and shifting the inherent frequencies of the different components away from the motor's pushing frequencies.

Considering the urgency of the pump requirements and the complexity of the task outlined in option 1, it became evident that modifying the base frame to enhance its natural frequency would be challenging due to the extensive labor involved in removing the entire skid.

The FE model was revised based on the ODS value of 12-13 Hz in motor pitching mode, and the grouted state was included in the input to anticipate a closely approximated result. The rebuilt finite element model, with the grouted design as input, demonstrated promising anticipated results as outlined in figure below.

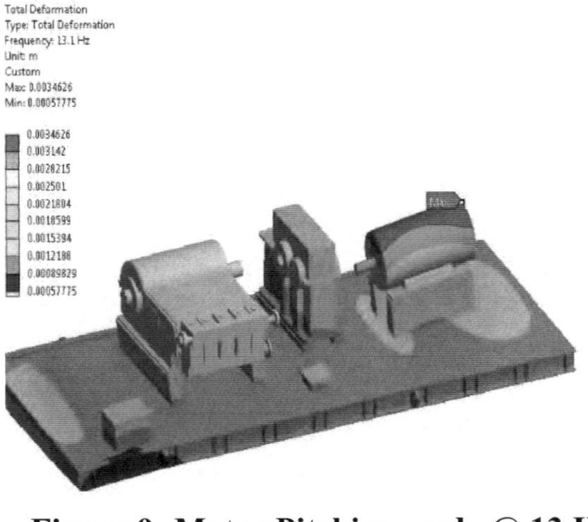

Figure 9- Motor Pitching mode @ 13 Hz

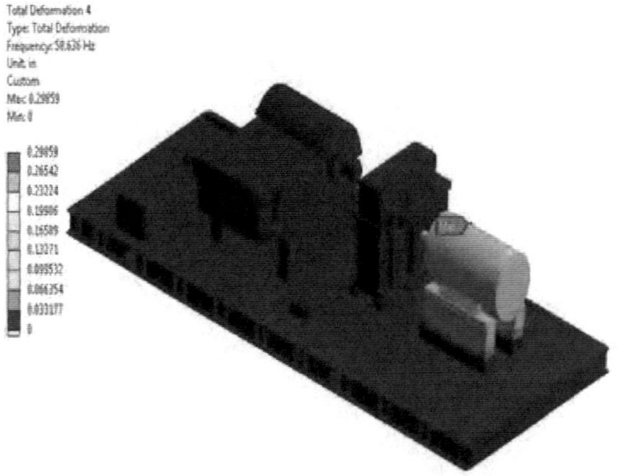

Motor pitching mode @ 59Hz

To ensure a monolithic structure of the existing foundation and no shrink grout filling, rods were inserted between base plate and foundation to recreate a rebar type structure and unified foundation. After the modifications at site, the repeat vibration survey showed much lower vibration throughout the operating regime of VFD.

How this issue could have been avoided

The compete unit was designed to be installed on a single metallic skid and with a grout-less design. At test bed, the unit was properly installed, but testing with the VFD was waived due to a schedule issue cited by unit supplier.

Case 3 – Pulsating Pipe Support Structural Vibration

Fluid pipes and accompanying vessels, such as separators, knockout pots, and scrubbers, constitute a full machinery unit of a plant. Pipelines and piping systems can induce pulsations and vibrations. In addition, the support structure of a pipe rack may shake even when the vibrations of the piping are deemed acceptable. This vibration may result in fatigue failure and structural damage.

The structures exposed to dynamic loads are engineered to ensure that their predominant natural frequencies do not coincide with the frequency of the applied forces, and modal mass participation is also considered to evaluate severity of vibration. EFRC and Energy institute guidelines provide the acceptable limits of piping and small-bore piping vibration. The German Standard DIN 4150 Part 3 mandates vibration velocity criteria for assessing the impact of vibration on structural integrity. A peak vibration velocity of 10 mm/s is regarded as a safe threshold for maintaining structural integrity for pulsating pipeline support.

A crucial stage during design engineering is to guarantee that structural vibration remains within permissible limits and does not coincide with the operating frequency and its harmonics, which necessitates doing a structural dynamic analysis utilizing commercially available tools. Following the static analysis of the structure, a forced harmonic analysis is performed with a frequency sweep of four times the operating frequency, determined by the number of cylinders in the unit (for four cylinders, a minimum sweep of four times the operational frequency is required). However, in this case, the structural design of the support structure was carried well before finalizing the machine in terms of configuration and stages. This delay was attributed to selecting a unit which has lower residual unbalance load to avoid pile foundation.

The subject equipment was later finalized as a 5 MW electric motor-driven reciprocating compressor, including a four-stage, six-cylinder configuration, operating at 331 RPM.

The gas pipe support structures for the reciprocating compressor were deemed sufficiently robust to accommodate small utility lines via L-shaped supports from the main pipe rack structure. The connections were executed at site without informing the structural design team.

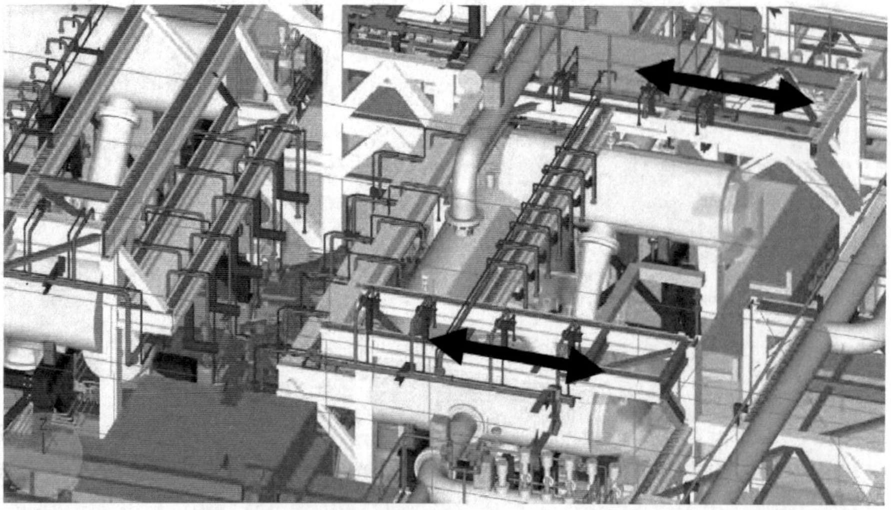

Figure 10 – Compressor gas pipe and utility piping vibration pattern

High vibrations were visually seen in the utility line support structure at the site whenever the reciprocating compressor was in operation. No vibratory movement was noticed while the reciprocating compressor was not operational. The operating frequency of the utility water pump was 25 Hz, and the vane pass frequency would have been a multiple of 25 Hz. The influence of vibration induced by the fire water pump was eliminated based on the relevant driving frequencies. As the vibration was within the visible range, it was determined that this vibration was not induced by any of the fault frequencies, which exceed the observable spectrum of human eyesight. The compressor piping support structure is comprehensive and depicted in Figure 11.

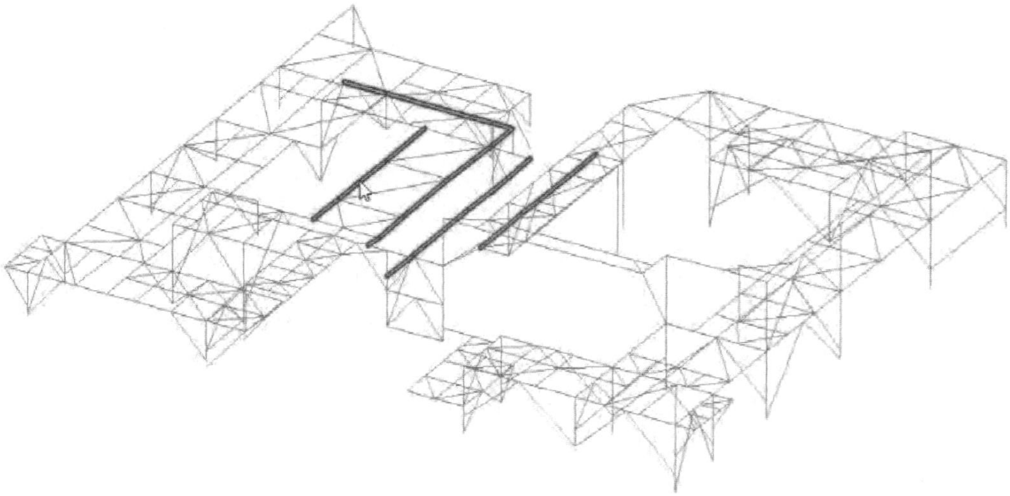

Figure 11 – schematic of gas pipe support structures for compressor unit.

A modal test verified that several parts exhibiting excessive vibration possessed a natural frequency within 1% of the compressor's working frequency. The FFT displayed 1X as its predominant peak. A hump indicative of a potential resonant state was seen. The remaining frequencies exerted minimal influence. The structural design program was re-executed, revealing that certain modes corresponded to the 1x operating frequency of the compressor; however, the entire structure did not exhibit vibrations at global resonance.

No immediate integrity threat to the compressor hydrocarbon piping elements was identified due to vibration, which was at low levels and deemed 'Acceptable' according to Energy Institute requirements. Fundamental frequency – 5.5Hz, FFT showed all Harmonics but higher harmonic has larger peak. 1X has comparatively lower energy content in suction and discharge piping .

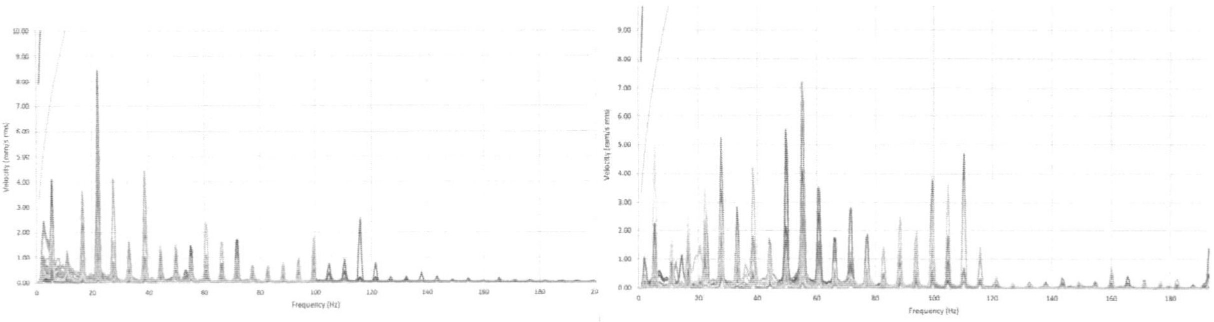

Figure 12 – Pulsating piping vibration spectra

Structural Vibration: The recorded structural vibration measurements were juxtaposed with the long-term vibration criteria outlined in DIN 4150-3:2016 and found to be in the high-risk zone. The vibration levels on the structure were over five times the recommended threshold. The surrounding items, which may be considered as sources of excitation energy, were vibrating at far lower levels, indicating that the structure had a resonant response. Several modes near 5.5 Hz are emphasized in the structural study conducted under "as installed and as built" conditions. To enhance verification and comprehension of the structural motions, an operational deflection shape analysis (ODS) was conducted using multi-channel analyzers, and animations were produced in WMV file format.

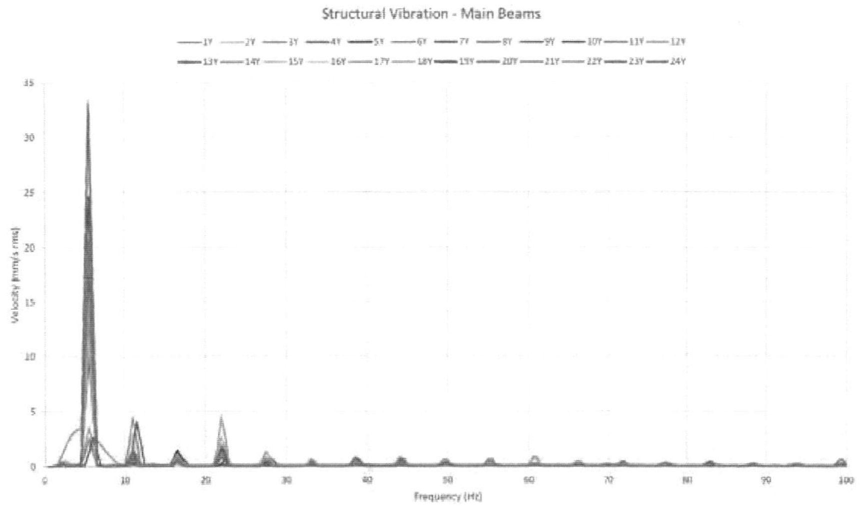

Figure 13 –vibration spectrum of structural beams

The ODS investigation provided a visual animation elucidating the direction of vibrations in some local structural parts. These provided a clearer comprehension of the reinforcing requirements for the construction. The primary source of the structural vibration was postulated to be structural resonance, mechanically stimulated by the 1X operational speed of the reciprocating gas compressor.

The time to analyze consumed a lot of work hours, mobilizing a third party with multi-channel analyzer, animation software, specialist services, issues of data transfer for conducting FE analysis to calculate time to fail. Handing over schedule to end user was impacted by 6 months.

How This Could Have Been Avoided

Analysis of structures are done in-house considering frame type structures in terms of user friendliness. FE based analysis is not considered even for pulsating line large support structure. Further review of the output file showed the presence of resonance condition but was overlooked considering lower modal mass participation and complete design was not revisited after ordering of compressor. To mitigate repeat occurrence, a design review check sheet has been developed for implementation in all future projects.

During the detailed engineering phase, Compressor RPM and frequency (1X,2X &4X – dominant frequencies based on compressor layout) have been indicated in the Table below for ready reference. The resonant frequencies of structure supporting the pulsating lines must maintain the separation margin. This has been shared with structural design engineer to ensure that the Structure Resonant frequency shall not fall between lower and upper range shown in red colour.

Table 1, Dominant Forcing and Separation Frequencies

Tag no.	Item Description	Compressor Operating RPM	1X (in Hz)	lower range	upper range	2X (in HZ)	lower range	upper range	4X (in Hz)	lower range	upper range
MB-1001 A/B	Make up Gas Compressor	330	5.5	4.4	6.6	11	9.35	12.65	22	19.8	24.2
NB-1015 A/B	Off Gas Compressor	368	6.13	4.91	7.36	12.27	10.43	14.11	24.53	22.08	26.99

These were also instructed as noted below:

1. Structures supporting pulsating lines (Gas compressor piping) shall not be connected to any other structures without prior analysis.
2. Supports of pulsating lines (Gas compressor piping) shall not be connected to the compressor platform
3. Structures support footings shall have adequate margin with respect to the skid.

Conclusion

The detailed engineering work should consider all major possible failure modes referenced in the repository. It must be understood what type of base frame design has been mandated in the

project specification, and in all cases, it must be understood that grouting should not be considered to dampen vibration.
The support structures must be analyzed for any local excitation / deflection. The workflow of detailed design engineering must have milestones to cover as predecessors and must use the type and rating of driver as installed at the skid to validate the functionality and possible vibratory responses.

Acknowledgements

The author wishes to tender sincere thanks for the support of VIATC 2025 program chair for accepting this paper for VIATC 2025 conference.

References

1. European Forum for Reciprocating Compressors (EFRC), "Guidelines for Vibrations in Reciprocating Compressor Systems," 4th Edition, March 2017.
2. Energy Institute, "Guidelines for the Avoidance of Vibration Induced Fatigue Failure in Process Pipework," 2008.
3. DIN, "Vibration in Buildings - Part 3: Effects on Structures," DIN 4150-3:2016
4. Visualizing Flying Horses and Vibrating Machines: Slow-Motion Amplified Video – Uptime Magazine March 2019 - Chad Paso - MSI Inc. NJ USA
5. Resonance and its effect on mechanical structures - Gene Vogel
   Resonance and its effect on mechanical structures - Resource Library - EASA | The Electro•Mechanical Authority
6. Modal Testing: A Practitioner's Guide - Peter Avitabile

NEW HYDROELECTRIC GENERATOR EXCESSIVE VIBRATION LEVELS DURING COMMISSIONING

Author: Bernard F. Boueri

VIATC TRAINING CONFERENCE & EXPO
NEWPORT NEWS, VIRGINIA | AUGUST 6-8, 2025

VIATC TRAINING CONFERENCE & EXPO

NEWPORT NEWS, VIRGINIA | AUGUST 6-8, 2025

NEW HYDROELECTRIC GENERATOR EXCESSIVE VIBRATION LEVELS DURING COMMISSIONING

Bernard F. Boueri
Senior Mechanical Engineer
Ontario Power Generation
bernard.boueri@opg.com

Bernard Boueri is currently a Senior Vibration Engineer in the Machine Dynamics Section at Ontario Power Generation (OPG). He has 20 years experience supporting rotating equipment commissioning, diagnostic and balancing activities. In recent years he was involved in the new commissioning and upgrades of the hydroelectric fleet at OPG. He was involved in the commissioning of the Upper and Lower Mattagami Projects which included the installation of 13 new units. Bernard is a graduate from the University of Florida. He is a licensed Professional Engineer in the province of Ontario and is a certified Category IV from the Vibration Institute.

Abstract

A hydroelectric unit was experiencing vibration trips during cold startup at the generator guide bearing. The author at the time identified that the vibration signature indicated that the unit is most likely misaligned. However, the OEM did not agree with the prognosis. The original proposal by The OEM was to increase the vibration trip levels at the generator guide bearing to alleviate the trips during cold startup. In addition, excessive vibration levels were also noted during unit loading through the rough zone and during shutdown. Initial inspection of the unit showed rub marks on the discharge ring. This paper will discuss the findings and vibration data collected by the author during and after commissioning and the root cause of the excessive vibration levels during cold startup and during the load transition.

Key Words: Hydroelectric, misalignment, natural frequency, rubs, impact.

Introduction

A new commissioned hydro generator propeller unit in 2014 with a speed of 100 RPM and rated power of 70 MW, was experiencing vibration trips during cold start-up at the generator guide bearing. The trips would occur during transition through the rough zone and at efficiency. The authors at the time identified that the vibration signature indicated that the unit is most likely misaligned. However, the OEM did not agree with the prognosis. The original proposal by the OEM was to increase the vibration trip levels at the generator guide bearing to alleviate the trips during cold start-up. The authors rejected the proposal because it did not address the root cause of the excessive shaft relative displacement levels.

In addition, high generator shaft relative displacement levels more than 40 mils p-p were also noted when the unit transitioned through the rough zone and during the rundown of the unit. The author suspected possible excitation of the shaft natural frequency a rotordynamic analysis of the unit was done to identify the unit critical speeds.

Instrumentation

The unit is protected with four permanently mounted proximity transducers, two proximity transducers 90° apart on each of the generator and turbine guide bearings as shown in Figure 1. The signals from the proximity transducers are captured through the buffered output of the Bently Nevada 3500 Rack (Figure 2). Temporary proximity probes were also installed on the oil bafflers directly opposite the permanent transducers at the generator guide bearing.

Temporary proximity probes were installed below the thrust bearing in line with the permanent proximity probes (Figure 3), these probes were located about 5 feet below the permanently installed proximity probes

Accelerometers were installed in the radial and axial directions to measure casing vibration levels. Accelerometers were installed on the turbine cone, scroll case door and draft tube door, to measure vibration levels due to the pressure pulsation and rope during the unit transition through the rough zone.

Figure 1: Generator and Turbine Transducer Setup

Figure 2: Bently Nevada 3500 Rack

Figure 3: Temporary Proximity Probes Below Thrust Bearing

Results

Figure 4 is the trend of the overall shaft relative displacement levels at the generator guide bearing during cold start up. As shown in Figure 4 the unit tripped twice on high overall shaft relative displacement levels when transitioning through rough zone. Highest levels recorded were about 30 mils p-p on the permanently installed proximity probes while the temporary proximity probes installed on the oil bafflers were about 41 mils p-p.

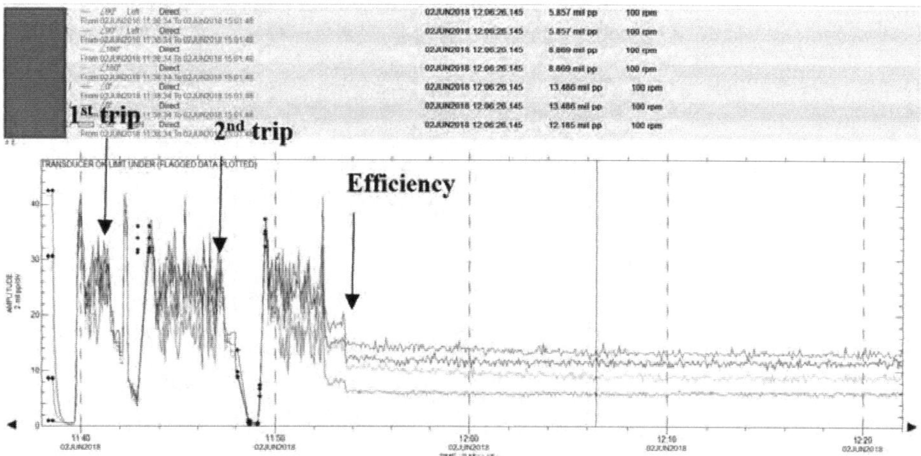

Figure 4: Generator Overall Shaft Relative Displacement Levels pre-outage

Table 1 is a summary of the generator shaft relative displacement levels when the unit is at efficiency. Note the difference in the overall shaft relative displacement levels between the temporary and permanent proximity transducers. The transducers should show similar shaft relative displacement levels. However, in the x- direction the difference in the shaft relative displacement levels between the temporary and permanent transducers were doubled. Note as well the significant vibration component at 2x RPM. These results are a possible indication that the shaft is not vertical. Note as well the 1x Phase difference between the permanent proximity transducers and those located below the thrust.

Table 1: Generator Shaft Relative Displacement Levels at Efficiency

	Overall (mils p-p)	1x RPM (mils p-p)	1x Phase (deg)	2x RPM (mils p-p)	2x Phase (deg)
Generator x	6.0	2.8	159	2.0	146
Generator y	9.8	7.8	85	1.0	5
Oil baffler x	12.5	5.0	340	0.9	167
Oil baffler y	14.2	7.0	257	2.0	316
Thrust x	2.5	1.0	43	0.7	324
Thrust y	4.0	3.0	291	0.7	240

Generator x, y are the permanent proximity transducers. Oil baffler are the temporary proximity transducers installed on the oil baffler 180° opposite the permanent transducers. Thrust are the temporary transducers installed in line with permanent transducers below the thrust (Figure 3).

Figure 5 is the shaft orbits and time waveforms at efficiency at all measured locations. Note the differences in orbit shapes at the generator guide bearings between the permanent transducers and the temporary transducers; possible indication that the journal is not vertical in the bearing.

The permanent proximity probes show a truncated time waveform and flat orbit, possible indication of a rub or heavy preload. The temporary thrust bearing orbit shows a Figure 8 shape indication of possible misalignment.

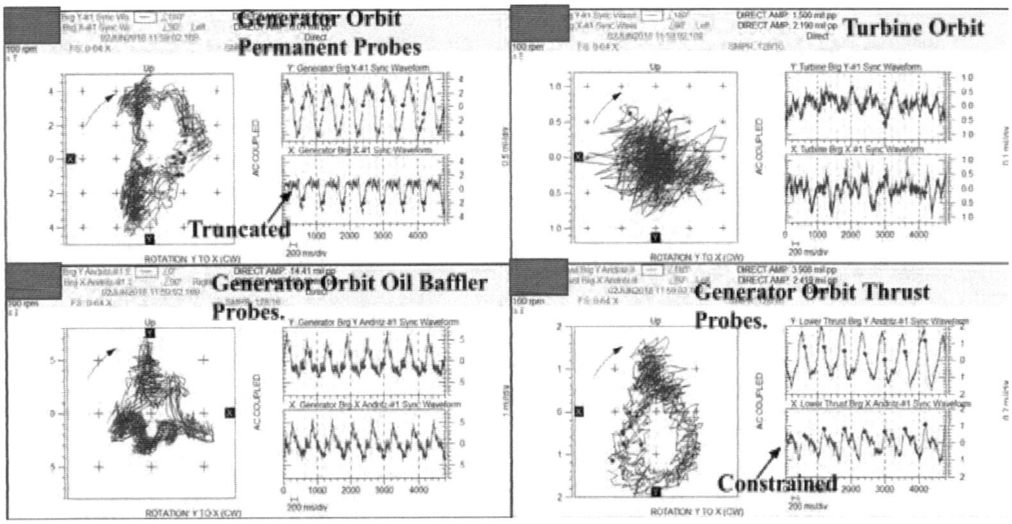

Figure 5: Generator and Turbine Shaft Orbits at Efficiency

The shape of the flat orbit and truncated time waveform prompted a closer look at the shaft behavior during start-up and loading of the unit. Figure 6 is the shaft centerline movement at the generator guide bearing and turbine guide bearing. The shaft centerline movement shows that during run-up the shaft moves 15 mils at the generator guide and 6 mils at the turbine guide. The data also shows that the generator shaft and turbine shaft are moving in opposite direction; also indication that the shaft is not vertical. Both generator and turbine bearings clearances were set at 20 mils diametrical. The static movement of 15 mils by the generator shaft consumes $\frac{3}{4}$ of the bearing clearances leaving only 5 mils for the dynamic movement (static movement is in x-direction). The shaft is free to vibrate in the y-direction as seen in the time waveform but restricted in the x-direction due to the generator shaft static movement that consumed most of the bearing clearance in that direction (Figure 7).

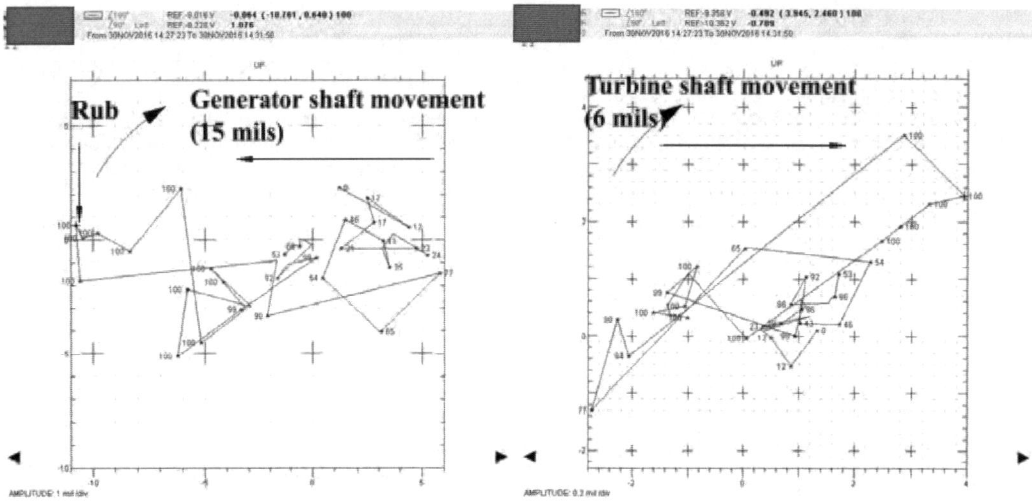

Figure 6: Generator and Turbine Shaft Centerline Movement during Run-up and Loading

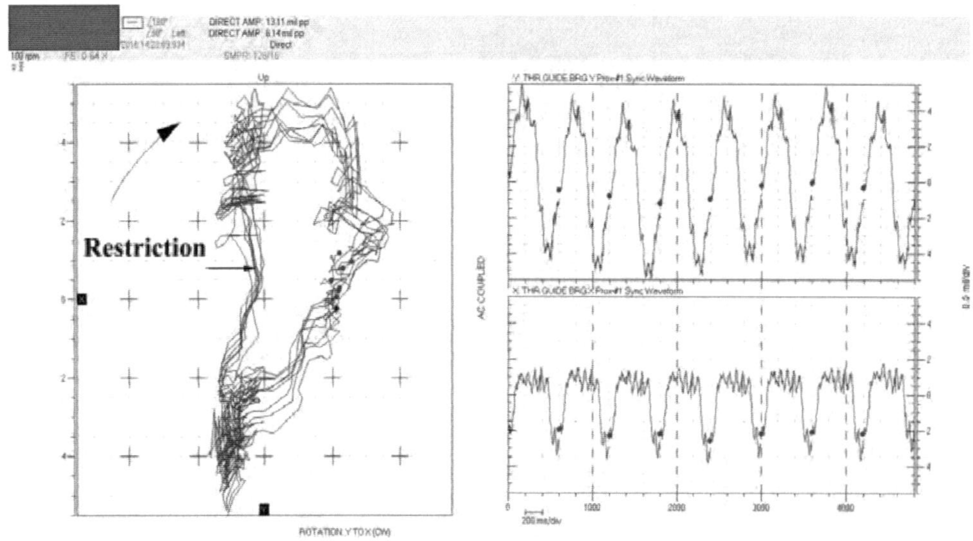

Figure 7: Generator Orbit and Time Waveform at Full Load

The casing time waveform also identified an impact at the generator guide bearing casing during transition from the rough zone to efficiency (Figure 8). These impacts could be due to generator shaft contacting stationary part causing the truncated time waveform.

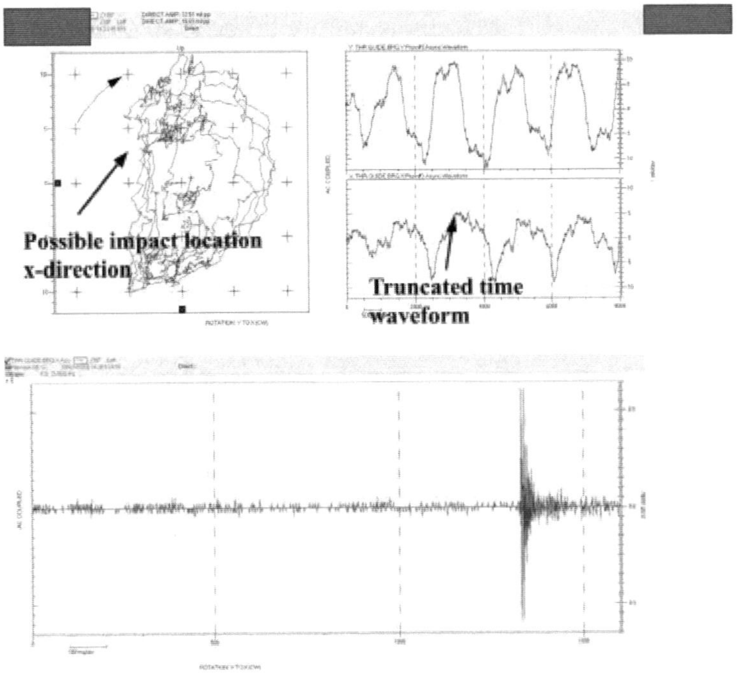

Figure 8: Generator Shaft Orbit and Impact

Figure 9 is the generator casing spectra prior and during the impact. Note the excitation of possible natural frequencies when impact occurs.

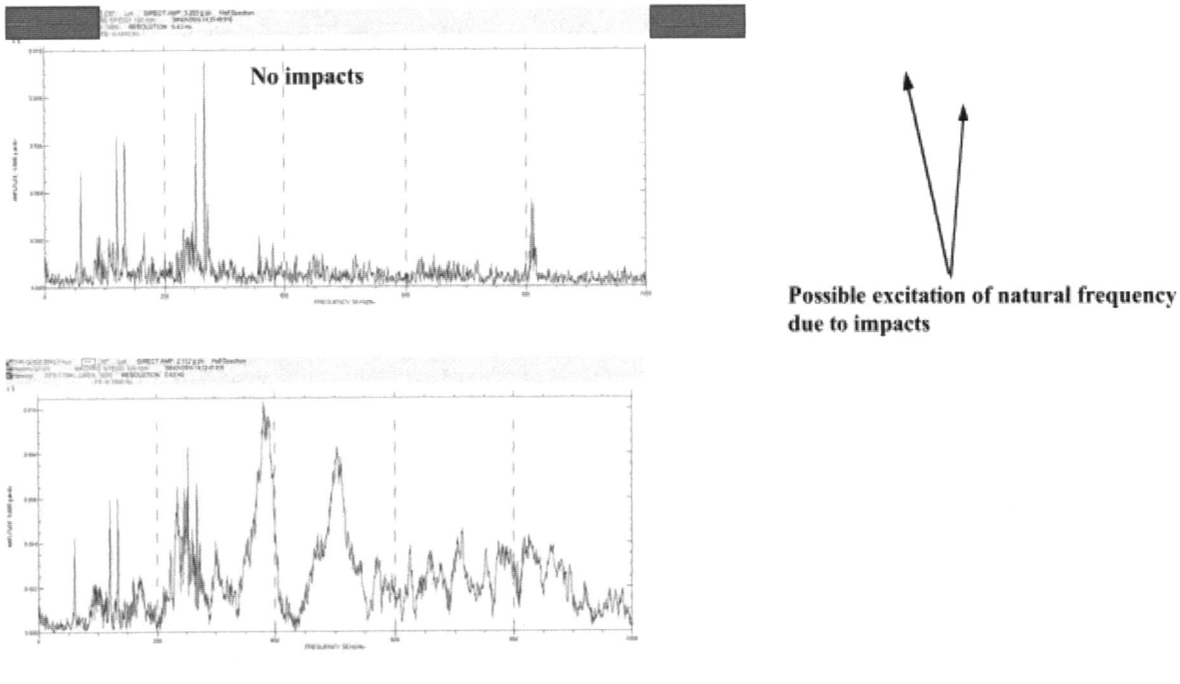

Figure 9: Generator Casing Spectra Prior and during

Based on the vibration data, the authors recommended inspection of both turbine and generator guide bearings and verification of the unit plumbness prior to any reduction in bearing clearances as recommended by OEM.

Inspection

Four generator guide bearing pads 90° apart were removed for inspection. Figure 10 are photographs of the inspected pads. NDT measurements did not reveal any Babbitt damage. However, it is evident from inspection that the upstream and downstream pads are experiencing different loading than the ones 90° away. This is consistent with the vibration data.

Figure 10: Generator Guide Bearing Pads

The OEM performed a verticality check of unit with generator and turbine guide bearings in place. The verticality of the unit was ***1.032 thou/ft (0.086 mm/m)***, acceptable tolerance ***0.24 thou/ft (0.02 mm/m)***. The measured value is more than double the tolerance. This is an indication that the unit was misaligned.

The out of tolerance verticality number prompted OPG to request the OEM for a proper alignment of the unit. Following removal of both generator and turbine guide bearings (shaft is free), the verticality check yielded ***0.95 thou/ft***. This slight improvement in the verticality is an indication that the generator bracket and turbine bracket were not aligned properly.

Turbine guide pads were also inspected. Figure 11 shows three of the twelve inspected pads. Downstream pads indicate a heavier load on pads; this is also consistent with vibration data.

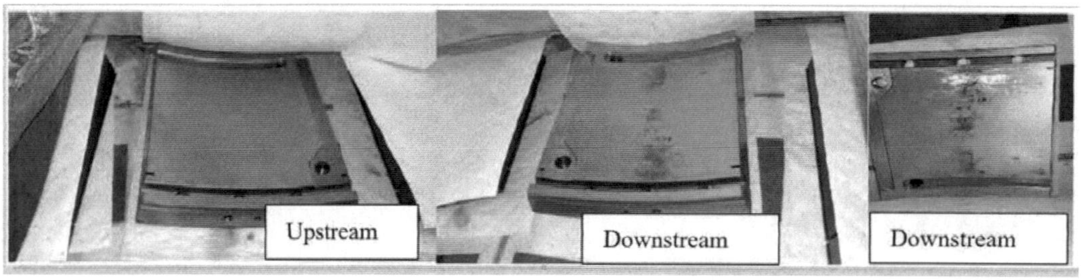

Figure 11: Turbine Guide bearing Pads

Inspection of the draft tube also indicated runner to discharge ring contact (Figure 12). The authors believe that the verticality and whirling of the unit could have contributed to the contact.

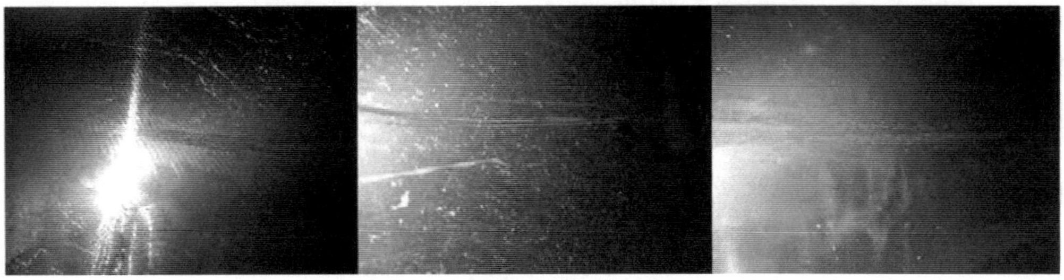

Figure 12: Runner to Discharge Ring Contact

Inspection also revealed that the generator bracket hold-down bolts were found to be loose (Figure 13), as per OEM these bolts were torqued to specification. OEM did not provide any explanation on how they would become loose. It was not clear at the time if the impacts experienced at the generator bracket could have been the cause.

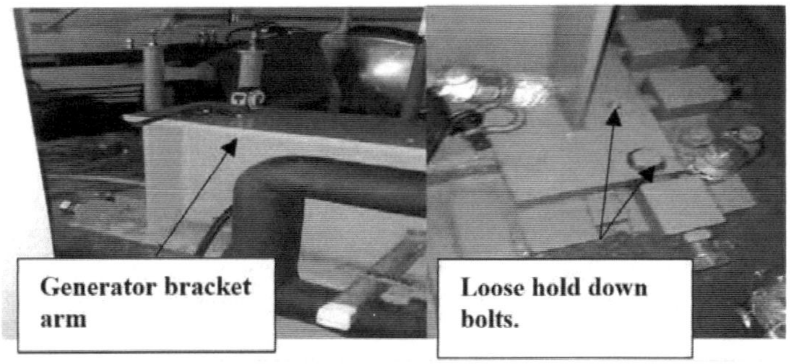

Figure 13: Generator Bracket Hold down Bolts

During the turbine inspection, it was noted that two carbon seal segments were found to be broken in half. The bottom section of the carbon seal had two broken segments and were found side by side on the downstream. The top carbon seal assembly had three vertical dowels broken off on the upstream side of the shaft. These dowels are to keep the segments from moving radially (Figure 14). These findings are also consistent with vibration data.

Figure 14: Turbine Broken Carbon Seals and Sheared Dowels

Inspection of the thrust bearing oil pot and pads revealed excessive debris in the oil and rust on the springs. This was identified in 2014 during commissioning of the unit (Figure 15).

Figure 15: Thrust Bearing Spring Bed

High Vibrations during Rough Zone and Shutdown

During testing excessive generator shaft relative displacement levels more than 40 mils p-p were measured during the transition of the unit through the rough zone, Figure 16 is the generator shaft orbit and spectrum. Note the triangular orbit shape. The spectrum shows a dominant vibration component at about 4.37 Hz and the second harmonic of 8.74 Hz.

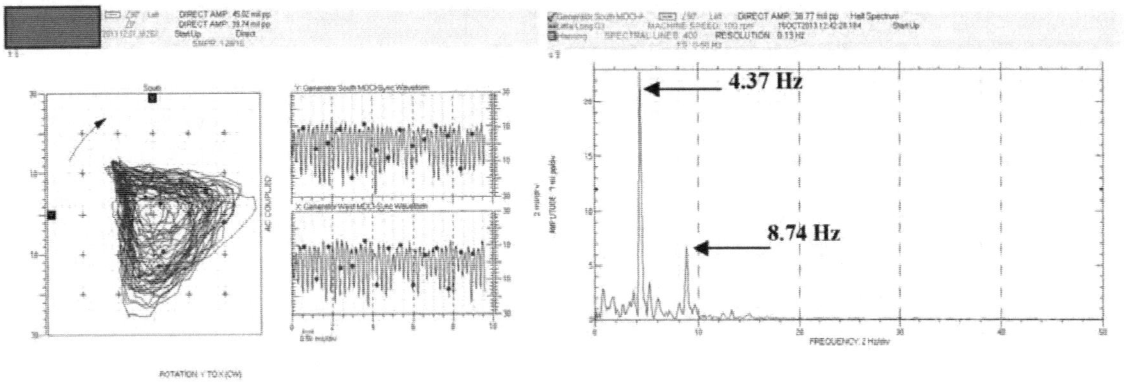

Figure 16: Generator Orbit and Spectrum during Rough Zone Transition

The level of the 4.37 Hz vibration component was about 25 mils p-p. Generator and turbine casing spectra (Figure 17) also show a dominant vibration component at 4.37 Hz. Note that the 4.37 Hz vibration component levels are about ten times higher on the generator than the turbine. The author suspected that 4.37 Hz to be the natural frequency of the shaft. The difference in the amplitude levels at the generator and turbine bearings is most likely due to the mode shape.

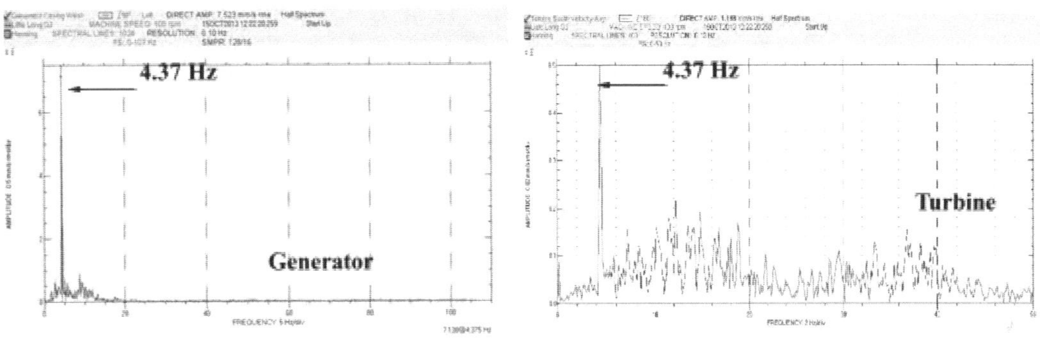

Figure 17: Generator and Turbine Casing Spectra during Rough Zone Transition

The author requested the Original Equipment Manufacturer (OEM) to provide a proper rotordynamics study to identify the critical speeds of the unit. Table 2 is the summary of the critical speeds of the unit.

Table 2: OEM Supplied Unit Natural Frequencies at 100 RPM

	Turbine in air no negative air gap stiffness	**Turbine in water no negative air gap stiffness**	**Turbine in water with negative air gap stiffness**
1st natural frequency	4.68 Hz	4.47 Hz	3.77 Hz
2nd natural frequency	8.03 Hz	7.18 Hz	7.05 Hz

The theoretical values in Table 2 seem to confirm site results. The OEM agreed that the hydraulic phenomenon during rough zone transition is exciting the first natural frequency of the unit. However, of interest is the second natural frequency being close to being a multiple of the first natural frequency. Which leads to the question that the vibration component at 8.74 Hz could also be exciting the second natural frequency of the unit.

Conclusion

The vibration data indicated that the unit was misaligned. Inspection and verification of the unit plumbness confirmed that the unit was more than 2 times the allowable tolerance. Inspection of both generator and turbine guide bearings showed that the guide bearing pads in the upstream and downstream locations were heavily loaded as compared to the remaining pads. The loading on the pads was consistent with the vibration data. Reducing the bearing clearances as per OEM recommendation prior to aligning unit would have most likely caused the bearing temperature to increase to unacceptable levels and could have damaged the bearings. Following proper alignment of the unit the cold trips at efficiency were eliminated.

Rotordynamics analysis confirmed that during rundown and rough zone transition the hydraulic phenomenon at the conditions is exciting the shaft natural frequency. To mitigate this issue in the future OPG has modified its specifications to alleviate this problem.

CONTROLLING PIPING/TUBING VIBRATION ON REFRIGERATION EQUIPMENT SYSTEMS

Author: Ray Kelm P.E.

VIATC TRAINING CONFERENCE & EXPO
NEWPORT NEWS, VIRGINIA | AUGUST 6-8, 2025

VIATC TRAINING CONFERENCE & EXPO
NEWPORT NEWS, VIRGINIA | AUGUST 6-8, 2025

CONTROLLING PIPING/TUBING VIBRATION ON REFRIGERATION EQUIPMENT SYSTEMS

Ray Kelm P.E.
President ZetaQuest Technologies, LLC
Ray.Kelm@ZetaQuest.com

Ray is the president of Kelm Engineering, LLC and ZetaQuest Technologies, LLC. Kelm Engineering is an engineering consulting firm in Houston focused on machinery dynamics including vibration, acoustics, noise, rotordynamics and other dynamic system analysis. ZetaQuest Technologies is a manufacturer of vibration damping hardware commonly used in piping systems and on other mechanical equipment. He received a BSME from Texas A&M University and an MSME from the University of Virginia. Ray is a registered professional engineer in Texas, a member of ASME and Vibration Institute, a member of the Texas A&M Turbomachinery Advisory Committee, and a past member of the Vibration Institute Board of Directors.

Abstract

Piping and tubing as provided on commercial refrigeration systems and other packaged skids can experience elevated vibration driven by dynamic forces generated by compressors and other equipment contained in the skid. Controlling vibration is historically done by adding additional pipe supports at locations where elevated vibration is identified.

With the prolific use of variable speed drive compressors, the traditional approach is less effective since the variable speed equipment will drive harmonic vibration over broad frequency ranges that eventually end up at natural frequencies. The resonant vibration resulting from harmonic forces at natural frequencies of the piping can produce rather high amplitude vibration that will result in pipe/tube fatigue failures.

Vibration damping hardware has been proven to reduce resonant vibration response to acceptable limits to reduce the risk of fatigue failures. A variety of examples are provided that demonstrate the effectiveness of this approach to controlling piping vibration.

Key Words: Damping, piping, refrigeration, tubing

Piping and Tubing Vibration

Piping vibration is caused by either forces acting on the piping or from transmitted vibration from machinery or other supports and connections to the piping. Elevated vibration (with enough amplitude to result in a fatigue failure) will almost always involve a natural frequency and resonance of something. For the purpose of this paper, the term piping and tubing will be used interchangeably for simplicity although tubing is generally thinner wall thickness than piping.

Dynamic forces acting on piping can occur from pressure pulsations in the piping that will react with the pipe at elbows, tees, closed ends, reducers, orifice plates or valves, or any other transition. Pressure pulsations can easily be generated by some machinery particularly including reciprocating pumps and compressors. Other equipment can also generate pulsations. These forces are usually found to act at some harmonic of the running speed of the machine such as 1xRPM, 2xRPM, etc.

High flow rates can also generate hydraulic loads on piping due to flow turbulence or other reactions internal to the piping. Conditions like pump cavitation or flow noise from control valves or other flow restrictions will generate rather broad band pulsations in the piping that can react similar to the harmonic pulsation, but with a broad frequency range that will excite natural frequencies in the piping.

The second common reason for elevated vibration is transmitted vibration from other equipment. If piping is connected to a pump and the pump is vibrating, the piping will vibrate as well. If the piping is vibrating more than the pump, it is probably due to vibration amplification due to natural frequencies on the piping.

Refrigeration System Layout

A typical refrigeration system for retail use will have two compression stages referred to as the low temperature (~-20°F) and medium temperature (~20°F). This is to accommodate both refrigeration cases (need about 36°F product temperature) and freezer cases (< 0°F).

The system must be designed to handle a wide range of cooling loads due to the various conditions in the store as well as ambient conditions since the refrigerant coolers (and the main equipment skids) are often mounted on the roof of the building it is serving. To achieve this requirement, there will be multiple compressors at each compression stage including one at each stage that is usually variable speed/load. This is so that the load on the system for each compression stage can vary from near zero to the full capacity of all compressors.

All the flow through the low temperature compressors must pass through the medium temperature compressors to get to the aftercooler. As a result, there is always more flow capacity (more compressors or larger capacity for each) on the medium temperature side.
In addition to the compressors there will also be an oil separator, a suction accumulator for each compressor suction line, filters, various cross exchangers, a liquid receiver and a variety of other control devices, instrumentation and relief devices.

All the equipment on the skid will require a host of refrigerant tubing of a variety of sizes and pressure ratings. Most refrigerants will operate at pressures that are acceptable for either Type K or L copper refrigerant tubing. However, current trends in refrigerants is to swap to environmentally friendlier products such as R-744 (CO_2), which has no negative impact for global warming.

R-744 operates with supercritical pressures/temperatures with a common rating pressure of 1,500 psig for the discharge of the medium temperature compressors and refrigerant aftercooler/receiver. Since this pressure rating exceeds typical Type K tubing pressure limits, it is necessary to change the tubing material to some product with higher pressure rating.

There are two obvious choices based on currently available tubing: stainless steel (304/316) or UNS C19400 which is a high strength modified copper that is infused with iron to increase material strength. The UNS C19400 can be preferred since it has similar assembly characteristics compared to Type K copper tube, although the brazing material is different and requires a slightly different process for acceptable assembly.

One drawback of the C19400 is that it has limited information regarding fatigue characteristics. Since it is a harder metal, the allowable dynamic stress is likely much lower than typical Type K tube. Per ASME code (add reference), the allowable dynamic stress may be closer to 5,000 psi where common copper tube may have a limit closer to 12,000 psi. As a result, it is certainly possible that the combination of higher static pressure (and higher stress due to pressure) combined with lower allowable dynamic stress can result in a higher risk for fatigue failures with the C19400 tubing.

Traditional Approaches

Most piping design guidelines recommend pipe routing that will avoid resonances at "excitable" frequencies. For a triplex reciprocating pump, that would normally result in a design goal to have the lowest piping system natural frequency at 120% of 6x pump speed. A 2 or 4 cylinder reciprocating compressor (like a refrigeration compressor) will generate moderate 2xRPM and 4xRPM as well as a variety of other harmonics.

The most common approach to addressing elevated piping vibration is to add supports. This is sort of a brute force approach to capture the vibration with more stiffness on the piping network. It is not uncommon to see attempts with people using come-a-longs, straps, wood blocks and a variety of other temporary methods with the hope of a longer-term solution yet to come that seems to rarely materialize.

The concept of adding a stiffener is not a flawed one. For a simply supported pipe with no lumped masses (valves, flanges, etc.), if you put a support in the middle, you will increase the natural frequency of the pipe by a factor of 4! That is based on the idea that a support can be added that is "rigid" and that it acts in both radial directions on the pipe at the midspan.

In reality, all supports are not created equal. It is a common problem that horizontal supports can be difficult to get stiff enough to be "rigid". This is since the "easy" supports will be something like a pipe jack or a hanger attached to something. These can add a lot of vertical stiffness but may not provide much in the horizontal direction.

As a result, fixing a high vibration problem with the traditional approach requires that you start by finding where it is shaking and just add enough supports until it stops vibrating. Unfortunately, this "adding support" effort should also involve review of static piping stress per ASME B31 codes to make sure that the piping network will not be overstressed or that the nozzle loads on connected equipment are too large and result in damage to other equipment. If the issue is not addressed well, the medicine can be worse than the disease and elevated vibration that is well controlled by additional support can result in quick fatigue failures due to elevated stresses both static and dynamic that combine to stress the material beyond safe limits.

The challenge of adding supports is particularly difficult for piping that operates at temperature well above or below ambient. Hot or cold service piping (as is common on refrigeration systems) will thermally move quite a bit. Having a very stiff and well supported piping network can easily aggravate fatigue issues due to elevated static stress combined with dynamic loading from the remaining vibration.

For example, a 6' long discharge header that runs at 200°F will thermally grow by about 0.070". If this header has and elbow on the end and a short run to something (oil separator, etc.), the bending stress in the tube near the elbow can be very high! A 1 5/8" OD tube with 12" run length past the elbow would have as much as 42,000 psi stress at the brazed fitting if there is not consideration for flexibility in any design modifications. 42,000 psi would well exceed the stress limits for the material in this case even with the higher strength of the C19400 material (20,900 psi yield strength).

Resonant Vibration

The worst pipe vibration will occur when a resonance condition exists. This condition happens when there is a mechanical natural frequency on the piping and there is some exciting source that has the same frequency. For example, a compressor discharge line with a 60 Hz natural frequency that is connected to a 2 cylinder compressor that operates at 1800 RPM will be resonant. So what happens at resonance?

Vibration of any mechanical system can be described by an equation of motion as detailed in equation 1 with the vibration displacement expressed in either displacement or velocity. The equation depends on the amplification factor, which is a function of exciting frequency, natural frequency and available damping.

$$X = \frac{F_0/k}{\sqrt{\left[1-(\omega/\omega_n)^2\right]^2+(2\zeta\omega/\omega_n)^2}}$$ Equation 1

$$X = \frac{F}{k}AF$$ Equation 2

$$AF = \frac{1}{\sqrt{\left(\left(1-\left(\frac{\omega}{\omega_n}\right)^2\right)^2+\left(2\zeta\frac{\omega}{\omega_n}\right)^2\right)^2}}$$ Equation 3

In the equations above the following are definitions:

- F_o (or F) is the dynamic force amplitude where Force = F sin (ωt)
- k is the modal stiffness
- m is the modal mass
- ω is the exciting frequency
- ω_n is the natural frequency where $\omega_n = \sqrt{\frac{k}{m}}$
- ζ is the damping ratio, or the percentage of critical damping, C_c
- $C_c = 2\sqrt{km}$

So at resonance (where $\omega = \omega_n$), the vibration will depend only on the damping level so that the equation reduces down to equation 2.

At resonance $AF = \frac{1}{2\zeta}$. Since damping ratios for piping systems tend to be rather low in the 0.1% to 2% range, the amplification factor can be very high with 0.1% damping producing an amplification factor of 500!

Typical amplification factors at resonance on piping will be at least a factor of 25 for most systems. Therefore, when the piping is resonant, it can shake as much as 25 times higher (or more) than it would without the resonant condition.

Compressor Vibration

Before going further with discussion about vibration limits on the piping, it must be clearly stated that the vibration on the compressors (usually the source for all vibration on the system) should be lower than the targets for the piping associated with them. Therefore, it is imperative that the compressor vibration be evaluated before addressing the piping system.

Compressors can be known to resonate as well. A simple bump test on the compressors can confirm if there is a natural frequency of the entire mounted compressor. This is not an uncommon situation with typical compressor mounting, particularly for any compressors that are mounted higher on the skid (as is typical for the low temperature compressors).

If there is a natural frequency at or below 4xRPM, this will be excited by some harmonic of the compressor speed as the speed is varied on variable speed compressors. Field testing on a variety of compressors has confirmed that a resonant compressor can have vibration as high as 7 in/sec pk at 2xRPM (for 2 cylinder compressor) when operated at the natural frequency. Figure 1 shows that when the compressor is operated at about 47 Hz VFD frequency (1,388 RPM), that the horizontal vibration is over 7 in/sec pk at 2xRPM!!

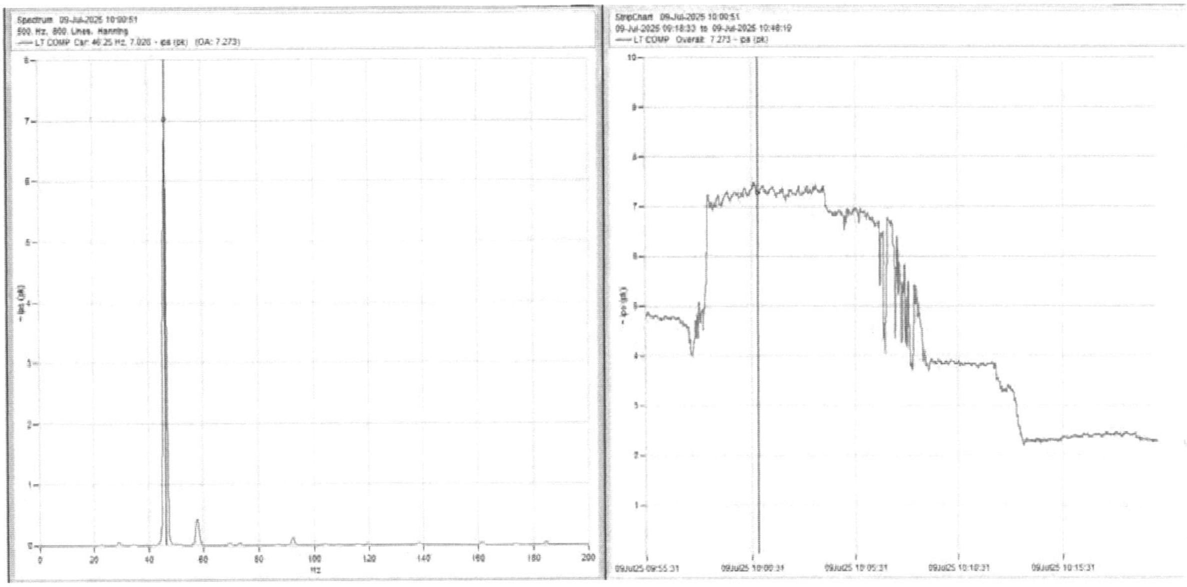

Figure 1 - Compressor Vibration When Resonant

Reducing compressor vibration will achieved by either adding enough structural support to the compressor structure to push the frequency well above 60 Hz or to add a vibration damper onto the compressor as detailed in figure 2. A damper tuned for the 46-50 Hz natural frequency was found to reduce the compressor vibration to less than 0.5 in/sec pk overall at resonance.

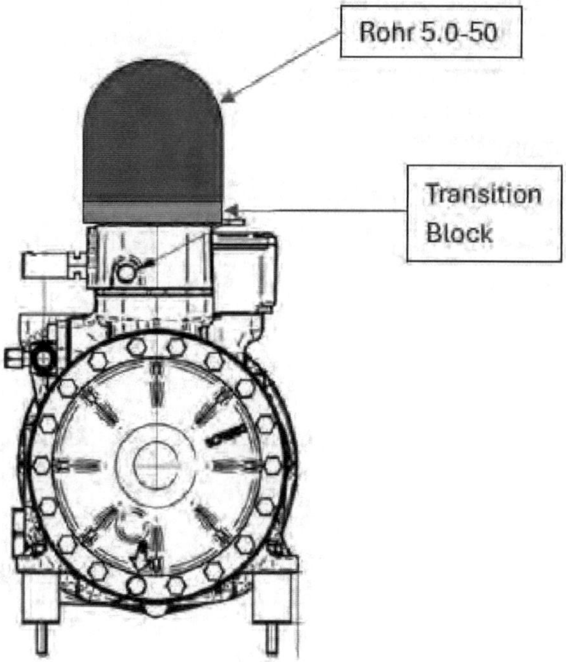

Figure 2 - LT Compressor with Damper Mounted

Compressor vibration that exceeds about 0.5 in/sec should be addressed prior to addressing the piping network as the vibration on piping connected to the compressors (and other structure) will vibrate at least as much as the compressors and/or structural frame vibration.

Acceptable Limits of Vibration

The next logical question is "what is the acceptable vibration level before I need to worry?" And, unfortunately, the answer to that question is not straightforward. In a simplistic summary, if the pipe has been shaking at the same level for a long period of time then the chances are that it can continue at the same level without significant risk of failure.

This is of course a simplistic statement because of some of the following considerations:

- Fatigue is cumulative and you may spend 90% of the time generating the crack and only 10% propagating it. So you can be very near failure and not yet failed.
- Many materials have defined endurance limits, or dynamic stress levels below which fatigue failures are not expected. These limits depend on a host of issues that can greatly reduce the actual tolerance of material to resist fatigue such as:
    - Corrosion fatigue where corrosion attack reduces fatigue strength
    - Stress corrosion cracking (chlorides, etc.)
    - Embrittlement issues (hydrogen or others)

- Fatigue is usually evaluated with consideration for both static (mean) stress and dynamic stress so that locations with high mean stress will experience fatigue failures more frequently than locations with low mean stress.
- Most materials don't really have endurance limits when considering very high cycle duty

Fatigue cycles that accumulate in piping systems can result in very high cycle loading. For instance, vibration at vane pass frequency on a pump that operates at 3,580 RPM with 5 vanes would accumulate almost 26 million cycles per day. A general rule for fatigue is that most fatigue failures will occur in less than 2 million cycles. So for this example, a classic fatigue failure (with dynamic stress higher than the endurance limit) would occur in about 1.86 hours.

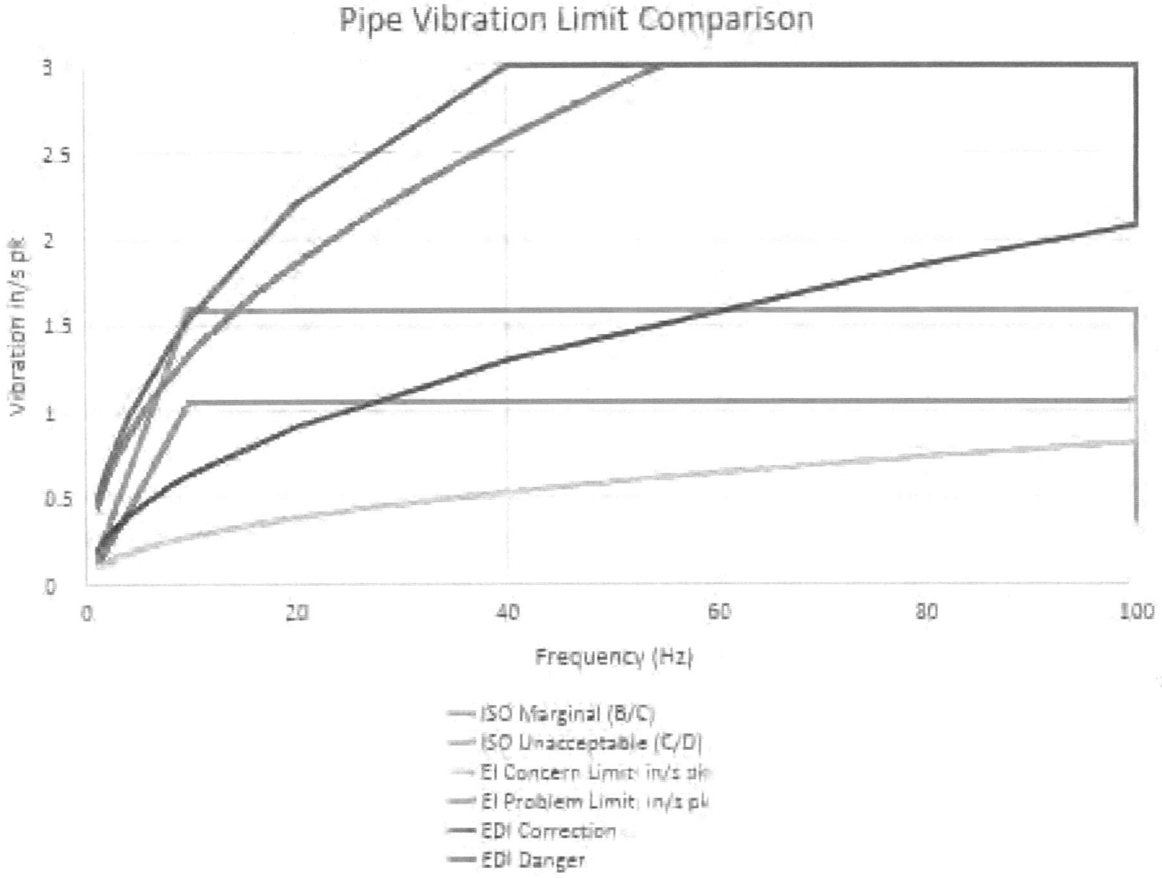

Figure 3 - Various Recommendations for Vibration Limits on Piping

So, when there are cases with very high vibration and relatively high frequencies, fatigue failures can occur quickly. For refrigeration systems, the vibration tends to be in the 30 -120 Hz range. So accumulating the 2 million cycles at 60 Hz would take about 9.3 hours.

It should be noted that the operating time estimates are based on total accumulated stress cycles. So if the input is intermittent or the amplitude varies for some reason (multiple machines on/off at different times), it may take a lot longer to accumulate enough cycles to experience an actual fatigue failure.

Acceptable vibration limits for operating measurements are suggested by a variety of codes and publications. Some of the available resources include (comparison of limits detailed in figure 3):

- Energy Institute (EI), "Guidelines for the Avoidance of Vibration Induced Fatigue Failure in Process Pipework"
- ISO 20816-8, "Mechanical Vibration – Measurement and evaluation of machine vibration – Part 8: Reciprocating Compressor Systems"
- Engineering Dynamics, Inc – Various publications

In general, a limit of about 1 in/sec pk is around the range where vibration will need to be addressed to reduce the risk of fatigue failures.
From a fatigue assessment approach, fatigue prevention is achieved by complying with ASME Section VIII, Division 2, Appendix 5 which is referenced by all the ASME B31 series piping codes. The approach suggested in the standard is based on some general assumptions that static (mean) stress is within normal design limits specified in the B31 codes and that fatigue can then be assessed using estimates of dynamic stress that can be determined from measured vibration amplitudes.

Vibration Limits Based on Fatigue Stress

To identify the actual vibration limit per ASME code, you are required to determine the relationship between vibration and stress at the location where fatigue is possible. This will normally be found at fittings such as brazed connections, etc.

Since piping vibration is commonly extreme when resonance of the piping is involved, it is helpful to generate some model of the piping network to calculate the natural frequency and mode shape to confirm that is the case. When natural frequencies are involved, vibration can be significantly amplified.

As an example, the pipe network shown in figure 4 was observed to vibrate at 2.4 in/sec pk at 58 Hz. An ANSYS finite element model was generated to calculate the natural frequency, with the model confirming that a natural frequency exists near 58 Hz with a mode shape that matches the field observation.

The same model was then used to apply a dynamic load to get the 2.4 in/sec pk vibration to calculate the maximum stress at the connections. Figure 5 shows the calculated stress, with the model showing 2,285 psi due to the bending loading from the vibrating pipe. Since the termination was a socket weld fitting, the actual stress is increased by the stress intensification factor of 2.1 based on ASME code. So the dynamic code stress was 4,799 psi. The code limit for carbon steel material (as was used for this pipe) is 7,000 psi.

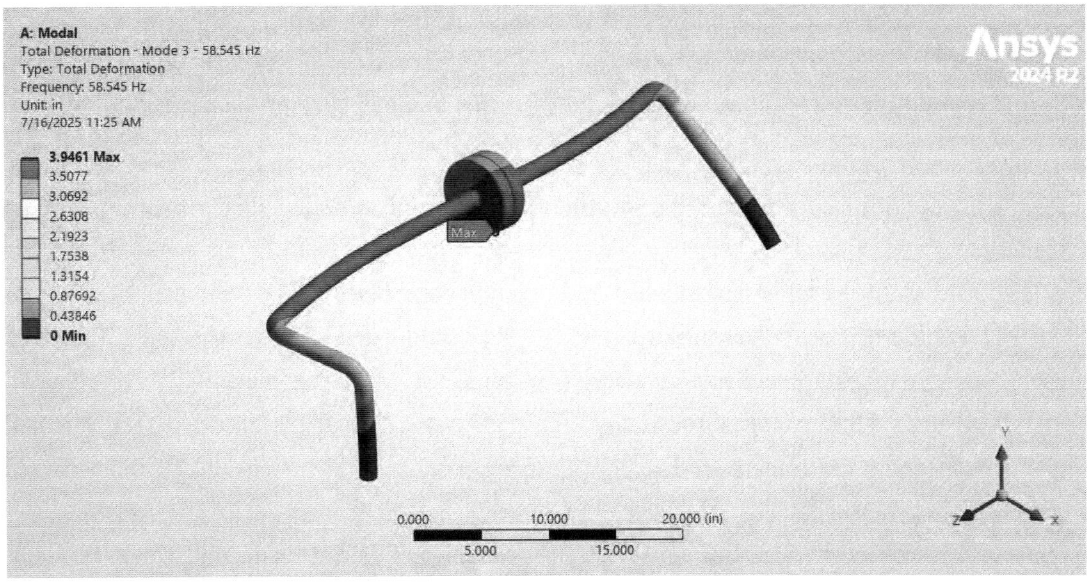

Figure 4 - Pipe Run Natural Frequency

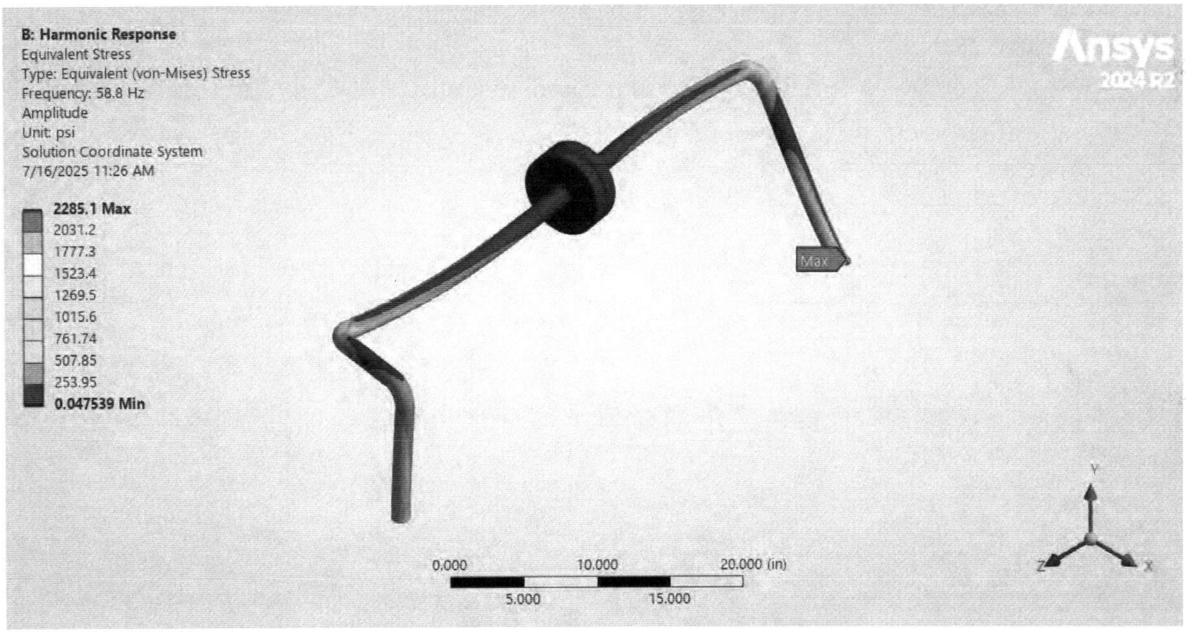

Figure 5 - Stress at Connection due to Measured Vibration

To further improve the fatigue margin, two Rohr 1.5-60 dampers were mounted on the line with a reduction in vibration to 0.24 in/sec pk.

Examples with Rohr Dampers

The option of using vibration dampers has some significant advantages over the traditional "stiffen it up" approach and some drawbacks as detailed below:
Advantages:

- Selection and mounting of the dampers is quick and easy with immediate feedback on whether it is successful to reduce vibration to an acceptable limit

- Most piping layouts can be addressed with the addition of clamp on devices
- Vibration targets of 1 in/sec or lower are easily achieved in most cases when applying damper(s) with combined mass of 10-15% of the modal (moving) mass on the piping

Disadvantages:

- Requires an additional device that should be documented on ISO's for future replacement if needed
- The "right" damper must be selected and oriented correctly
    - The dampers are manufactured with pre-set "target" frequencies with the frequencies spaced for coverage of a wide range of applications
    - If the system natural frequency changes (other supports get loose or additional stiffeners are added), a different damper may be required to match the new natural frequency
- Clamp on damper devices have no benefit if the piping is not resonant
- For more complicated layouts where the vibrating mode is not the 1st mode, the added weight of the damper could shift lower modes so that they are resonant when they previously weren't
    - A vibration test is always recommended to confirm that vibration is acceptable after mounting the dampers

The typical mounting of the dampers is shown in figure 6.

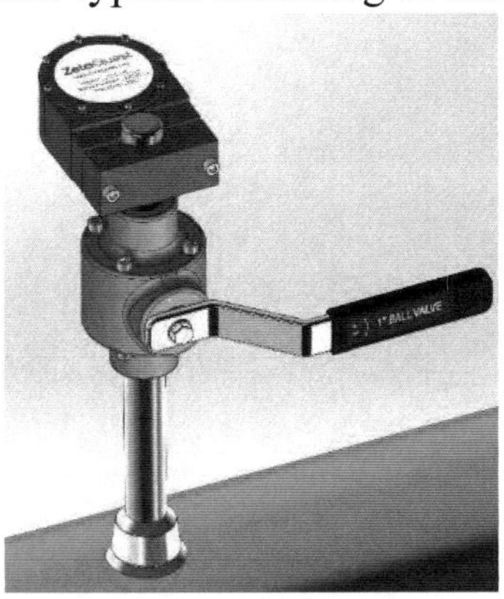

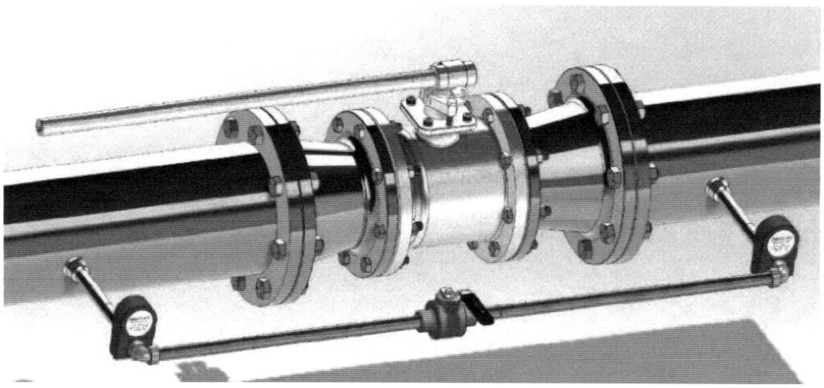

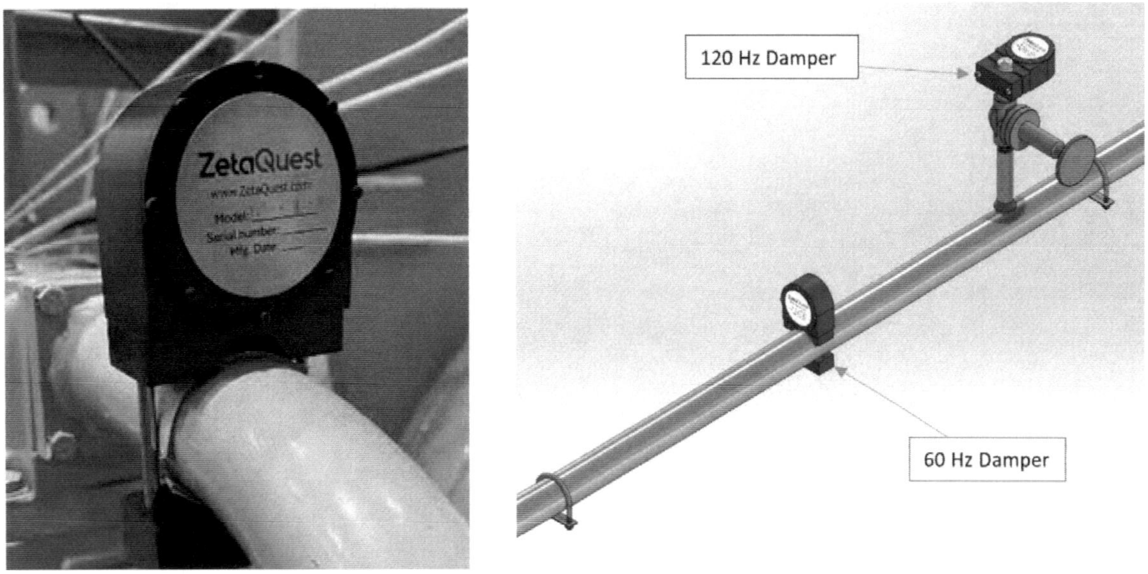

Figure 6 - Damper Mounting

A tubing run with 1 5/8" XHP (C19500) tubing is shown in figure 7.

Figure 7 - Tubing Layout Example

The vibration was found to be at 58 Hz with an amplitude of 2.1 in/sec pk. A damper was added with 60 Hz target frequency as shown in figure 7 with the results detailed in figure 8. The maximum vibration as speed varied was reduced to less than 1 in/sec pk which was the intended goal.

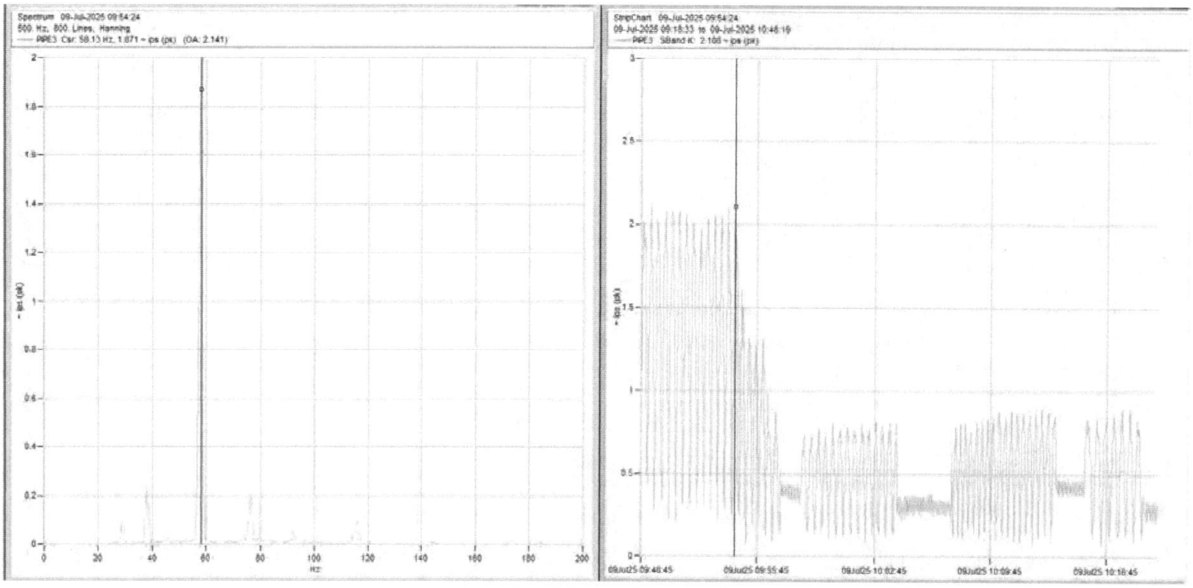

Figure 8 - Results of Adding a Damper

Conclusions

Piping/tubing vibration can be very high when the layout results in natural frequencies near exciting frequencies. Equipment with variable speed equipment will be particularly susceptible to high vibration issues since harmonic vibration from machinery will often amplify significantly at resonances.

Traditional modifications of stiffening the piping can improve vibration particularly if the natural frequencies can be shifted away from harmonic vibration from the associated machinery. If variable speed equipment is used, damping may be a preferred option so that the issue is not shifted from one frequency to another by adding some stiffness.

The addition of vibration damping hardware is easy to implement and provides immediate feedback on whether the damping devices selected will produce the desired results. If not, an additional damper can be used at potentially a different frequency or location to achieve the goal. Periodic inspection of piping should be done to identify locations with vibration in excess of 1 in/sec pk. When elevated vibration is identified, it should either be confirmed to be acceptable by reviewing the dynamic stress or it should be reduced using some method. Available methods of improving vibration are currently limited to either reducing the forces generating the vibration (always the first choice), adding support stiffness (often a good 2nd choice), or by adding enough damping so that the vibration is acceptable even with the resonance (may be the easiest option with quick results!).

SHAFT MISALIGNMENT: THE HIDDEN DRIVER BEHIND VIBRATION AND EQUIPMENT FAILURE

Author: Terry C. Southall, MA. Ed

VIATC TRAINING CONFERENCE & EXPO
NEWPORT NEWS, VIRGINIA | AUGUST 6-8, 2025

SHAFT MISALIGNMENT: THE HIDDEN DRIVER BEHIND VIBRATION AND EQUIPMENT FAILURE

Terry C. Southall, MA. Ed
Global Brand & Distribution Manager
Hamar Laser Instruments, Inc.

Terry C. Southall has provided training and skills support within operations, maintenance, and reliability for over 12 years, working for companies such as Honeywell Aerospace, Georgia-Pacific Wood Products, Reliability Solutions Training LP, People and Processes, Inc., and most recently at Hamar Laser Instruments, Inc. Terry has a M.A. in Adult Education and Training as well as a B.S. in Management. Terry also served and is a protected veteran of the U.S. Air Force.

Abstract

Despite significant progress in condition monitoring and predictive maintenance, shaft misalignment remains one of the most common, and commonly underestimated, contributors to equipment failure. It often manifests through elevated vibration, accelerated bearing and seal wear, and chronic reliability issues that erode performance and shorten asset life. Yet in many facilities, alignment is still treated as a one-time setup task rather than a critical, ongoing component of machine health.

This paper examines both the mechanical and financial impact of misalignment and explains why precision alignment should be a core part of modern maintenance strategies. It outlines how misalignment produces specific vibration signatures, how these can be distinguished from other common faults, and how the use of outdated methods has created a gap between problem detection and resolution. A real-world case study underscores the measurable benefits of adopting accurate, repeatable alignment practices, demonstrating why alignment deserves renewed attention in reliability-focused programs.

1. Understanding the Consequences of Shaft Misalignment

The Silent Threat Behind Rotating Equipment Failures

Shaft misalignment occurs when the rotational centerlines of two coupled shafts are not collinear. This condition leads to unbalanced forces, cyclic loading, and excessive stresses on components such as bearings, seals, and couplings. Unlike more conspicuous mechanical issues, misalignment often develops gradually and may initially appear as minor inefficiencies or wear patterns. Over time, however, it creates a domino effect of degradation across multiple components in the drivetrain.

The symptoms of misalignment are well documented: elevated vibration amplitudes (especially at 1× and 2× running speed), axial movement, coupling wear, and thermal growth discrepancies. What complicates the diagnosis is that these symptoms can also be attributed to other faults, such as unbalance or resonance, making root cause identification difficult without precise measurement tools.

Why Legacy Alignment Practices Fall Short

Many maintenance teams still rely on traditional tools such as dial indicators, straightedges, or feeler gauges for shaft alignment. While useful for rough alignment procedures, these tools lack the resolution and repeatability needed to achieve precision alignment, especially under dynamic operating conditions. Compounding the issue, these legacy methods do not account for soft foot, thermal growth, or baseplate distortion, which can result in misalignment being reintroduced even after adjustments are made.

In many industries, shaft alignment is treated as a set-and-forget task, performed during installation or major overhauls but not integrated into routine condition-based maintenance programs. This lack of ongoing validation leads to a slow but steady degradation in alignment, performance, and ultimately, reliability.

2. Misalignment and Vibration: The Missing Link

How Misalignment Drives Specific Vibration Patterns

Vibration analysis is a primary diagnostic tool in predictive maintenance. However, without correlating the vibration signatures to alignment condition, teams may misidentify the root cause. Misalignment typically produces high vibration readings at 1× RPM (due to angular misalignment) and 2× RPM (due to offset misalignment), along with elevated axial vibration. In severe cases, harmonics and sidebands may appear, especially if the misalignment has induced looseness or coupling deterioration.

Correctly identifying these patterns as misalignment rather than imbalance or mechanical looseness is critical. An incorrect diagnosis can lead to unnecessary balancing procedures, ineffective part replacements, or repeat failures. Worse, the real issue remains unaddressed, eroding both confidence and system reliability.

Restoring Alignment as a Reliability Pillar

Modern laser alignment systems eliminate the ambiguity and guesswork of traditional alignment approaches. These systems provide high-resolution, real-time measurements and correct for factors such as soft foot, dynamic movement, and thermal growth. When combined with a standardized procedure, such as incorporating alignment into pre- and post-maintenance routines, organizations can significantly reduce misalignment-related downtime and improve Mean Time Between Failures (MTBF).

The transition from reactive to proactive alignment also supports broader reliability-centered maintenance (RCM) objectives. Proper alignment not only extends component life but also enhances the effectiveness of adjacent PdM technologies such as vibration, oil, and thermal analysis.

3. Case Study: Pump Train Failures in a Pulp & Paper Mill

A pulp and paper facility operating in the southeastern United States was experiencing frequent, unplanned outages caused by repeated failures in a six-pump train system. The system, responsible for circulating process water throughout the mill, suffered from increasing vibration levels, with several pumps requiring unscheduled bearing and coupling replacements every 4–6 weeks.

Initial assessments using standard vibration analysis revealed elevated 1× and 2× vibration frequencies, along with signs of axial misalignment. Maintenance crews had historically used dial indicators for alignment but had not revisited alignment settings in over a year. Compounding the problem was thermal expansion in the piping system that altered shaft positions once the pumps reached operating temperature.

To investigate further, the facility opted for a full precision laser alignment procedure during a scheduled shutdown. Technicians identified significant angular misalignment at three of the pump couplings and moderate offset misalignment in the remaining units. Additionally, several soft foot conditions were discovered and corrected.

Once alignment was completed with consideration for thermal movement and corrected soft foot conditions, the pumps were brought back online. Vibration readings fell within acceptable standards, and no bearing or coupling failures occurred for over eighteen months following the intervention.

Results:

- 60% reduction in alignment time through the use of real-time laser feedback and Machine Train Mode™ alignment features.
- 40% increase in MTBF, reducing replacement costs and labor hours.
- Improved energy efficiency due to reduced shaft loading and friction.
- Enhanced reliability confidence, prompting leadership to integrate alignment into future PdM planning.

Conclusion

Shaft misalignment remains a critical, but too often underestimated, source of equipment failure and operational inefficiency. As this article demonstrates, aligning shafts is not just a mechanical necessity; it's a reliability strategy. By understanding the link between alignment and vibration, embracing high-precision measurement tools, and integrating alignment into the maintenance lifecycle, organizations can dramatically improve performance, reduce costs, and build a stronger foundation for asset health.

References

1. Bloch, H. P., & Geitner, F. K. (1999). Machinery component maintenance and repair (3rd ed.). Gulf Professional Publishing.
2. Mobley, R. K. (2002). An introduction to predictive maintenance (2nd ed.). Butterworth-Heinemann.
3. Electric Power Research Institute (EPRI). (2014). Shaft alignment best practices for rotating equipment (Report No. 3002002855).

IMPACT TESTS OF A SYSTEM WITH RATE-DEPENDENT STIFFNESS AND DAMPING

Author: Jianke Wang, Ph.D., P.E.

VIATC TRAINING CONFERENCE & EXPO
NEWPORT NEWS, VIRGINIA | AUGUST 6-8, 2025

VIATC TRAINING CONFERENCE & EXPO
NEWPORT NEWS, VIRGINIA | AUGUST 6-8, 2025

IMPACT TESTS OF A SYSTEM WITH RATE-DEPENDENT STIFFNESS AND DAMPING

Jianke Wang, Ph.D., P.E.
Staff Engineer, MS Directional, a company of Pattern-UTI Energy
jianke_wang@yahoo.com

Jianke Wang is a staff engineer for MS Directional, a company of Patterson-UTI Energy. He earned his MS and PhD both on the vibration control by design of high-speed rotating machinery and has over 20 years industrial experiences in vibration control and analysis. He is a subject matter expert on Machine Design & Materials of NCEES PE Mechanical Engineering PAKS and Exam Committees and Drilling Equipment of NCEES PE Petroleum Engineering PAKS Committee. He has been a certified Vibration Analyst IV since 2007.

Abstract

Three different methods (i.e., half-power, phase shift, and logarithmic decrement) are often used to evaluate system natural frequencies and damping ratios. For a system with constant stiffness and damping, the readings of these three methods are close and support each other. However, for a system with progressive stiffness and damping, it has been found that logarithmic decrement method gets significantly different readings from the method of phase shift. For logarithmic decrement method, the fundamental natural frequency reading increases as the number of time waveforms (i.e., cycles) scoped increases. These discrepancies are explained based on how frequency and phase results are obtained through Fast Fourier Transformations and the physical meanings of the rate-dependent stiffness and damping. A set of impact test results of a downhole high-pressure and high-temperature shock isolator with pressure and temperature compensations is presented as an example.

Key Words: Impact Test, Nonlinear System, Hysteresis, Progressive Damping

1. Introduction

Measurement While Drilling (hereinafter referred to as MWD) tools are critical to the success of long-reach directional drilling in oil and gas industry. The main challenges facing in an 8000-10000 feet deep underground with another 8000-16000 feet long horizontal reach are high ambient pressure (e.g., 20000 psi) and high ambient temperature (up to 392 °F) and extremely high shock (up to 1000 g with 0.5 milli-second duration, 1 g = 386 in/sec^2) with high recurring frequencies (e.g., 47 counts per second). To minimize the shock transmitted to MWD tools for the protection of expensive electronics and sensors, a pressure and temperature compensated shock isolator with progressively increasing damping was developed. Figure 1 shows a patent-pending shock isolator. A progressive spring-damper mechanism is built in along the axial direction of the shock isolator. It has a 1000g (0.5 milli-second duration) axial shock capacity.

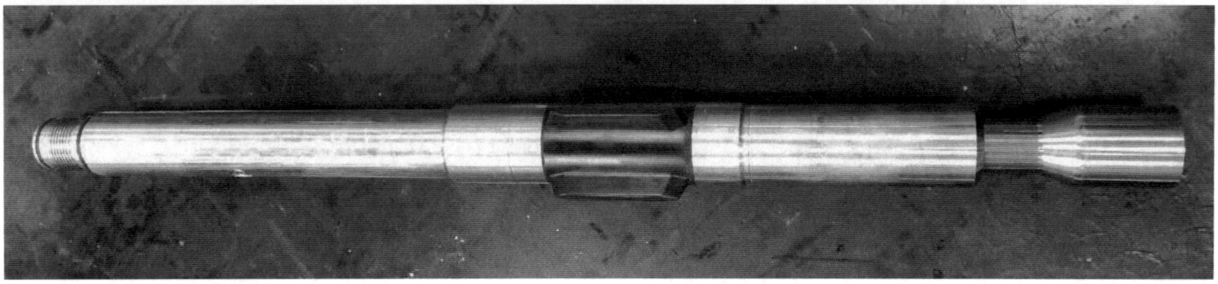

Figure 1 Patent-pending shock isolator with progressive spring-damper mechanism

For the shock isolator to work, there are three critical system parameters, which are system fundamental natural frequency, damping, load capacity. System natural frequencies and damping ratios can be measured by impact test (often called as bump test in industrial practices).

To conduct a successful impact test, Wang (2011) listed a set of comprehensive recommended practices on instrument selection and how to conduct impact test [1]. These recommended practices are followed in this research.

With impact tests done, three different methods (i.e., half-power, phase shift, and logarithmic decrement) are often used to evaluate system natural frequencies and damping ratios [1-4]. Since there are two natural frequencies close to each other for shock isolator, half-power method is not suitable. Phase shift and logarithmic decrement methods are applied to evaluate the natural frequencies and damping ratios of the new shock isolator. Logarithmic decrement method is one of the best methods to measure the damping of a system, whose fundamental natural frequency is dominant [3,4].

2. Compression Tests

A series of timed compression tests have been done to quantify the stiffness and load capacity of the new shock isolator. Figure 2 shows the test setup.

In Figure 2, a through-hole (donut-shaped) load cell is used to measure compression force. An intelligent digital handheld display is used to display and record the compression test data including instant peak/valley and steady state readings. Its sampling rate is set at 100Hz. A height gauge with an absolute encoder built-in is used to monitor compression displacement. The resolution of the height gauge is 0.0005" and its accuracy is ±0.0015". The base of hydraulic press is set level. Shock isolator and load cell sit upright. A special washer separates load cell and hydraulic press plunger to assure that a purely vertical compression load is transmitted to the assembly of load cell and shock isolator. Height gauge is set to zero when the initial compression load reading is around 1 lb. For each load step, effort was put to monotonically compress as slowly as possible to minimize any inertia load.

Figure 2. Timed compression test setup.

There are three main observations during the compression tests:

1. While compressing down, initial reading is the highest (peak) and then goes down and levels out.
2. While self-retracting (in other words, slacking) up, initial reading is the lowest (valley) and then goes up and levels out.
3. There is not any detectable change in the overall length of shock isolator before and after compression tests.

Figure 3 shows the loading and then unloading history of how compression load changes with the increasing/decreasing displacement of shock isolator. During both loading and unloading processes, the force reading approaches steady state at around 15-20 seconds after a new compression displacement is set.

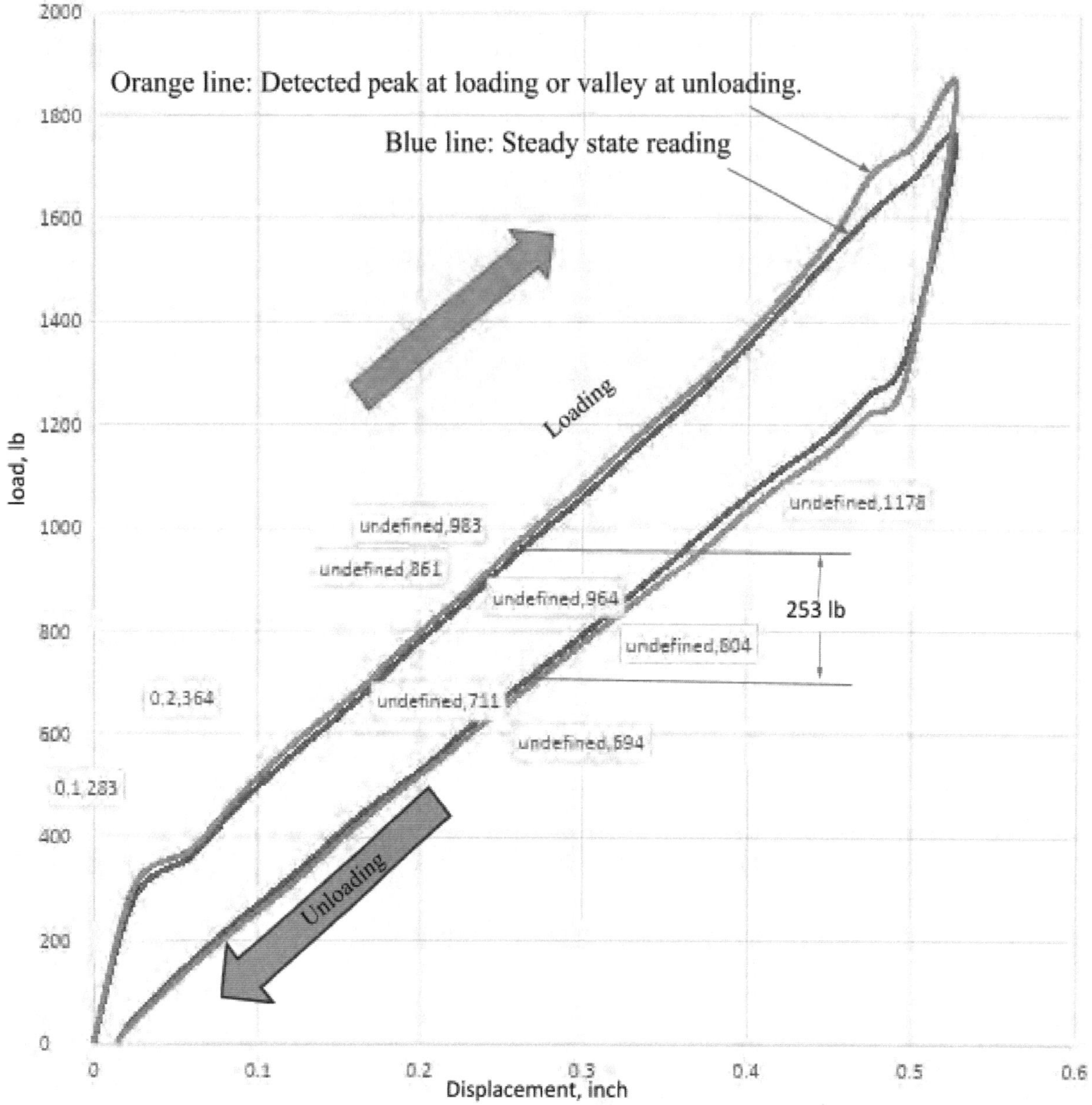

Figure 3 Hysteresis Phenomenon discovered in shock isolator compression test (loading and then unloading) (Blue line: steady state reading; Orange Line: peak (during loading process) or valley (during unloading process) value detected).

In Figure 3, a significant hysteresis phenomenon shows up between the loading and unloading compression processes. Similar hysteresis phenomenon has been seen often in sensor performance [5]. Simply put, loading and unloading are going through a different path. It takes more force to compress to the same displacement than self-retract (slack) and then hold the compression down. The force difference (i.e., the hysteresis loop width) can be as wide as 253lb at the compression displacement of around 0.266" (0.265" at loading and 0.268" at unloading).

Figure 3 also shows that there is an obvious difference between the peak value detected during loading process and the steady state reading. For example, at the compression displacement of 0.265", the peak force detected is 983 lb while the steady state reading is 964 lb. Similarly, but in the opposite direction, there is a clear force difference between the valley value detected during unloading process and the steady state reading. During unloading, the steady state reading (711 lb) at 0.268" displacement is 17 lb more than the valley value detected.

3. Impact Test Setup

Following the recommended practices listed in Reference [1] for bump test, the following instruments are selected and used in this research: a six-channel FFT vibration analyzer with cross-channel functions such as transfer function and coherence features; an instrumented hammer (a hammer with a force sensor embedded inside) with multiple tips of different sizes and hardness; and a triaxial accelerometer. For successful impact test, the careful selection of a suitable hammer with a suitable tip (hardness and size) is especially important and multiple trial tests are often necessary [1]. Figure 4 shows a test setup for the impact test of shock isolator.

Figure 4 Test setup for the impact test of shock isolator.

Different placement combinations of the accelerometer and instrumented hammer and shock isolator have been tried and the most effective combination is shown in Figure 4. In the final test configuration, shock isolator sits on a steel table corner (8" from each of the right corner edges).

The steel table is rigidly mounted on lab floor. The accelerometer shown in Figure 4 is a tri-axial ICP accelerometer, which is glued on an adaptor mounted on the top of the shock isolator. Hammer hits a location on the same adaptor with 180° across the accelerometer. Hammer force sensor is connected to the Channel (1) of the FFT vibration analyzer. The X, Y, and Z axes of the accelerometer are connected to the Channels (2), (3), and (4) of the vibration analyzer, respectively. As shown in Figure 4, accelerometer X axis (with cable tail) points to the shock isolator axial center line perpendicularly; Y axis is 90° away from X axis in the same orthogonal plane; Z axis is parallel with the shock isolator center axis, which is in the same direction as the primary progressive spring-damper mechanism. During the final tests, all channels use uniform window (also called as rectangular window); hammer hits with a white tip (medium hardness) and an extension weight; Frequency span was set at 800Hz with 6400 FFT lines; a pre-trigger of 3g is used; a linear averaging of 4 is adopted to reduce noises; clipping reject function is turned on. Trial tests were conducted at the beginning to assure that hammer impact response dies out over the time record.

Since the Z axis (Channel 4 of vibration analyzer) of the accelerometer is in the same direction as the progressive spring-damper mechanism of the shock isolator, the primary focus of the following analysis will be on the impact force input on Channel (1) and the impact response on Channel (4).

4. Impact Test Results and Discussions

4.1 Impact Test Results and Calculations

Figure 5 shows a set of impact test results on the shock isolator shown in Figure 4.

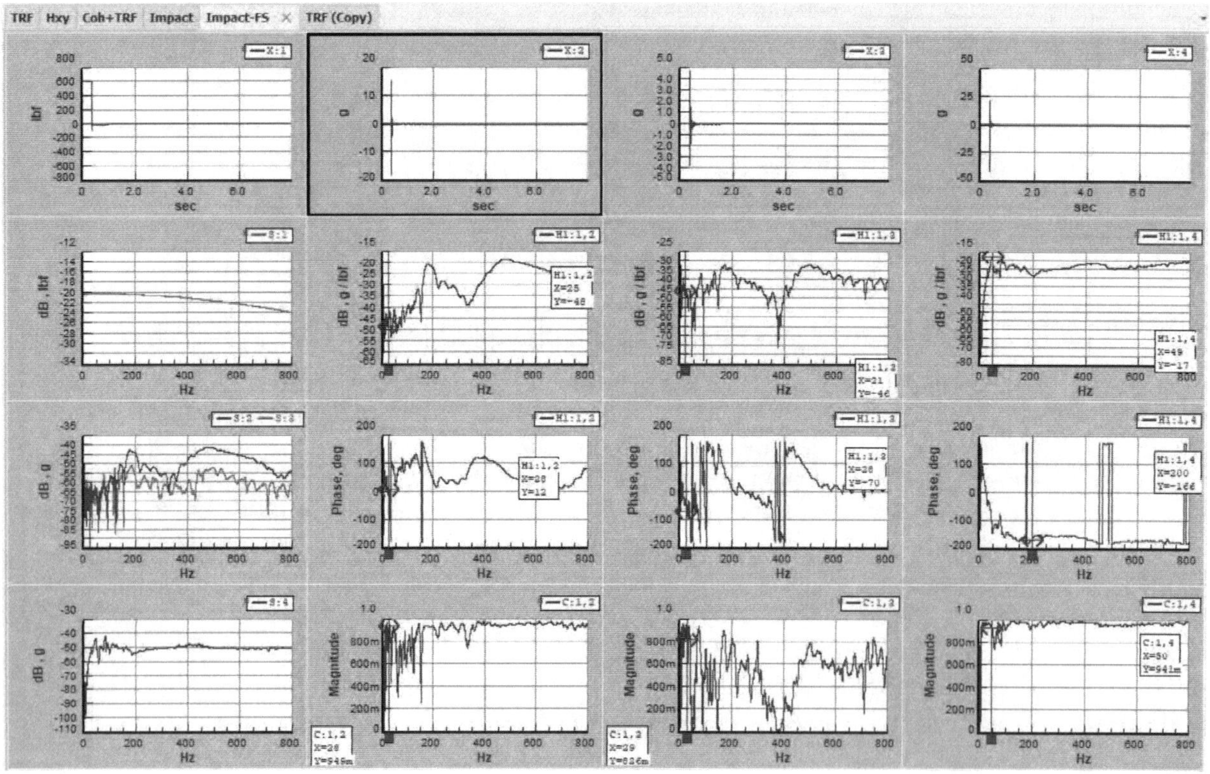

Figure 5 Impact test results of the shock isolator shown in Figure 4.

In the sub-titles of Figure 5, "X" stands for the time waveforms of impact force input (i.e., 1: impact force on Channel 1) or the impact response in each of the three accelerator channels (2: X, 3: Y, and 4: Z). While following the same suffix protocols of ": 1/2/3/4" as the time waveforms "X", "S" indicates linear spectrum. "H" denotes the transfer function between two signals. "C" is the coherence between two signals. Since Figure 5 shows a single spike (no double-hit) in the time waveform plot of the impact force and a flat impact force spectrum (less than 2dB drop for the interested frequency span of 0-500Hz) and a coherence of more than 0.8 at the frequency spike and all signal inputs (both impact force and its responses) start and end at zero, these impact test results are valid [1]. For clarity, Figure 6 shows a close-up view of the coherence (C:1,4) between impact force and its axial shock response.

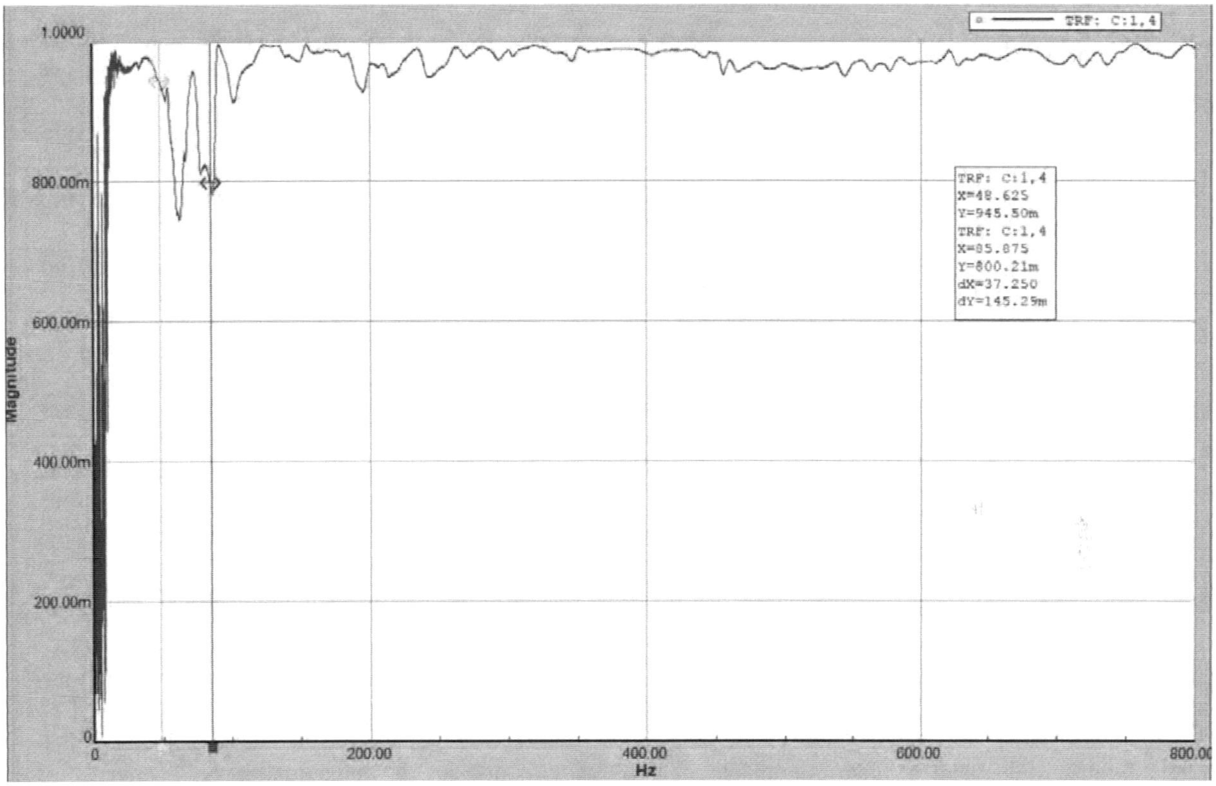

Figure 6 Close-up view of the coherence between impact force and its axial shock response.

The transfer function between impact force on Channel 1 and its impact response on Channel 4 (in the axial direction of shock isolator) is shown in Figure 7.

Figure 7 shows that the system fundamental natural frequency f_n is 48.625Hz and there is a 110° phase shift cross this natural frequency. As shown in Figure 6, the coherence (C:1,4) between the impact force and its axial shock response at f_n (48.625Hz) is 0.945, which is more than the recommended minimum of 0.8 [1]. Though Figure 7 also shows a secondary natural frequency of

85.875 Hz, it has a phase shift about 50° with a low coherence so that it barely can be seen in all time waveforms.

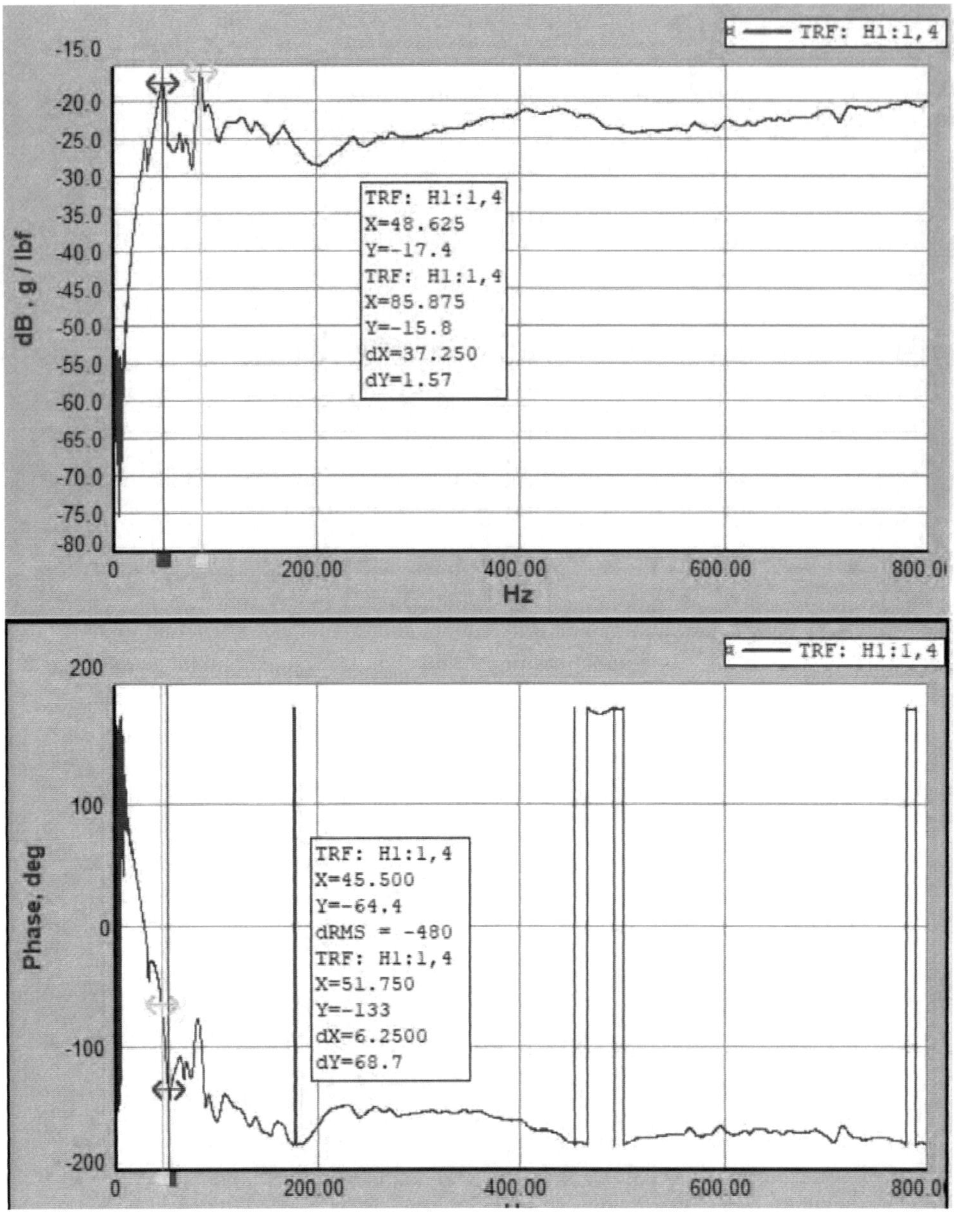

Figure 7 Transfer function between impact force and shock response in axial direction.

Phase shift method is used to calculate system damping at the fundamental natural frequency of f_n (48.625Hz) with a phase shift of 110° as shown in Figure 7. Equation 1 is used to calculate resonance amplification factor Q:

$$Q = \frac{\pi f_n}{360} \times \frac{\Delta\phi}{\Delta f} \tag{1}$$

where Δf is a frequency span around the fundamental natural frequency f_n (48.625Hz). $\Delta\phi$ is the phase shift across the frequency span Δf, where the phase decreases linearly as frequency increases. Δf and $\Delta\phi$ are given by Equations (2 and 3) respectively.

$$\Delta f = |f_1 - f_2| \tag{2}$$

$$\Delta\phi = |\phi_1 - \phi_2| \tag{3}$$

where f_1 is the leading frequency and f_2 is the trailing frequency; ϕ_1 is the leading phase and ϕ_2 is the trailing phase correspondingly, | | means the absolute value of the result. Both frequency span Δf and phase shift $\Delta\phi$ are in absolute value.

Plugging Equations (2 and 3) into Equation (1) yields the amplification factor Q (4.661), then Equation (4) defines damping ratio ζ:

$$\zeta = \frac{1}{2Q} \tag{4}$$

In Figure 7, the readings of f_n, f_1, f_2, ϕ_1, ϕ_2 have been called out. Table 1 shows a summary of the transfer function plot readings, qualification factor Q and damping ratio ζ derived using Equations (1-4).

Table 1 Transfer function plot readings and calculated damping ratio.

Fundamental natural frequency f_n, Hz	48.625
Coherence at f_n	0.945
Leading frequency f_1, Hz	45.5
Leading phase ϕ_1, °	-64.402
Trailing frequency f_2, Hz	51.75
Trailing phase ϕ_2, °	-133.06
Amplification factor Q	4.661
Damping ratio ζ	0.107

To easily read the first cycles (waveforms) of impact response in time domain, Figures 8-12 show a close-up view of the impact response in shock isolator axial direction (X: 4) shown in Figure 5. Figures 8-10 are in the direct reading of acceleration by the accelerometer while Figures 11-12 are in velocity, which is derived by one integration of the acceleration readings shown in Figures 8-10.

As shown in Figures 8-12 obviously, the period of impact response decreases as time goes. Decreasing cycle period means increasing cycle frequency since cycle frequency is the reciprocal of cycle period.

During the impact test, the displacement of a damper piston within the shock isolator increases or decreases as time goes. Therefore, the time-dependent period and frequency can also be explained as displacement-dependent.

All time waveforms of the shock responses shown in Figures 8-12 show that the fundamental natural frequency is dominant. Therefore, logarithmic decrement method (often called as Log Dec in industrial practices) is also used to calculate system fundamental natural frequency and its damping ratio.

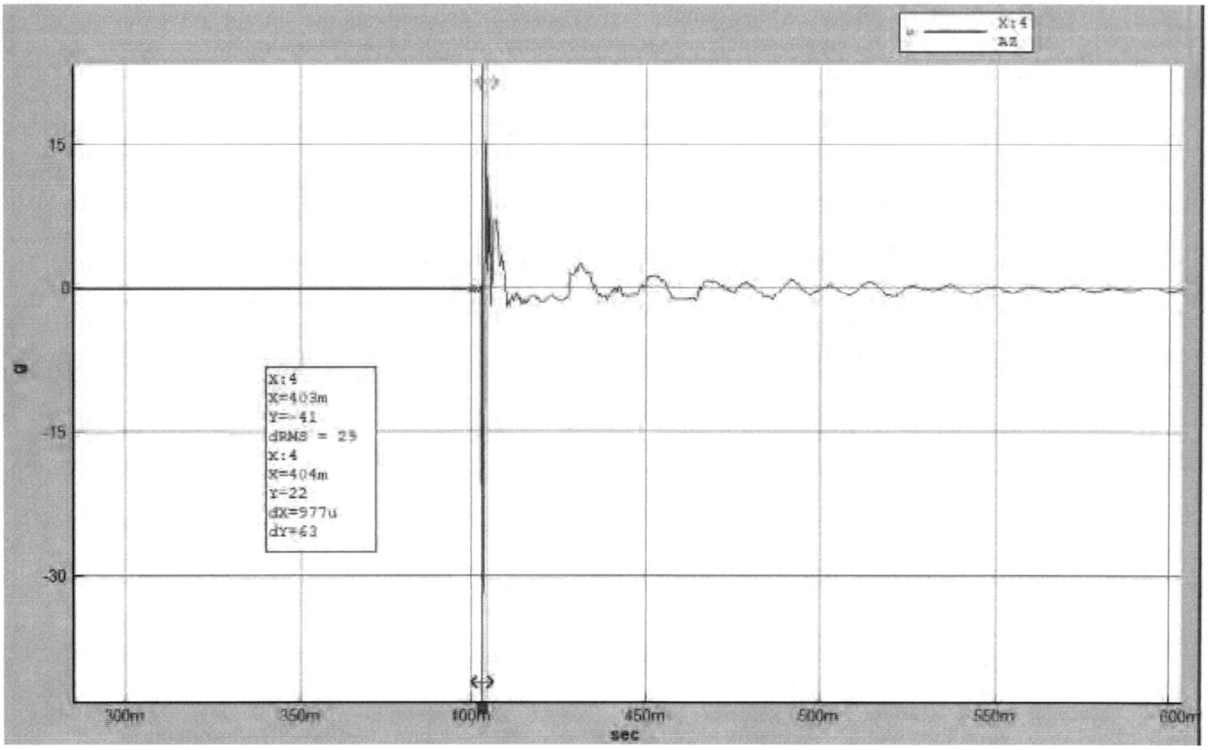

Figure 8 Close-up impact response of shock isolator in its axial direction.

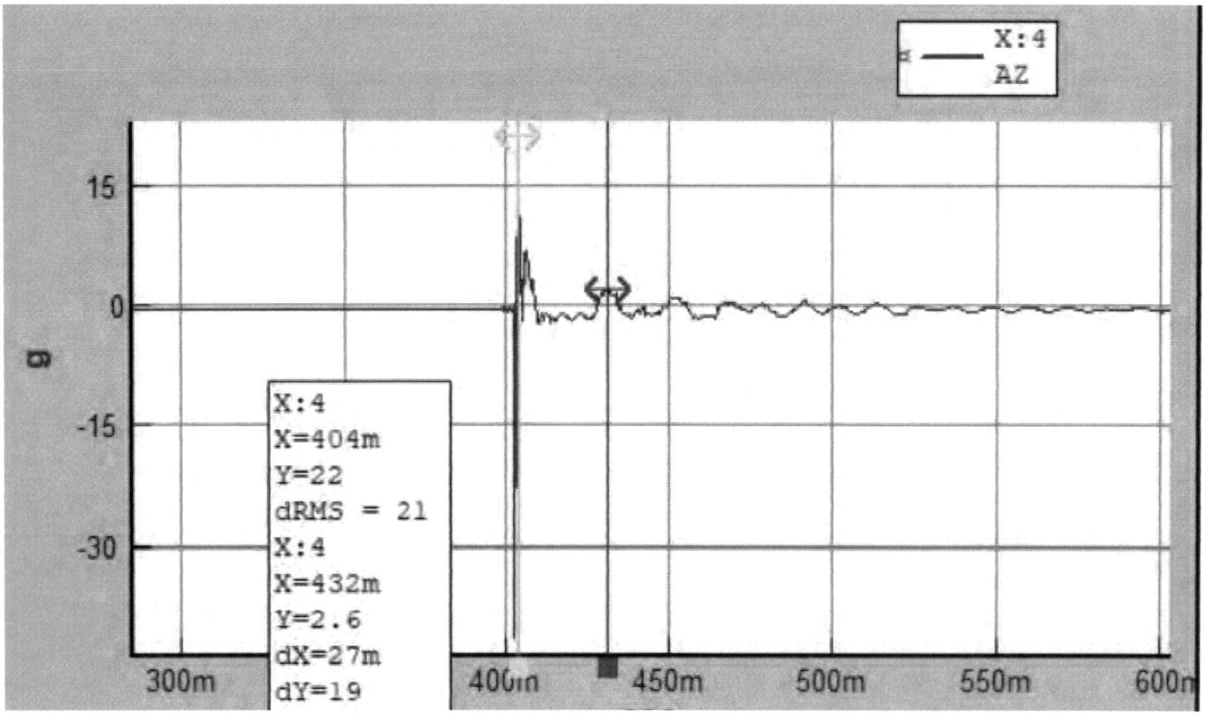

Figure 9 Close-up impact response of shock isolator in its axial direction.

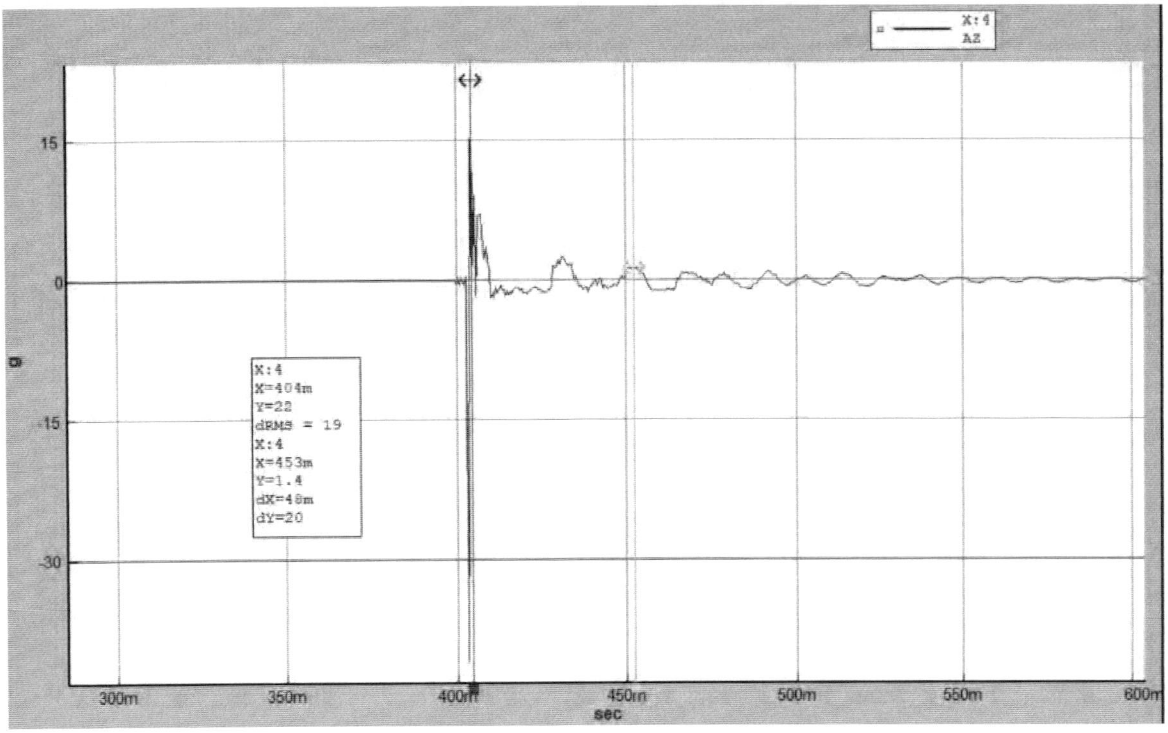

Figure 10 Close-up impact response of shock isolator in its axial direction.

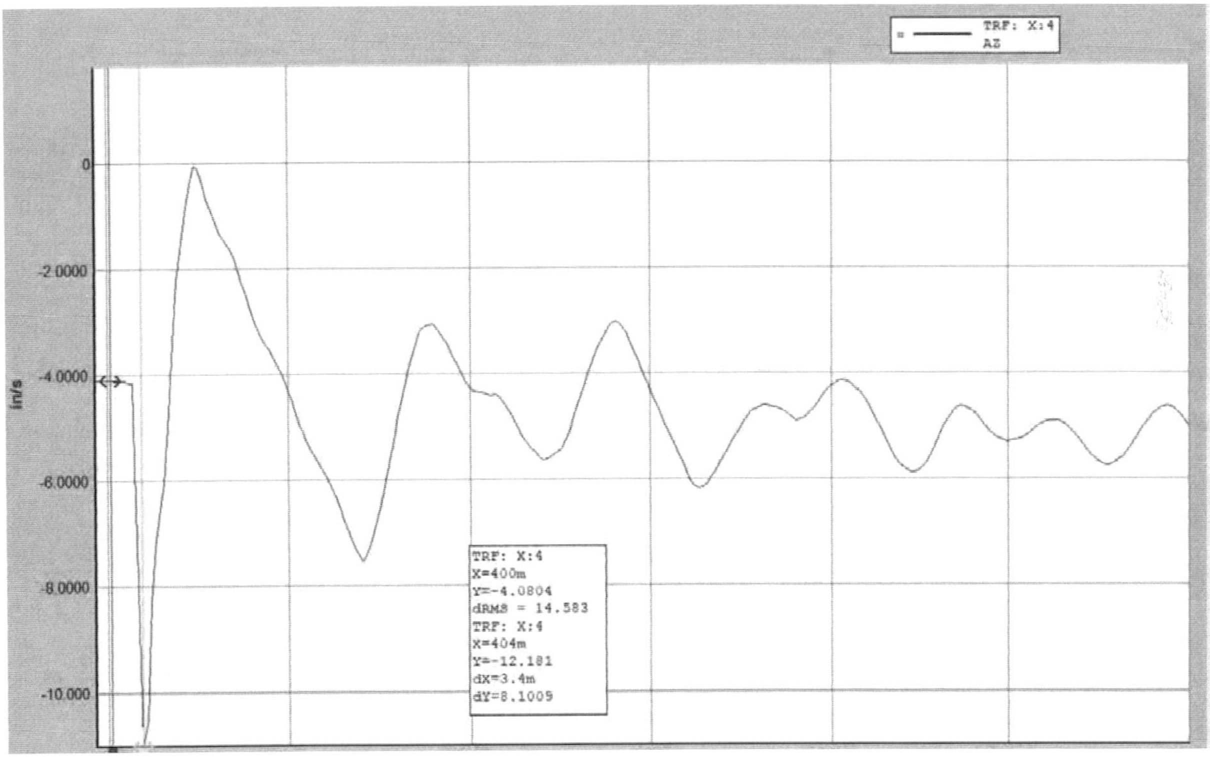

Figure 11 Close-up impact response of shock isolator in its axial direction.

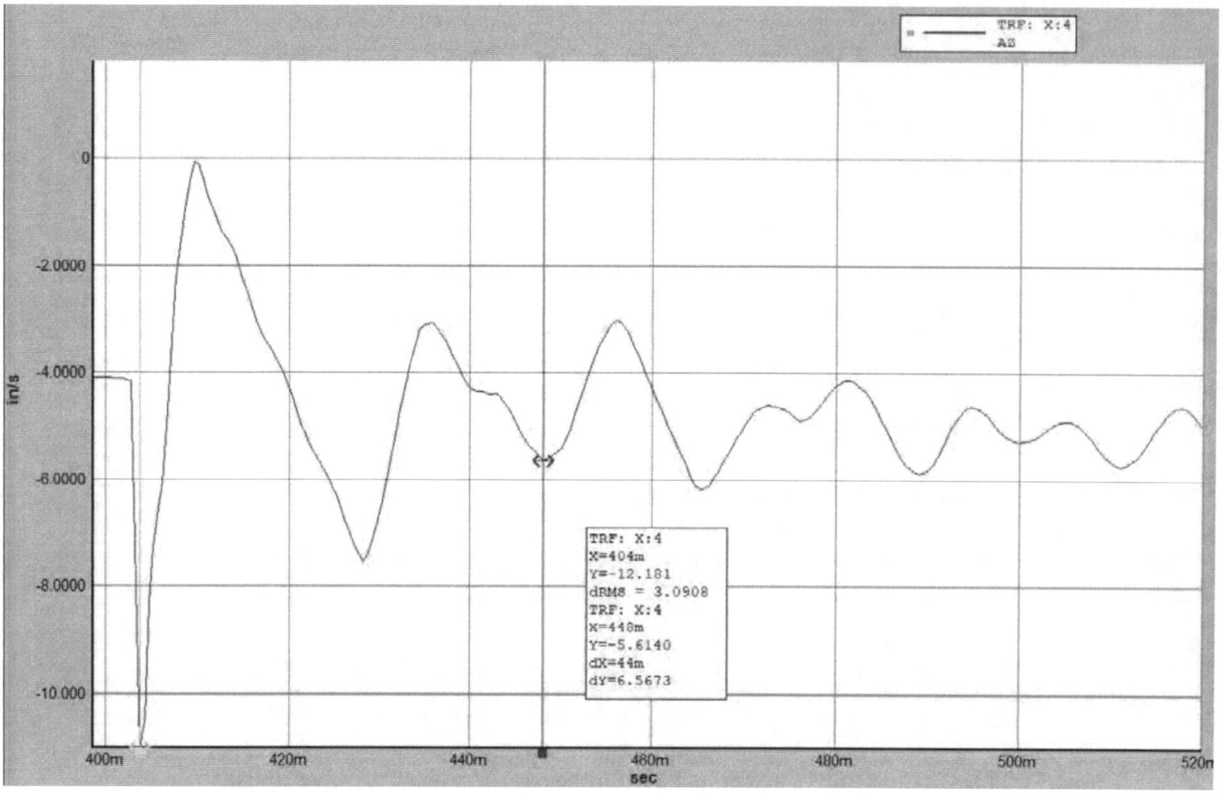

Figure 12 Close-up impact response of shock isolator in its axial direction.

Logarithmic Decrement δ is based on the time waveforms of the impact test response in axial direction. It is given by Equation (5).

$$\delta = \frac{1}{n} ln\left(\frac{X_0}{X_n}\right) \tag{5}$$

where X_0 is the shock response amplitude of the initial cycle to be counted, X_n is the shock response amplitude of Cycle n, which is the total number of the cycles counted. ln is natural logarithm. X_0, X_n, and n can be read directly from the impact response waveforms shown in Figures 9-12, and then Logarithmic Decrement δ can be derived using Equation (5).

With the obtained Logarithmic Decrement δ, system damping ratio ζ is given by Equation (6):

$$\zeta = \frac{1}{2\pi}\delta \tag{6}$$

where a trivial second order term of $\sqrt{1-\zeta^2}$ is often neglected on the right side of Equation (6) for simplicity. The maximum error of calculated damping ratio ζ because of this simplification is 6% overestimated in this research.

Table 2 shows a summary of three cases. Each of the cases uses either acceleration or velocity signature to derive the Logarithmic Decrement δ and then system damping ratio ζ. The fundamental natural frequency is the reciprocal of the cycle period read directly from Figures 8-12. Since the velocity signature shown in Figures 11-12 is derived by an integration of the acceleration measurement shown in Figures 8-10, there is a velocity drift, which is an integration constant (numerical integration error) of -4.0804 in/sec at the first cycle as shown in Figure 11. The start and end velocity amplitudes shown in Table 2 have been corrected approximately with the initial drift of -4.0804 in/sec. Since the second cycle has more accumulated shift as shown in Figures 11-12, finer correction for each cycle has a lower corrected second velocity amplitude and then higher damping ratio would be obtained (not shown in this paper).

Table 2 Results of three cases using logarithmic decrement method.

Impact response signature used	Acceleration	Acceleration	Velocity
Figure showing the impact response	Figure 9	Figure 10	Figure 12
Number of cycles n	1	2	2
Initial shock spike peaking time T_0, second	0.404	0.404	0.404
Initial cycle amplitude X_0, g or in/second	22	22	-8.1006
Cycle n shock spike peaking time T_n, second	0.432	0.453	0.448
Cycle n amplitude X_n, g or in/second	2.6	1.4	-1.5336
Fundamental natural frequency f_n, Hz	35.71	40.82	45.45
Logarithmic Decrement δ	2.136	1.377	0.832
Damping ratio ζ	0.340	0.219	0.132

As shown in Figure 5 and zoomed-in Figures 8-10, the impact response amplitude decays 88% off (22g => 2.6g) in first cycle and then the rest of 12% slowly decays away. It matches with the design of the progressive spring-damper mechanism of the patent-pending shock isolator. The damping of the progressive spring-damper mechanism includes hydraulic inertia damping (both bidirectional and unidirectional), viscous damping, and Coulomb friction damping. It progressively increases as system compression progresses.

For a system with a dominant natural frequency, its damped system natural frequency ω_d is given by Equation (7).

$$\omega_d = \sqrt{1-\zeta^2}\,\omega_n \tag{7}$$

where ω_n is the undamped natural frequency.

Equation (7) shows, for a given undamped natural frequency ω_n, the higher the system damping ratio ζ, the lower the damped system natural frequency ω_d will be. Since cycle period is the reciprocal of damped system natural frequency, the higher the system damping ratio ζ, the longer the cycle period will be correspondingly. This theory explains the observations of the reducing time periods shown in Figures 8-12 and the derived results summarized in Table 2.

As quantified in Table 2, the system fundamental natural frequency retrieved from the acceleration signature of the first two impact response cycles is 40.82 Hz, which is higher than that (35.71 Hz) derived from the first impact response cycle. System damping ratio calculated follows the opposite trend: the system damping ratio retrieved from the first two impact response cycles is 0.219, which is lower than that (0.340) from the first impact response cycle. These observations match exactly with the definition of damped natural frequency by Equation (7).

In summary of the impact test results of the new shock isolator with progressive damping, the initial response amplitude is high, so is the damping ratio and then the damped natural frequency is low and corresponding cycle period is long. As time goes, response amplitude reduces significantly, so does the damping ratio and then natural frequency increases and cycle period decreases.

In addition, Figure 8 shows the first maximum acceleration and deceleration. Table 3 shows them and the ratio between the first maximum acceleration and deceleration.

Table 3 First maximum acceleration and deceleration and their ratio.

First acceleration peaking time T_A, second	0.403
First acceleration amplitude X_A, g	-41
First deceleration peaking time T_D, second	0.404
First deceleration amplitude X_D, g	22
Ratio, $\left\|\frac{X_A}{X_D}\right\|$	1.864

4.2 Discussions

For linear or reasonably linear system, whose natural frequency is not a function of time, its frequency response is a property solely of itself and independent of excitation (or stimulus). However, for nonlinear system, its frequency response is not only a property of itself but also the excitation (or stimulus) applied. The results vary with the amplitude and frequency of the excitation. As what Reference [6] said, "The frequency response of linear networks has proven to be a very powerful tool and so naturally people have tried to extend it to non-linear analysis, particularly since other nonlinear analysis tools have proved intractable." Different test signals with dynamic signal analyzers are "best" (optimum) in certain senses. The linear or non-linear "network" in Reference [6] is referred as to linear or nonlinear system in this research.

In signal analysis, Fourier Transform is used to transfer a signal from its time domain to frequency domain. The inverse Fourier Transform can inversely transfer the signal back to time domain from its frequency domain. Equation (8) is the equation of the Fourier Transform and Equation (9) is the equation of the inverse Fourier Transform.

$$F_x(f) = \int_{-\infty}^{\infty} x(t)e^{-i2\pi ft}dt \tag{8}$$

$$x(t) = \int_{-\infty}^{\infty} F_x(f)e^{i2\pi ft}df \tag{9}$$

where $i = \sqrt{-1}$ is the imaginary unit, $e^{-i2\pi ft} = \cos cos\, 2\pi ft \; - i \sin sin\, 2\pi ft$ and $e^{i2\pi ft} = \cos cos\, 2\pi ft \; + i \sin sin\, 2\pi ft$ are the applications of Euler Formula, t is the time variable, f is the frequency variable, $x(t)$ is a time-dependent signal, $F_x(f)$ is the frequency domain representation — spectrum of the signal $x(t)$. There are four governing theorems of Fourier Transform: Existence of the Fourier Transform, Linearity of the Fourier Transform, Fourier Transform of the Derivative of signal $x(t)$, and Convolution Theorem [7]. These four governing theorems are used extensively in dynamic signal analyzers with dual channels to compute transfer functions and coherences between different signals based on power spectrum and cross power spectrum.

Whether a signal $x(t)$ is continuous or not, to computer Fourier Transform of the signal $x(t)$ digitally, the signal $x(t)$ must be sampled and recorded discretely within a defined time record no matter how frequently the data sampling is. Therefore, the digitalized signal $x(t)$ and its derived Discrete Fourier Transform approximate the actual signal and its spectrum although more frequent sampling results in more accurate digital representation of the actual signal (often continuous or piece-wise continuous). Equation (10) presents a simplified Discrete Fourier Transform $S_x(m\Delta f)$.

$$S_x(m\Delta f) \approx \frac{T}{N}\sum_{n=0}^{N-1} x(n\Delta t)e^{\frac{-i2\pi mn}{N}} \quad \text{for } m = 0, \pm 1, \pm 2, \cdots, \pm M - 1 \tag{10}$$

where T is the length of time record, N is the total number of samples in the time record T, M is the number of spectrum lines, Δt is time interval between time samples, Δf is the frequency spacing between lines.

With the obtained spectrum by Equation (10) and based on the Convolution Theorem of Fourier Transform, power spectrum G_{xx} and cross power spectrum G_{xy} are defined by the next two equations correspondingly.

$$G_{xx}(f) = S_x(f)S_x(f) \tag{11}$$

$$G_{yx}(f) = S_x(f)S_y^*(f) \tag{12}$$

where $S_x(f)$ is the spectrum of signal $x(t)$ that is often an input signal in application, $S_y(f)$ is the spectrum of signal $y(t)$ that is often a response signal induced by the input signal $x(t)$, and * means the complex conjugate of the function.

With the power spectrum $G_{xx}(f)$ and cross power spectrum $G_{xy}(f)$ given by Equations (11 and 12), the transfer function $H_{yx}(f)$ from input signal $x(t)$ to output signal $y(t)$ and coherence $\gamma_{yx}^2(f)$ between them are defined by Equation (13) and Equation (14) correspondingly [6].

$$H_{yx}(f) = \frac{\overline{G_{yx}(f)}}{\overline{G_{xx}(f)}} \tag{14}$$

$$\gamma_{yx}^2(f) = \frac{\overline{G_{yx}(f)}\,\overline{G_{yx}^*(f)}}{\overline{G_{xx}(f)G_{yy}(f)}} \tag{15}$$

where $\overline{\ }$ denotes the *average* of the function and * means the complex conjugate of the function. The important fact that the derived transfer function is the ratio of the *averages* of power and cross power spectrums explains that the natural frequency and damping ratio retrieved from the transfer function plots through either half-power or phase shift method (e.g., the results shown in Table 1) are an average over the whole signal record. For linear system with constant natural frequencies and damping ratios, averaging or not does not make any difference. But for a nonlinear system with rate-dependent stiffness and damping, there will be difference between the parameters derived from transfer function (e.g., the results shown in Table 1 by phase shift method) and those derived from time waveforms directly (e.g., the results shown in Table 2 by logarithmic decrement method).

Fast Fourier Transform (FFT) is a fast algorithm to compute the discrete Fourier Transform (DFT) shown in Equation (10-15). Since it is not plausible to transform the data in a continuous manner, FFT transforms a finite and digitized time domain data sample into frequency spectrum while assuming the data before and after the data sample are continuous (i.e., repeating throughout time [6]). In short, as a fast approximation, FFT only provides limited samples of Fourier Transform and transforms a finite time record of the data. Just like sweep filter analyzer, FFT is treated as a batch processer and every data point in the data sample buffer is important. If the different frequency regions of a spectrum were to be accurate in their relative amplitude, then the amplitude and frequency of the signal must not vary much over the sample time [6, 8]. If system frequency changes, there will be a "fat" peak shown up in the spectrum [8]. How "fat" the peak is depends on how far the signal frequency varies.

The time waveforms with decreasing period (i.e., increasing frequency) of the impact test results shown in Figures 8-12 and the time-dependent hysteresis loop of the compression test results shown in Section 2 have proven that the progressive spring-damper mechanism built in the shock isolator is a nonlinear system. The decreasing time period (i.e., the increasing natural frequency) comes along with a decreasing cycle amplitude, which transfers into a deceasing damping ratio by the progressive spring-damper mechanism of the new shock isolator.

For the new shock isolator with a progressive spring-damper mechanism built in, the higher the initial displacement amplitude, the higher the initial system damping ratio is, and correspondingly the lower the damped system natural frequency is (in other words, the longer the response cycle period is). Since the transfer function is an average by its definition per Equation (14), it explains why, in Section 4.1, the fundamental system natural frequency read from transfer function plots (Table 1) is an average that is higher than those read from the first one or two cycles of time waveforms (Table 2), and the damping ratio read from transfer function plots (Table 1) is also an average while lower than those read from the first one or two cycles of time waveforms (Table 2). The initial high damping (first cycle or cycles) is critical for the shock isolation performance of the shock isolator, which isolates extremely high shock (up to 1000g with the duration of 0.5 milli-second) with high recurring frequencies (e.g., 47 counts per second). Therefore, for the nonlinear shock isolator system, the phase shift method used to evaluate system damping ratio based on the frequency response in this research is only taken as a reference. It is not as accurate as the logarithmic decrement method, which is based on the raw time waveforms.

5. Conclusions

For a shock isolator with progressive damping, the period of the impact response in time domain decreases continuously, the damped natural frequency increases correspondingly, and the system damping ratio decreases continuously along with the decreasing cycle amplitude.

Logarithmic decrement method, which is based on the time domain data, can be used to evaluate the transient damping ratio of a nonlinear system and how the damping ratio progresses. So is the transient damped natural frequency.

Per the definition of transfer function, phase shift method based on the derived transfer function plots reads an averaged damped natural frequency and damping ratio over the whole-time record. For the shock isolator with progressive damping, the averaged damped natural frequency retrieved from transfer function plots is higher than those retrieved from the first cycle or cycles of the time waveforms; the averaged damping ratio retrieved from transfer function plots is lower than those retrieved from the first cycle or cycles of the time waveforms using logarithmic decrement method.

For a nonlinear system, it has been shown that logarithmic decrement method is more suitable to evaluate the transient system damping ratio and damped natural frequency.

Acknowledgements

The author wishes to thank MS Directional (a Patterson-UTI Energy company) Engineering Department for their help throughout the compression and impact tests, and for permission to share the important scientific findings during the tests.

References

[1] Wang, J.K., *"Recommended Practices for a Bump Test"*, Vibrations, Vol 28, No 3, September 2011.

[2] Wang, J.K., *"Virtual Prototyping plus Physically Tested Validation Delivers State-of-the-Art Shale Shakers"*, Proceedings of NAFEMS World Congress 2011, Boston, USA, May 23-26, 2011.

[3] Baxter, N.L., *"Empirical Comparisons of Damping Equations"*, Advanced Vibration Analysis, Vibration Institute, 2006.

[4] Eshleman, R.L., *"Resonance and Critical Speed Testing"*, Machinery Vibration Analysis: Diagnostics, Condition Evaluation, and Correction, Vibration Institute, 2002.

[5] Wang, J.K., et. al., *"Jointly Developed Novel Torque Transducer Ensures Jack-up Rig Integrity",* SPE-176480-MS, Proceedings of SPE/IATMI Asia Pacific Oil & Gas Conference and Exhibition, Bali, Indonesia, October 20-22, 2015.

[6] Hewlett-Packard, The Fundamentals of Signal Analysis, Application Note 243, 1994.

[7] Kreyszig, E., et. al., "Chapter 11 Fourier Analysis", Advanced Engineering Mathematics, 10th Edition, John Wiley & Son, 2011.

[8] Frarey, J.L., *"Fourier Transform — The Door to Frequency Domain Analysis"*, Advanced Vibration Analysis, Vibration Institute, 2006.

OPTIMIZING VIBRATION PROTECTION SYSTEM TO REDUCE EQUIPMENT SPURIOUS TRIP

Author: Juhyung Yoo

VIATC TRAINING CONFERENCE & EXPO
NEWPORT NEWS, VIRGINIA | AUGUST 6-8, 2025

VIATC TRAINING CONFERENCE & EXPO
NEWPORT NEWS, VIRGINIA | AUGUST 6-8, 2025

Optimizing Vibration Protection System to Reduce Equipment Spurious Trip

Juhyung Yoo
Sr. Maintenance Engineer, Saudi Aramco

Juhyung Yoo P.E. has over 18 years of experience in sound and vibration measurement and analysis of rotating equipment. He is Sr. Maintenance Engineer with Saudi Aramco where he is responsible for conducting vibration analysis and field vibration testing for rotating and stationary equipment. Juhyung is a registered professional engineer in the States of Texas and is a certified ISO Category IV Vibration Analyst, Motion Amplification Analyst, Ultrasound Level 1, MLA Level I, and Project Management Professional.

Abstract

Unexpected machine shutdowns caused by faulty vibration sensors and related systems can be both costly and disruptive. This paper presents a method to reduce such false shutdowns by implementing dual (2oo2) and multi (2oo4) voting logics. We discuss the drawbacks of single voting (1oo2) systems, illustrate how dual voting help prevent spurious trips, and explain how multi-voting addresses severe directional vibration issues. Additionally, we discuss machine selection criteria and various system improvements (e.g. latching relocation, BN 3500 rack configuration). After implementing these changes on the equipment, zero spurious trips has been reported and it has enhanced reliability and plant uptime.

Key Words: AND/TRUE voting, API 670, BN3500 Configuration, Faulty Probe System, Latching, 'Not OK' Alarm, Spurious (Unexpected) Trip, Vibration Protection System, Voting Logic (Single, Dual, Multi)

1. Introduction

In industrial plants, rotating equipment such as pumps, compressors, and turbines use vibration sensors for protection. A **spurious trip** occurs when a faulty sensor reading falsely indicates high vibration, causing an unnecessary trip (shutdown). These false trips can result in production loss, unplanned maintenance, and operational inefficiencies.

Many plants have adopted single voting logic (1oo2), which shuts down a machine if **one** of two sensors detects high vibration above the trip setpoint. Although this approach is protective, it also makes the systems prone to false trip from only one faulty sensor. This paper proposes and demonstrate dual (2oo2) and multi-voting (2oo4) strategies, along with several design improvements, to reduce false shutdowns while maintaining strong equipment protection.

2. Problem Statement

2.1 False Trips from Sensor Malfunctions

Many plant shutdowns classified as "Vibration Trips" were, in fact, triggered by sensor or instrumentation malfunctions rather than true high vibration caused by mechanical defects. Figure 1 shows an example distribution of vibration-related trips, where instrumentation issues dominate.

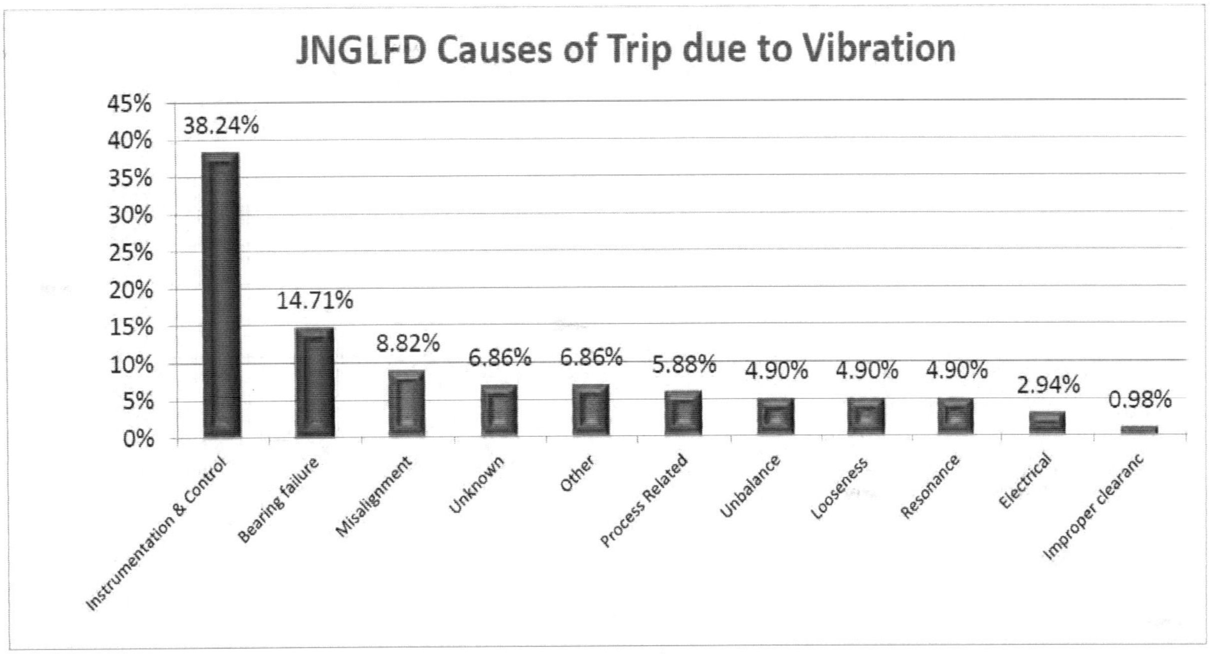

Figure 1: Vibration-Related Trips in the Plant (for the past 5 years)

2.2 Impact of Unnecessary Shutdowns

A critical incident in 2022 involved a machine trip due to a failed vibration sensor. Although the plant avoided major losses thanks to equipment redundancy, the unplanned shutdown could have led to significant production loss. This incident underscored the importance of reliable vibration protection strategy, as single defective sensor could cause an entire unit to shutdown unexpectedly.

Table 1: Trip History Caused by Vibration Probe Malfunction in Critical Equipment in 2022

Equipment	Spurious Trip Date	Cause of Trip
R84-K-101	06/19/2022	High Vibration Vib. Probe Malfunction
R93-K-251	03/11/2022	High Vibration Vib. Probe Malfunction
R93-K-301	03/10/2022 04/11/2022	High Vibration Vib. Probe Malfunction
R93-K-401	03/04/2022	High Vibration Proximity Cable Looseness

2.3 Considering 2oo2 Voting Logic

After reviewing false trip incidents, we recognized that default single voting (1oo2) configuration made equipment vulnerable to spurious trips caused by a single defective sensor system. We therefore explored dual voting (2oo2) logic, which mandates both sensors detect high vibration before triggering a trip. This approach can significantly reduce the risk of unnecessary trips stemming from one sensor's malfunction.

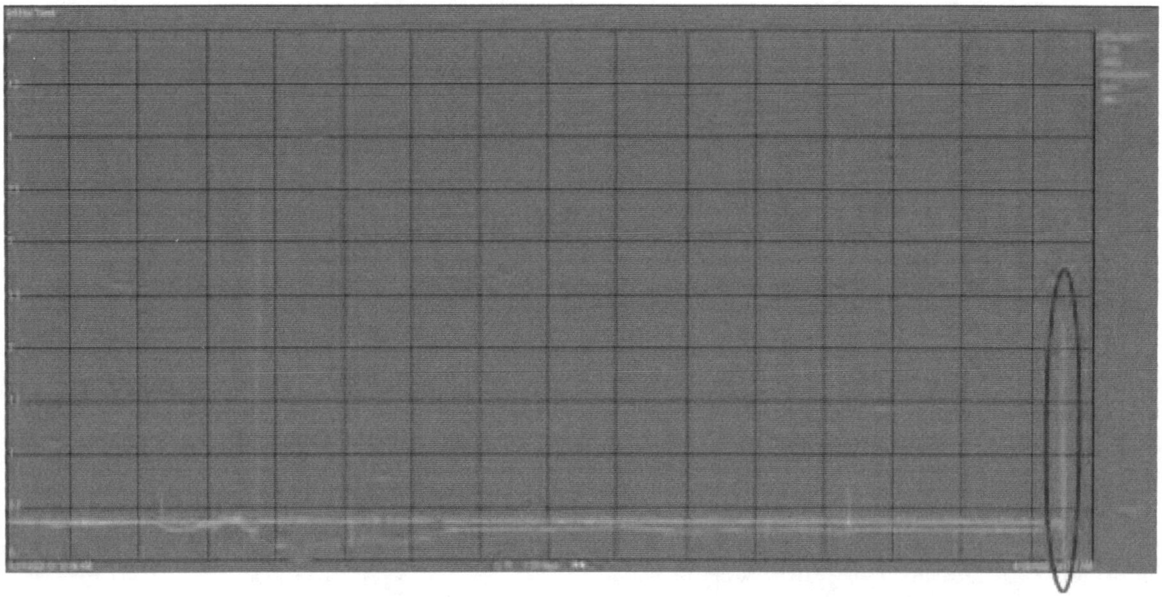

Figure 2: False Trip caused by a Defective Sensor

However, while 2oo2 addresses many false trip concerns, it may still miss highly directional vibration that strongly affects only one sensor. This is particularly an issue when vibration occurs strongly in one direction (e.g., horizontal or vertical), while remaining within acceptable limits on the other axis. In real scenarios, highly directional vibration (showing severe elliptical orbit shapes) can arise from various factors such as preload, misalignment, and/or certain bearing designs. If only one sensor reads this high vibration while the other remains within the trip setpoint, a 2oo2 voting logic might keep running despite real vibration problem. Consequently, we realized that a more robust approach might be necessary for equipment susceptible to these directional vibration issues.

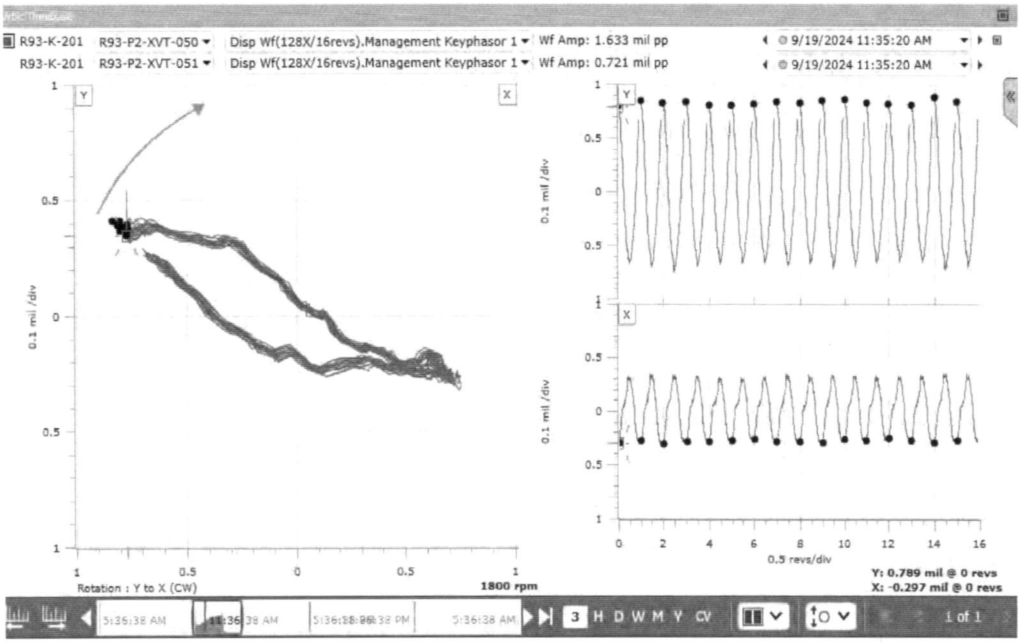

Figure 3: Elliptical Orbit Shape

2.4 Extending to 2oo4 Voting Logic

Because of limitation of 2oo2 in highly directional scenarios, we investigated 2oo4 voting logic. Under 2oo4 voting logic, the machine trips only if two out of four sensors detect high vibration. By considering four vibration sensors across different orientations, 2oo4 logic can:

1. Tolerate the failure or false reading of one sensor without trip.
2. Better capture directional vibration that might be missed if only one axis shows abnormal behavior.

2oo4 voting logic provides stronger protection for machines with the issues such as preload, misalignment, etc. that can produce highly directional vibration and prevent unexpected trip caused by vibration sensor malfunction. Below table compares three different voting logics.

Table 2: Comparison of Three Different Voting Logics

	1oo2	2oo2	2oo4
Trip the Machine for Real Vibration	Yes	Some times	Yes
Avoidance of spurious trips (instrument failure)	No	Yes	Yes
Applicability	Regardless of Orbit Shape	Not Applicable to Directional Vibration	Regardless of Orbit Shape

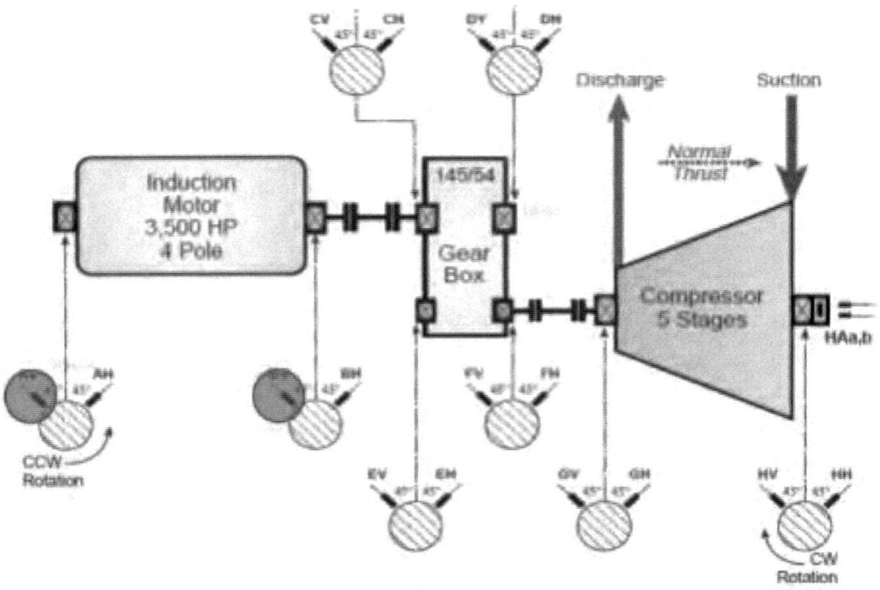

Figure 4: One of Trip Conditions under 2oo4 Voting Logic

3. Study Approach and Equipment Selection

3.1 Internal Standard Confliction

Saudi Aramco's internal documents present varying requirements.

- **34-SAMSS-625** Machinery Protection System (API 670 Based): Recommends dual voting as a standard.

- **SAEP-415** Saudi Aramco Engineering Standard: Requires single voting (1oo2) unless a strong justification is provided.

To reconcile these, we conducted a detailed analysis of each machine, aiming to justify where single, dual or multi-voting is genuinely needed.

3.2 Key Evaluation Factors

- **Production vs. Protection Priority**
  - Production Priority Equipment
    - Unexpected trip causes large financial or production impact
    - Prefer 2oo2 or 2oo4 to minimize false trip
  - Protection Priority Equipment
    - Minimal or no production impact caused by unexpected trip
    - Equipment damage concerns are very high (e.g. Spare part)
    - Maintain 1oo2 voting to protect equipment
- **Dual Voting Applicability**

  Machines must have two vibration probes per bearing to configure dual voting. The machine equipment single vibration probe per bearing maintains single voting logic.
- **Bearing Types**

  Lemon bore or elliptical bearings can produce high directional vibration easily, favoring 2oo4 voting logic.
- **Historic Failure Mode**

  By reviewing failure record of equipment, sort the equipment experienced directional vibration issues and implement 2oo4 voting logic.

3.3 Equipment Excluded from the Study

Table 3: Equipment Out of Scope of Study

Equipment Out of Scope	**Reasons**

Machines with vibration systems other than BN 3300 and 3500	• Many of these systems in our plant are not configurable for dual voting
Machines with a single vibration probe per bearing	• Dual voting considers two (2) vibration probes
Vertical pumps driven by motors	• Vibration measured from motor top bearing is usually late to respond to pump failures • Vertical machines/structures usually have significantly different dynamic properties in the two directions

3.4 Equipment Selection for Dual/Multi Voting Logics

Below picture shows how we select each machine. We considered dual voting applicability, circuit fault alarm ('not ok' alarm) configurability, possibility of directional vibration issue, bearing types, production and protection priority, etc.

Equipment	Tag name	Chennel Type	Voting Logic (Proposed)	Voting Logic (Current)	Dual Voting Applicability	Circuit Fault Alarm Configuration (Yes / No)	Common Alarm(through Rack or Module)	Potential significant elliptical shaft orbit (due to misalignment/preload) (Yes / No)	Bearing Lemon-type design (Yes / No)	VMS System-1 functionality (Yes / No)	Production vs. Machine Protection Priority
R93-KM-101	R93-P1-XVT-050	Radial Vibration	2oo2	1oo2	Applicable	Yes	Yes	No	No	No	Production
	R93-P1-XVT-051	Radial Vibration	2oo2	1oo2		Yes	Yes				
	R93-P1-XVT-052	Radial Vibration	2oo2	1oo2		Yes	Yes				
	R93-P1-XVT-053	Radial Vibration	2oo2	1oo2		Yes	Yes				
R93-KG-101	R93-P1-XVT-054	Radial Vibration	2oo2	1oo2	Applicable	Yes	Yes	No	No	Yes	
	R93-P1-XVT-055	Radial Vibration	2oo2	1oo2		Yes	Yes				
	R93-P1-XVT-056	Radial Vibration	2oo2	1oo2		Yes	Yes				
	R93-P1-XVT-057	Radial Vibration	2oo2	1oo2		Yes	Yes				
	R93-P1-XDT-058	Thrust Position	NA	NA		Yes	Yes				
	R93-P1-XDT-067	Thrust Position	NA	NA		Yes	Yes				
	R93-P1-XGT-066	Acceleration	NA	NA		Yes	Yes				
R93-K-101	R93-P1-XVT-059	Radial Vibration	2oo2	1oo2	Applicable	Yes	Yes	No	No	Yes	
	R93-P1-XVT-060	Radial Vibration	2oo2	1oo2		Yes	Yes				
	R93-P1-XVT-061	Radial Vibration	2oo2	1oo2		Yes	Yes				
	R93-P1-XVT-062	Radial Vibration	2oo2	1oo2		Yes	Yes				
	R93-P1-XDT-064	Thrust Position	2oo2	NA		Yes	Yes				
	R93-P1-XDT-065	Thrust Position	2oo2	NA		Yes	Yes				

Equipment	Tag name	Chennel Type	Voting Logic (Proposed)	Voting Logic (Current)	Dual Voting Applicability	Circuit Fault Alarm Configuration (Yes / No)	Common Alarm(through Rack or Module)	Potential significant elliptical shaft orbit (due to misalignment/preload) (Yes / No)	Bearing Lemon-type design (Yes / No)	VMS System-1 functionality (Yes / No)	Production vs. Machine Protection Priority
R81-KT-108D	R81-90-XVT-839	Radial Vibration	2oo2	1oo2	Applicable	No	Yes	No	No	Yes	Production
	R81-90-XVT-841	Radial Vibration	2oo2	1oo2		No	Yes				
	R81-90-XVT-838	Radial Vibration	2oo2	1oo2		No	Yes				
	R81-90-XVT-843	Radial Vibration	2oo2	1oo2		No	Yes				
	R81-90-XVT-842	Thrust Position	NA	NA		No	Yes				
R81-GM-103A	R81-93-XVT-117	Acceleration	NA	NA	NA	No	Yes	No	No	Yes	Protection
R81-G-103A	R81-93-XVT-116A	Radial Vibration	NA	1oo2	NA	No	Yes	No	No	Yes	
	R81-93-XVT-116B	Radial Vibration	NA	1oo2		No	Yes				
R81-GM-103B	R81-93-XVT-119	Radial Vibration	NA	NA	NA	No	Yes	No	No	Yes	Protection
R81-G-103B	R81-93-XVT-118A	Radial Vibration	NA	1oo2	NA	No	Yes	No	No	Yes	
	R81-93-XVT-118B	Radial Vibration	NA	1oo2		No	Yes				

Figure 5: Equipment Selection Criteria for Applying Voting Logics

From the eligible set, 24 out of 70 equipment trains were selected. These machines largely comprised high capacity pumps and compressors that could incur major losses if shutdown unnecessarily.

4. Solutions Implemented

4.1 Optimized Dual/Multi Voting Logic

API 670 states that trip relay shall be activated when both vibration channels violate the shutdown (danger) setpoint at the same time (HiHi & HiHi). However, we introduced three conditions to trip a machine as below.

- HiHi & Hi (Hi & HiHi)
- Hi & Not Ok (Not Ok & Hi)
- Not Ok & Not Ok

If sensors show 'Not Ok' and 'Hi', it is considered as a real mechanical issue and the system recognize 'Not Ok' as 'HiHi' alarm to protect equipment. This modified trip logics helps minimize false trips and protect equipment from real machical issues.
Table 4 and 5 show detailed trip condition cases and logics for 2oo2 and 2oo4.

Table 4: Trip Conditions for 2oo2 Voting Logic

	1	2	3	4	5
DE_X	HiHi	Hi	Hi	Not Ok	Not Ok
DE_Y	Hi	Not Ok	Hi Hi	Hi	Not Ok
NDE_X	Hi Hi	Hi	Hi	Not Ok	Not Ok
NDE_Y	Hi	Not Ok	Hi Hi	Hi	Not Ok

The 2oo2 logic configured in relay modules in BN 3500 system is as below.

For DE: $((DE_X\_HH + DE_X\_NO) * (DE_Y\_H + DE_y\_NO)) + ((DE_Y\_HH + DE_Y\_NO) * (DE_X\_H))$

For NDE: $((NDE_X\_HH + NDE_X\_NO) * (NDE_Y\_H + NDE_Y\_NO)) + ((NDE_Y\_HH + NDE_Y\_NO) * (NDE_X\_H))$

Table 5: Trip Conditions for 2oo4 Voting Logic

	1	2	3	4	5	6	7	8	9	10	11	12
DE_X	HiHi	HiHi	HiHi	Hi	Healthy	Healthy	Hi	Healthy	Healthy	Hi	Healthy	Healthy
DE_Y	Hi	Healthy	Healthy	HiHi	HiHi	HiHi	Healthy	Hi	Healthy	Healthy	Hi	Healthy
NDE_X	Healthy	Hi	Healthy	Healthy	Hi	Healthy	HiHi	HiHi	HiHi	Healthy	Healthy	Hi
NDE_Y	Healthy	Healthy	Hi	Healthy	Healthy	Hi	Healthy	Healthy	Hi	HiHi	HiHi	HiHi
	13	**14**	**15**	**16**	**17**	**18**	**19**	**20**	**21**	**22**	**23**	**24**
DE_X	Not Ok	Not Ok	Not Ok	Hi	Healthy	Healthy	Hi	Healthy	Healthy	Hi	Healthy	Healthy
DE_Y	Hi	Healthy	Healthy	Not Ok	Not Ok	Not Ok	Healthy	Hi	Healthy	Healthy	Hi	Healthy
NDE_X	Healthy	Hi	Healthy	Healthy	Hi	Healthy	Not Ok	Not Ok	Not Ok	Healthy	Healthy	Hi
NDE_Y	Healthy	Healthy	Hi	Healthy	Healthy	Hi	Healthy	Healthy	Hi	Not Ok	Not Ok	Not Ok
	25	**26**	**27**	**28**	**29**	**30**						

DE_X	Not Ok	Not Ok	Not Ok	Healthy	Healthy	Healthy
DE_Y	Not Ok	Healthy	Healthy	Not Ok	Not Ok	Healthy
NDE_X	Healthy	Not Ok	Healthy	Not Ok	Healthy	Not Ok
NDE_Y	Healthy	Healthy	Not Ok	Healthy	Not Ok	Not Ok

The 2oo4 logic configured in relay modules in BN 3500 system is as below. Under 2oo4 voting logic, there are thirty (30) cases to trip a machine and the logic looks much more complicated than 2oo2 voting logic and difficult to maintain the logic safely. Therefore, special caution is required during and after implementing 2oo4 voting logic to ensure proper operation and maintenance.

$(DE_X\_HH * (DE_Y\_H + NDE_X\_H + NDE_Y\_H)) + (DE_Y\_HH * (DE_X\_H + NDE_X\_H + NDE_Y\_H)) + (NDE_X\_HH * (DE_X\_H + DE_Y\_H + NDE_Y\_H)) + (NDE_Y\_HH * (DE_X\_H + DE_Y\_H + NDE_X\_H)) + (DE_X\_NO * ((DE_Y\_H + DE_Y\_NO) + (NDE_X\_H + NDE_X\_NO) + (NDE_Y\_H + NDE_Y\_NO))) + (DE_Y\_NO * ((DE_X\_H) + (NDE_X\_H + NDE_X\_NO) + (NDE_Y\_H + NDE_Y\_NO))) + (NDE_X\_NO * ((DE_X\_H) + (DE_Y\_H) + (NDE_Y\_H + NDE_Y\_NO))) + (NDE_Y\_NO * (DE_X\_H + DE_Y\_H + NDE_X\_H)$

4.2 Normal AND Voting and True AND Voting

We adopted Normal AND voting, which revers to 1oo1 logic if one probe is 'Not Ok'. This approach prevents an out-of-service (or malfunctioning) sensor from disabling the trip function entirely. In contrast, True AND voting would keep dual voting active even with a faulty sensor, potentially missing the 'Not Ok & Hi' case.

4.3 Latching Configuration Improvements

Originally, all vibration probes were configured with latching, causing them to remain in an alarm state until manual reset. This configuration allowed engineers to recognize and track the equipment's alarm status and aligned with API 670 as well as internal standard. However, during one incident involving a steam turbine, a vibration probe on the NDE bearing detected a high vibration spike and triggered an alarm. Before operators could manually reset it, another vibration probe on DE bearing detected high vibration activating trip relay and trip the steam turbine. An investigation revealed a defected rack module as the primary factor behind this false trip. Additionally, it became clear that the existing latching configuration could be one of the factors of false trip. Consequently, probe level latching was removed and latching is currently only applied at the relay to minimize spurious trips.

4.4 BN Rack Module Reconfiguration

Vibration probe connections were changed from a **slot-based** arrangement (all probes of one machine in a single slot) to a **channel-based** arrangement (each slot has probes from different machines). Under dual and multi voting logic, this modification allows users to replace faulty probe or module without equipment shutdown or bypassing. Therefore, this arrangement significantly improves maintenance efficiency and safety.

Figure 6: Bently Nevada 3500 Rack

Slots

		Motor	Gear LS	Gear HS	Comp.	Thrust
		2	3	4	5	6
Channel	1	NDE_X	NDE_X	NDE_X	NDE_X	A
	2	NDE_Y	NDE_Y	NDE_Y	NDE_Y	B
	3	DE_X	DE_X	DE_X	DE_X	
	4	DE_Y	DE_Y	DE_Y	DE_Y	

Figure 7: Slot-Based Arrangement

			Slots				
			2	3	4	5	6
Channel	Motor	1	NDE_X	DE_X	NDE_Y	DE_Y	A
	Gear LS	2	DE_X	NDE_X	DE_Y	NDE_Y	
	Gear HS	3	NDE_X	DE_X	NDE_Y	DE_Y	
	Comp.	4	DE_X	NDE_X	DE_Y	NDE_Y	A

Figure 8: Channel-Based Arrangements (Right)

5. Conclusion

This paper outlined how transitioning from a 1oo2 single voting system to optimized dual (2oo2) and multi (2oo4) voting drastically reduced false trip caused by faulty vibration signals. Major takeaways include:

- Balancing Production Priority and Protection Priority: High production-loss related equipment benefits from dual (2oo2) / multi voting (2oo4); low production-impact but high maintenance risk equipment stays on single voting (1oo2)

- Normal AND Voting and 'Not OK' Configuration: Effective at minimizing spurious trips from defected probes.

- Relay-Level Latching: Eliminates spurious trip caused by stacked sensor alarms.

- BN Rack Reconfiguration: Allows straightforward maintenance without risking equipment shutdown.

As a result, these innovative approaches have yielded zero spurious trips for the equipment, reinforcing both operational continuity and equipment safety.

Acknowledgements

The author would like to thank Vibration Institute Annual Training Conference Advisory Committee for the opportunity to present the paper. Thanks especially to Juaymah NGLF Department (Saudi Aramco) management for their supports to implement optimized voting system, as well as reliability unit, automation unit, maintenance and operation team members who contributed valuable data and insights during study and implementation.

References

[1] API Standard 670 (5th Edition), *"Machinery Protection System"*, American Petroleum Institute, Washington, D.C., November, 2014

[2] 34-SAMSS-625 (Material System Specification), *"Machinery Protection System"*, Instrumentation Standards Committee, Saudi Aramco, October 20, 2022

[3] SAEP-415 (Engineering Procedure), *"Vibration Limits for Existing Rotating Machinery, Piping, and Structure"*, Vibration and Lubrication Standards Committee, Saudi Aramco, July 5, 2022

[4] Robert C. Eisenmann, Sr. and Robert C. Eisenmann, Jr., *"Machinery Malfunction Diagnosis and Correction: Vibration Analysis and Troubleshooting for the Process Industries"*

[5] Bently Nevada 3500 Monitoring System Training Manual

AN OVERVIEW OF LEVEL 1 PIPING VIBRATION CRITERIA

Authors: Gyorgy Szasz, PhD, Scot McNeill, PhD, PE

NEWPORT NEWS, VIRGINIA | AUGUST 6-8, 2025

AN OVERVIEW OF LEVEL 1 PIPING VIBRATION CRITERIA

Gyorgy Szasz, PhD
Stress Engineering Services, Inc.
13610 Westland East Blvd, Houston, TX 77041
gszasz@stress.com

Scot McNeill, PhD, PE
ExxonMobil
22777 Springwoods Village Pkwy, Spring, TX 77389
scot.i.mcneill@exxonmobil.com

Dr. Szasz has over 25 years of experience with hands-on vibration testing, measurement and analysis. He has performed an extensive amount of field work globally in various power plants (including hydro, fossil and nuclear), refineries, pipeline stations, LNG facilities, offshore platforms, etc. He is a VI certified category IV vibration analyst, member of ASME, president of the VI board of directors, and a former member of the ASME Technical Committee on Vibration and Sound. He received his Ph.D from Auburn University.

Dr. McNeill is a lead Noise and Vibration Engineer and SME at ExxonMobil in Houston, Texas. He has extensive experience with flow induced vibration of fixed equipment in upstream, midstream and downstream applications. Scot has widely contributed to udating industry standards, such as: the Energy Institute Guidelines, API 579, and EPRI MRP-482. Scot received his B.S. and M.S. degrees in engineering mechanics from the University of Wisconsin at Madison and his Ph.D. in mechanical engineering from the University of Houston.

ABSTRACT:

Piping vibration poses significant challenges in industrial systems, potentially leading to fatigue failures and operational hazards. Level 1 assessment criteria serve as preliminary screening tools to identify vibration issues before conducting detailed analyses. This paper provides an overview of piping vibration screening criteria (later referred to as Level 1 piping vibration criteria), including those outlined in ASME OM-3, the Energy Institute Guidelines, the API 579-1/ASME FFS-1 Part 15 draft, and the EPRI MRP-482 report. Ideally, the criteria from the various sources would converge over time to a unified set. While progress has been made toward this goal, it has not quite been reached yet. The paper aims to familiarize vibration analysts with currently available criteria for piping vibration assessment.

KEY WORDS: Piping vibration, Level 1 assessment, vibration screening, ASME OM-3, API 579-1/ASME FFS-1 Part 15, Energy Institute Guidelines, EPRI MRP-482.

Historical Overview

Before the formalization of screening criteria, engineers relied heavily on experience and qualitative judgment. Design changes were often reactive, based on failures. Vibration problems were usually noticed after leaks, fatigue failures, or operational complaints.

By the 1980s, empirical data and industry failures led to the development of initial screening methods, but many of these were region specific (USA, Europe, Russia). All of these methods were targeting bending mode vibration, shell mode vibration was not considered.

In the US, Buddy Wachel was a pioneering engineer and consultant in the field of fluid-induced vibration and pulsation analysis, particularly known for his work with reciprocating compressor systems. He was affiliated with Southwest Research Institute (SwRI) and later co-founded Engineering Dynamics, Inc. (EDI), and co-authored several foundational texts and technical papers still cited in industry today. EDI also contributed significantly to the development and application of the API RP 618 which is a source of vibration and pulsation guidelines for reciprocating compressors.

In Europe the United Kingdom Offshore Operators Association (UKOOA) developed piping vibration limits for offshore oil and gas operations and published its "Guidelines for the Avoidance of Vibration Induced Fatigue Failure in Process Pipework" [1]. It introduced vibration velocity RMS versus frequency based screening curves, applicable to pipe spans and small bore connections. In 2007 the UKOOA was dissolved and its technical work (including vibration guidelines) was transferred to the Energy Institute (EI). In 2008 the EI published the 2nd edition of the Guidelines and Shell's Design and Engineering Practice (DEP) adopted its vibration limits.

The screening curves in the 2nd edition of the EI Guidelines, the 'Concern' curve in particular, have proven to have an outsized influence on subsequent screening criterion development. The developers for the 3rd edition of the EI Guidelines, API 579-1/ASME FFS-1 Part 15, and EPRI MRP-482 all heavily referred back to this curve for the updated screening methods.

Wachel Curves (SwRI/EDI) circa 1980

The piping vibration screening curves approximately reproduced in Figure 1 are widely attributed to J.C. (Buddy) Wachel. One of the earliest known publications is the 1981 paper titled "Piping Vibration and Stress," presented at the Vibration Institute's Machinery Vibration Monitoring & Analysis Seminar in New Orleans [1]. While Wachel was the principal figure in their creation, his work was supported by collaborations with colleagues during his tenure at Southwest Research Institute (SwRI) and later at Engineering Dynamics, Inc. (EDI) [2]. Wachel introduced these curves as a practical tool to assess the severity of piping vibrations, particularly in high-energy process systems like refineries and power plants. His methodology provided a graphical means to evaluate vibration levels based on measured vibration amplitude and frequency, aiding engineers in determining the need for further analysis or corrective actions. Wachel's work significantly influenced piping vibration screening curves that followed in later years.

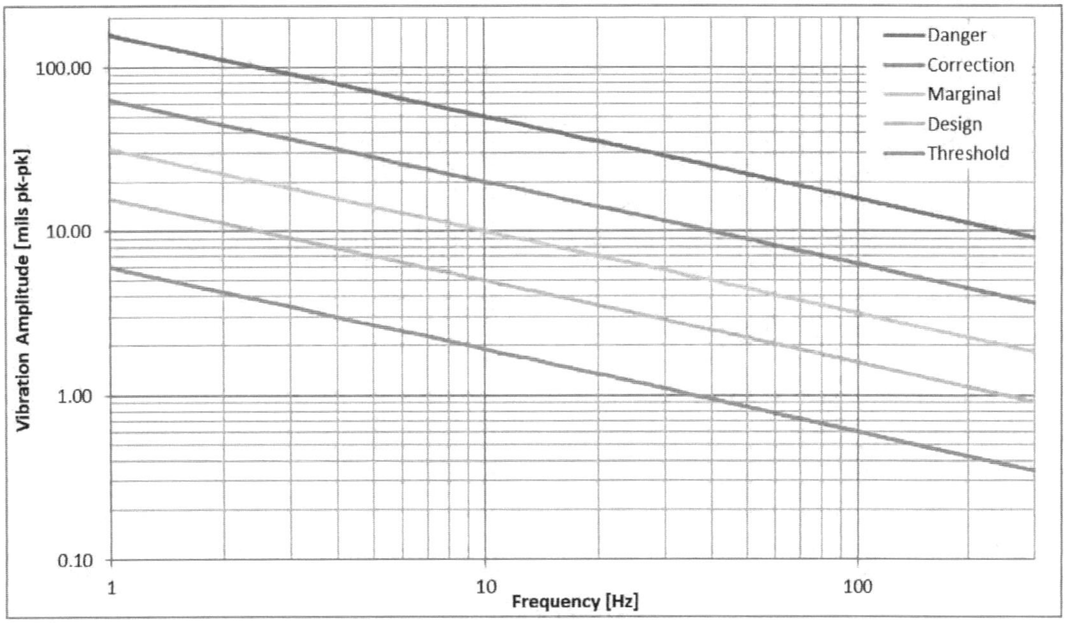

Figure 1: EDI Piping Vibration Screening Curves

ASME OM3 - 1982

A 0.5 in/s (12.7 mm/s) peak (0-peak) vibration velocity screening criterion was first introduced in the original ASME OM-3 standard, titled "Requirements for Preoperational and Initial Start-Up Vibration Testing of Nuclear Power Plant Piping Systems," published on September 30, 1982. This standard provided general requirements for assessing piping system vibration in nuclear power plants, including steady-state and transient vibration testing, acceptance criteria, and recommendations for corrective action when required. The 0.5 in/s peak velocity was established as a conservative threshold to identify piping systems that might require more detailed analysis to ensure structural integrity and operational safety. This criterion, however, proved to be extremely conservative in practice and too many safely operating high energy piping circuits in industry will fail it, forcing the practitioner to move to a higher level of assessment, therefore it is not considered a very practical screening tool. The standard was updated in 2009 [3], but this limit has not been increased.

EdF Piping Vibration Screening Limit - 2005

A 12 mm/s RMS (0.472 in/s RMS) piping vibration criterion was widely adopted by Électricité de France (EdF), and it serves as a conservative screening threshold for assessing piping vibration in nuclear power plants. This benchmark was established through an extensive study applying the ASME ANSI-OM3 standard to 181 different piping configurations. The analysis revealed that 99.7% of these configurations, which met static and seismic design requirements, also exhibited allowable vibration velocities exceeding 12 mm/s RMS. Consequently, the 12 mm/s RMS threshold is considered a reliable and conservative criterion for preliminary vibration assessments in nuclear piping systems. This criterion remains the only true single value piping vibration limit in industry, though it is best applied to circuits exhibiting vibration at 10 Hz or greater. It is noteworthy, the nuclear industry often utilizes stainless steel piping (higher fatigue limit) and has more rigorous welding standards compared to other industries, for Class 1 piping in particular. This criterion was first formally published in 2005 at the ASME Pressure Vessels and Piping (PVP) Conference [4].

Energy Institute Guidelines (2nd Edition, 2008)

The EI guidelines for avoidance of vibration induced fatigue failure in process pipework provide curves for vibration screening assessment of piping systems [5]. The screening curves apply to both the main pipe and small bore connections. The problem and concern level screening curves for vibration velocity in terms of mm/s RMS are provided in the equations below and depicted in Figure 2.

$$V_{Concern} \geq 10^{\frac{(log(f)+0.48017)}{2.127612}} \quad (1), \quad V_{Problem} \geq 10^{\frac{(log(f)+1.871083)}{2.084547}} \quad (2)$$

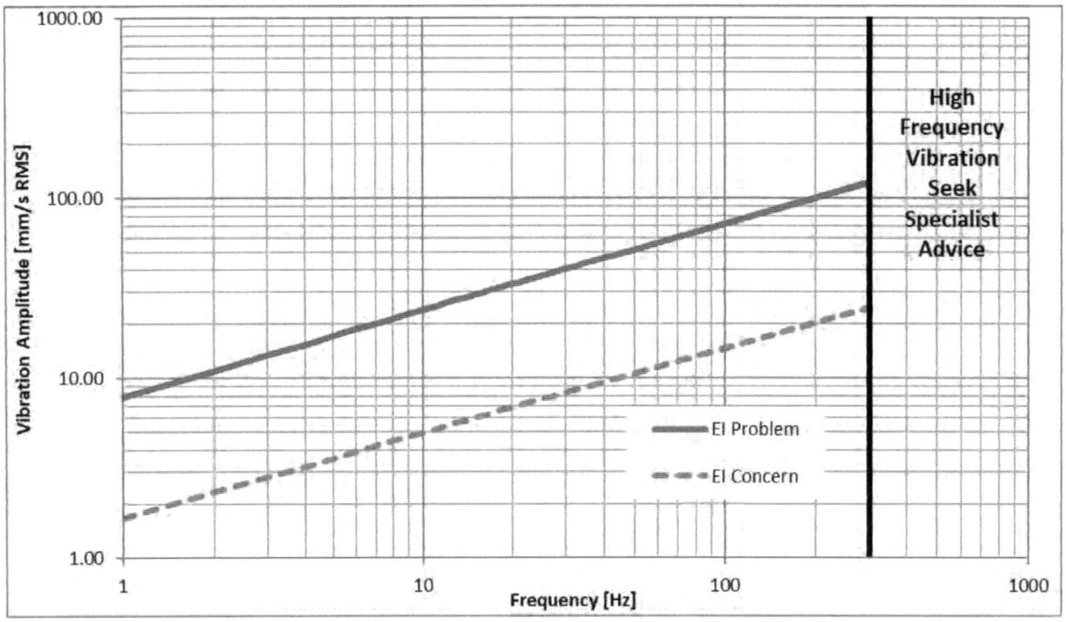

Figure 2: EI Piping Vibration Screening Curves

In the reference, it is stated that the measured RMS levels and peak frequency of the measured response should be used in the assessment. The frequency in equations (1) and (2), f, refers to the dominant peak frequency in Hz. Noting that the finite element analysis (FEA) data that was used to derive the EI screening curves was in the form of overall RMS levels, it stands to reason that the overall RMS velocity along with the frequency of the dominant spectral peak of the measured data should be plotted on the screening curve chart (as a single point) for vibration screening assessment of piping response between 1 and 300 Hz. Further reasoning for this approach can be found in [5].

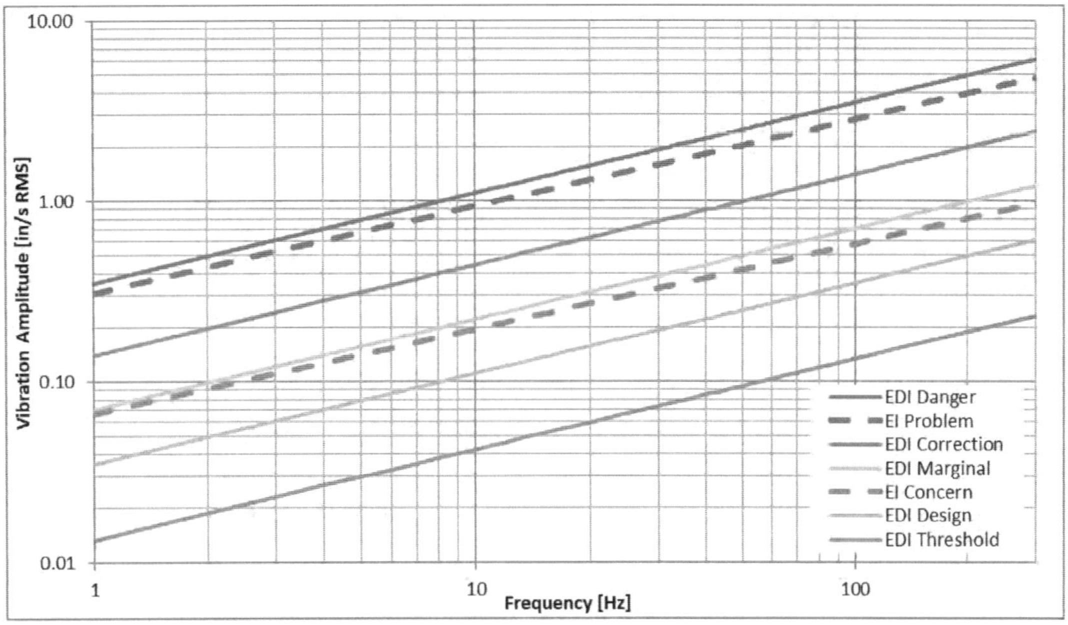

Figure 3: EI vs EDI Piping Vibration Screening Curve Comparison

While at first glance the EI curves look strikingly different from the Wachel/EDI curves, this is largely due to being expressed in different units. When a conversion is made from displacement to velocity and peak-to-peak to RMS (using a perfect sinewave assumption, crest factor = 1.41) the EI curves nest fairly well into EDI curve set as shown on Figure 3.

A methodology for extending the applicability of the EI screening curves to random vibration and multi-mode vibration was published in [6] in 2017.

EPRI Vibration Fatigue Handbook (MRP-482) - 2023

The Electrical Power Research Institute (EPRI) first addressed piping vibrations in their Fatigue Handbook (MRP-235) [7], which was originally intended to cover pressurized water reactor (PWR) piping, but gained acceptance in a wider circle within the power industry. This document was updated and split into three separate volumes in 2023. The volume addressing piping vibration became the Vibration Fatigue Handbook (MRP-482) [8]. The revision included the introduction of the API nomenclature of Level 1, Level 2 and Level 3 assessments with respect to determining fitness for service (FFS). Since MRP-235 did not have any pipe geometry independent screening criteria (Level 1 criteria), all assessments were based on piping geometry

specific calculations. In the FFS sense that process is typically regarded as a Level 2 assessment, therefore a new Level 1 assessment needed to be slotted-in below that.
It would have been convenient to simply adopt the Concern curve from the EI guidelines, however, an opportunity was seen to offer a more useful tool to the practitioner by addressing two issues: (i) in practice it was found that the Concern line tends to be overly conservative at low frequencies, and (ii) the application requires the calculation of a less than straightforward formula. While the single value limit from EdF seemingly addresses these issues, it was not meant to be used for low frequency vibrations, and it becomes quite conservative at high frequencies.

A still relatively straightforward approach was therefore investigated to apply field experience that approximately 1/3, ½ and 1 in/s RMS may be suitable screening limits for low, medium, and high frequency piping vibration respectively. When the concept was evaluated against a population of piping geometries, it required some adjustments, and the values in Table 1 were established. Note that most piping vibration is in fact random in nature (amplitude, frequency, or both) and the limits for equivalent life (as compared to periodic vibration in terms of fatigue damage) are reasonably similar to the original target values. The values for periodic vibration were based on a sinusoidal response as can be deduced from the peak to RMS ratio (crest factor) of square root two.

Table 1 Level 1 Piping Vibration Screening Limits from EPRI MRP-482

Frequency	Peak	RMS (Periodic)	RMS (Random – Equivalent Life)	RMS (Random – Infinite Life)
Hz	in/sec	in/sec	in/sec	in/sec
4-10	1.00	0.71	0.33	0.22
10-100	1.25	0.88	0.42	0.28
100-300	2.50	1.77	0.83	0.56

While these numbers are good for individually assessing piping circuits, algorithmic application to a large set could result in odd results because of the discreet cutoff frequencies between frequency ranges. For example, 1.1 in/s peak would pass at 10.01 Hz but it would fail at 9.99 Hz. To lessen the impact of hard step changes, an exponential smoothing function was applied to the transitions to arrive at the curves shown in Figure 4. These curves are only recommended to be used for algorithmic application or to avoid non-sensical situations as mentioned above.

The new curves are shown to fall in the general range of existing screening curves in Figure 5. The equivalent life limits are recommended for most cases. The significantly more conservative infinite life limits are best suited to most critical applications where the risk tolerance is extremely low, or piping with components resulting in high stress concentrations.

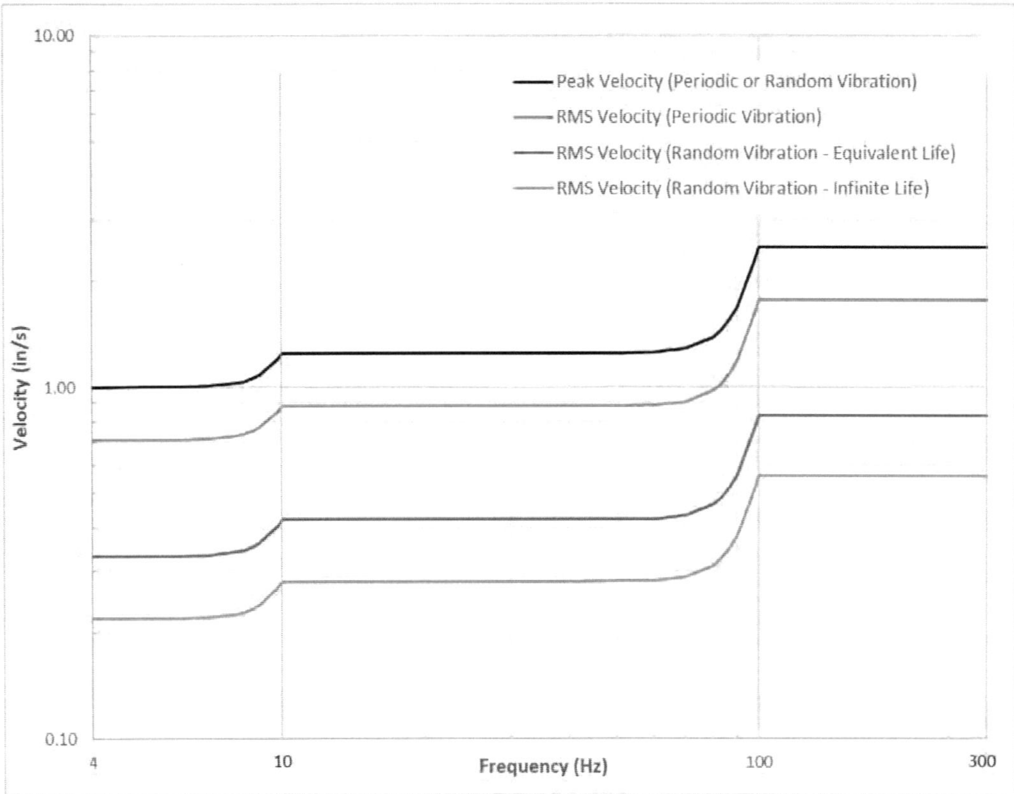

Figure 4: EPRI MRP-482 Piping Vibration Screening Curves

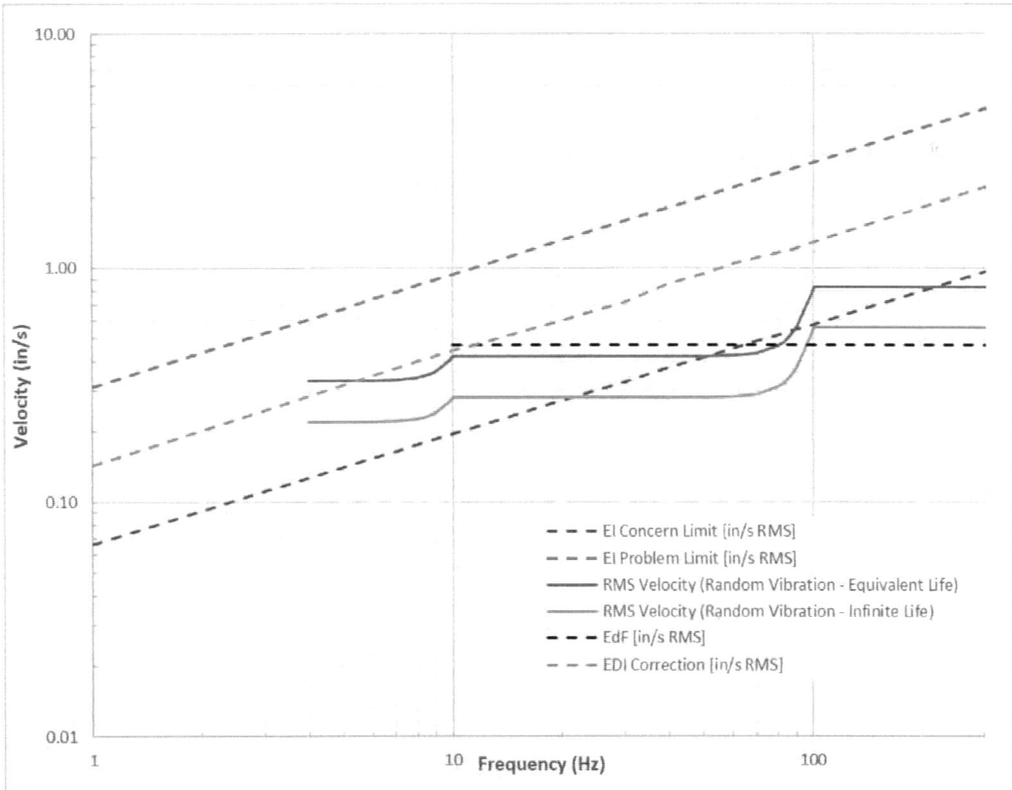

Figure 5: EPRI MRP-482 Piping Vibration Screening Curve Comparisons

API-579-1/ASME FFS-1, Part 15

The new Part 15 of API 579-1/ASME FFS-1 is another attempt to create a universal set of piping vibration guidelines similar to the EI and EPRI documents. While it only exists in draft form and not yet available in the public domain, it is very close to publication. In this document the Level 1 screening criteria will be frequency independent, however, there will be different values provided for different types of piping and fittings. There will be detailed guidelines provided to help the user to determine which screening limit may be applicable to their piping. A preview of the proposed limits is provided in Table 2.

Table 2 Level 1 Piping Vibration Screening Limits from the API-579 Part 15 Draft

Configuration	Overall Vibration Velocity RMS
Main Line – Low Stress Concentration (Lateral Bending). No Concentrated Masses.	22.8 mm/s (0.90 in/s)
Main Line – Low Stress Concentration (Lateral Bending) Including Concentrated Masses.	16.1 mm/s (0.63 in/s)
Main Line – High Stress Concentration (Lateral Bending).	5.0 mm/s (0.20 in/s)
Small Bore Connections (SBCs) in any configuration and all Schedule 10 Main Lines, Threaded Fittings, and Fabricated Tees)	5.0 mm/s (0.20 in/s)
SBC Welded-on Fitting (Branch Butt weld) (Straight Cantilever Configuration)	7.3 mm/s (0.29 in/s)
SBC Welded-on Fitting (Branch Butt weld) (Bend Cantilever Configuration)	10.0 mm/s (0.39 in/s)
SBC Welded-on Fitting (Branch Socket weld) (Straight Cantilever Configuration)	5.2 mm/s (0.21 in/s)
SBC Welded-on Fitting (Branch Socket weld) (Bend Cantilever Configuration)	7.1 mm/s (0.28 in/s)
SBC Welded-In Contour Insert Fitting	9.3 mm/s (0.37 in/s)
SBC Welded-on Contoured Body Fitting	12.0 mm/s (0.47 in/s)
Main Line (Shell Flexural). No Welded Discontinuities, No Branch Connections.	36.0 mm/s (1.42 in/s)
Main Line (Shell Flexural). Welded-On Fitting Branch Connections.	10.0 mm/s (0.39 in/s)
Main Line (Shell Flexural). Fabricated Tee Branch Connections.	18.0 mm/s (0.71 in/s)
Main Line (Shell Flexural). Welded-In Contour Insert Fitting Branch Connections.	21.0 mm/s (0.83 in/s)

Note that since these values are from an unpublished draft they are *for illustration only*, no attempt should be made to use them for engineering assessment. There is also insufficient information provided in this paper to determine the applicability of the numbers provided to the reader's piping circuit. The reader is strongly urged to wait until the API document is finalized and published, and use the information in the published version.

Consolidated Piping Vibration Criteria

As the reader may infer from this overview paper, there are still several piping vibration criteria, and the reality is that it is an objectively difficult process to assign indisputable hard values to the boundaries of acceptability. Regardless of the authors and the publication, decisions must be made during the process of setting up the criteria, and some of these decisions will be inevitably experience based.

For someone who is only interested in developing a high level feel for what is acceptable versus excessive piping vibration, the color scheme in Figure 6 is offered to give an indication of the approximate comfort level of an experienced analyst when first confronted with piping vibration, prior to performing any detailed analysis or assessment. The level of attention will tend to increase as the vibration level moves from green to yellow and red. Vibration levels beyond 1.0 in/s RMS can still be screened as acceptable if high frequency and sinusoidal or shell mode, but Figure 6 is focused on bending mode random vibration, which constitutes the vast majority of piping vibration problems in industry.

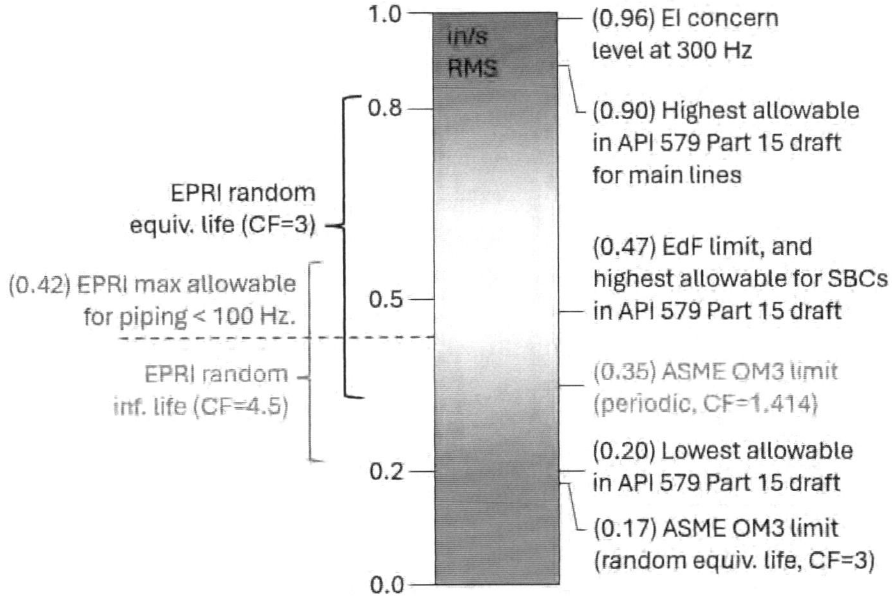

Figure 6 General Piping Vibration Assessment Color Chart

Note that all of the vibration criteria mentioned in this paper assume well behaved systems that exhibit periodic or gaussian vibration with a normal distribution. The limits are not applicable to special cases with high crest factors (CF>5) or high kurtosis (ku>3), for example due to impacting or slug flow.

Conclusions

Level 1 piping vibration assessment criteria serve as vital tools for early detection of potential vibration-related problems. By comparing ASME OM-3, API 579-1/ASME FFS-1 Part 15, EdF, and the Energy Institute Guidelines, this paper highlights the strengths and limitations of each approach. Engineers should consider the specific requirements of their applications when selecting an assessment criterion to ensure effective piping vibration management and system integrity.

Acknowledgements

The authors wish to thank Tom Damiani and Bob McGill at EPRI for supporting their work in this field.

References

1. Wachel, J.C., "Piping Vibration and Stress," https://engdyn.com/wp-content/uploads/2020/04/27-piping_vibration_and_stress_-_jcw.pdf

2. Wachel, J.C., Morton, S. J, Atkins, K.E., "Piping Vibration Analysis" 19th Turbomachinery Symposium, 1990.

3. American Society of Mechanical Engineers, OM-2009 Part 3 – Vibration Testing of Piping Systems. ASME. 2009.

4. Caillaud, S., Pons, Y., Moussou, P., and Gaudin, M. "A 12 MM/S RMS Screening Vibration Velocity for Pipes Using ANSI-OM3 Standard and Regulatory Design Rules," ASME 2005 Pressure Vessels and Piping Conference, 2005.

5. Guidelines for the Avoidance of Vibration Induced Fatigue Failure in Process Pipework. Energy Institute. 2008.

6. Breaux, L., McNeill, S., and Szasz, G., "A Screening Method for Random Piping Vibration" Proceedinsg of the Vibration Institute Annual Training Conference, 2017.

7. EPRI, Fatigue Handbook (MRP-235), 1990

8. EPRI, Vibration Fatigue Handbook (MRP-482): Guidelines for Management of Vibration-Induced Fatigue, 2023.

THE USE OF CURRENT SIGNATURE ANALYSIS FOR ELECTRICAL AND MECHANICAL CONDITION OF ELECTRIC MACHINES AND DRIVEN EQUIPMENT

Author: Howard W Penrose, Ph.D., CMRP, CEM, CMVP

VIATC TRAINING CONFERENCE & EXPO
NEWPORT NEWS, VIRGINIA | AUGUST 6-8, 2025

VIATC | TRAINING CONFERENCE & EXPO

NEWPORT NEWS, VIRGINIA | AUGUST 6-8, 2025

THE USE OF CURRENT SIGNATURE ANALYSIS FOR ELECTRICAL AND MECHANICAL CONDITION OF ELECTRIC MACHINES AND DRIVEN EQUIPMENT

Howard W Penrose, Ph.D., CMRP, CEM, CMVP
President, MotorDoc® LLC
howard@motordoc.com

President MotorDoc LLC, a Veteran Owned Small Business, Chair CIGRE-USNC A1 (motors and generators), IEEE PES grid working group, ACP clean energy systems standards chair, Convenor CIGRE A1.78 (ESA/MCSA project), IEEE PES standards group, NEMA rail electrification standards committee, global leader in ESA/MCSA research and development, certified CMRP, CEM, CMVP, data science/AI/machine learning from Stanford University, University of Michigan, AWS and IBM. Life member MENSA.

Abstract:

A concerted effort over the past decade has been made to explore the full capabilities and limitations of electrical and current signature analysis systems. In this paper we will discuss the use of two different methods of electrical and current signature analysis and the practical study performed in collaboration with repair facilities, equipment OEMs and end users. The results involved a series of methods and evaluation techniques for electrical and mechanical conditions and the identification of primary fault drivers, or cause analysis. Several case studies from the study will be involved including insulation breakdown in an electric machine to ground-noise mapping to identify the cause of shaft currents.

Key Words: MCSA, current signature analysis, RHI, supra-harmonics, reliability

Introduction

Electrical and current signature analysis (ESA/CSA), sometimes misnamed as motor current signature analysis (MCSA) or motor current analysis. However, ISO 20958[1] specifically separates them as electrical and current signature analysis with definitions. ESA is typically three phases of voltage and current while CSA is typically one phase of current, per the standard, and MCSA is not defined. CSA is also defined in IEEE 1068-2015[2] for customer and motor repair, and IEEE 1415-2006[3] (which has been administratively withdrawn) for electric motor troubleshooting. ESA has been defined in the ACP Gearbox Playbook[4] as a method for evaluating bearings and gears in wind turbine gearboxes. The CIGRE A1.78 project[5] was proposed in an effort to identify the full present capabilities of the technology for application in power generation, transmission transformers, and large electric machine systems. EPRI and the US Navy (NAVSEA) have studied the application of MCSA and ESA since 1985 with technologies developed at Oak Ridge National Labs (ORNL).

The original concept of ESA/CSA as envisioned by the ORNL researchers was the application of the air gap of an electric motor or generator as a vibration transducer.[6] The understanding of how the radial and torsional motions effects impact in the same way the magnetic field, and, as a result, the current waveform, was first identified in a 2025 IEEE paper, "The Application of Electrical and Current Signature Analysis…"[7] by this author. This ground breaking paper identified how ESA/CSA technologies can identify the entire powertrain all at the same time.

Since 2020, with the advent of rapid growth of data centers, and their impact on the electrical system initiated studies into supra-harmonics, discussions have arisen around frequencies between 2 kHz and 150 kHz, as outlined in IEC 61000-4-30.[8] Due to specific issues with the values associated with the current values as the frequencies increase no definition for how to manage supra-harmonics exist. As part of a study on the causes of bearing failures in wind turbine generators, neutral/ground systems, and inverter applications, MotorDoc proposed the Relative Harm Index (RHI), which was published at the IEEE Electrical Insulation Conference as, "Evaluation of Wind Turbine Electrical and Mechanical Defects Due to Nacelle Circulating Currents,"[9] which is the result of almost 5 years of field research. The paper also identifies that the issue is a common problem in commercial and industrial applications.

About ESA and CSA

The use of the magnetic field is effectively the same as using a vibration transducer to detect acceleration, velocity and displacement. However, the standard for analyzing the data is to view it in decibels down from 0 dB, represented as -dB levels, that are in peak pairs. The reason is that the carrier frequency is required in order to perform a condition analysis versus systems, such as vibration, where you are counting 'up' from a zero value. The information is then contained in both the positive and negative sides of the sine wave (Figure 1) as sub frequencies. In most instances these values are identical at the positive and negative sides of the sine waves resulting in similar peaks on either side of line frequency.

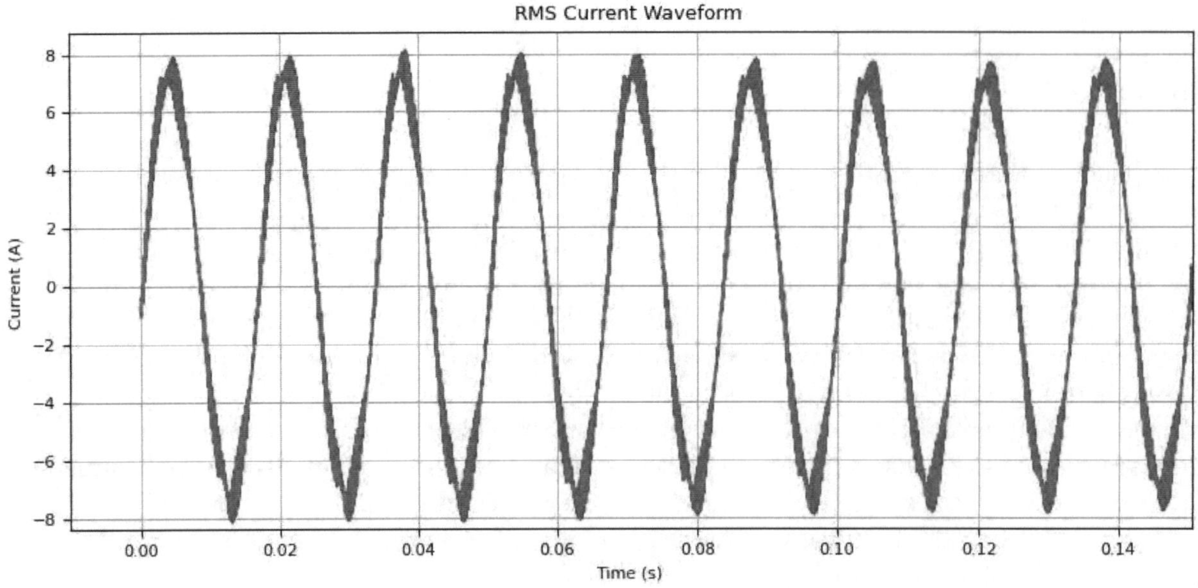

Figure 1: One phase of current waveform with sub frequencies.

While Figure 1 is of current, voltage is managed in the same manner. Other values that can be pulled including torque and watts are calculations based on the RMS values resulting in singular peaks. Filters, averaging, bin sizes, fmax, and resolution are similar to vibration. However, it is more important to ensure that the resolution is good enough as all conditions and components in the machine and driven equipment are shown in the FFT spectra, as well as power quality issues in ESA, at the same time.

For the purposes of this paper we will not go into the details of ESA/CSA theory which are well covered in the previously described literature. Instead we will focus on use cases and the discovery of the relative harm index (RHI) and it's implications in relation to electric machine bearings, insulation systems, and electronics, let alone other electrical reliability issues.

The Relative Harm Index – RHI™

The discovery of the RHI was the result of a study on circulating currents in turbine nacelles that impact generator, gearbox and main bearings, as well as generator windings and gearbox gears. The investigation discovered that the design of the towers and the grounding system does not prevent nacelle circulating currents which were found to exceed 150 Amps in certain conditions.

Ground leads are primarily for lightning protection and personnel protection with limited ability for ground fault protection (Figure 2). The ground leads are smaller than the phase conductors and high frequency harmonics present in the ground currents cause the impedance of the ground conductors to be excessive.

In the study, ground currents were gathered from multiple locations including across shaft grounding brushes. This work is well established in the literature which resulted in the development of the RHI.

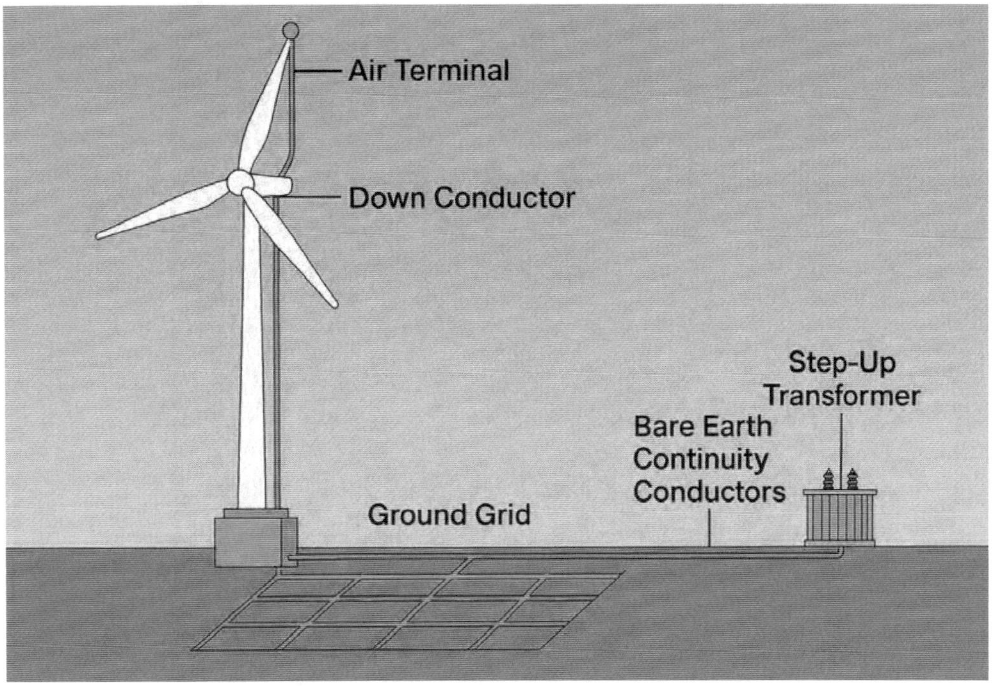

Figure 2: Ground system set up for lightning protection from nacelle.

Figure 3: Generator ground cable tested at the generator.

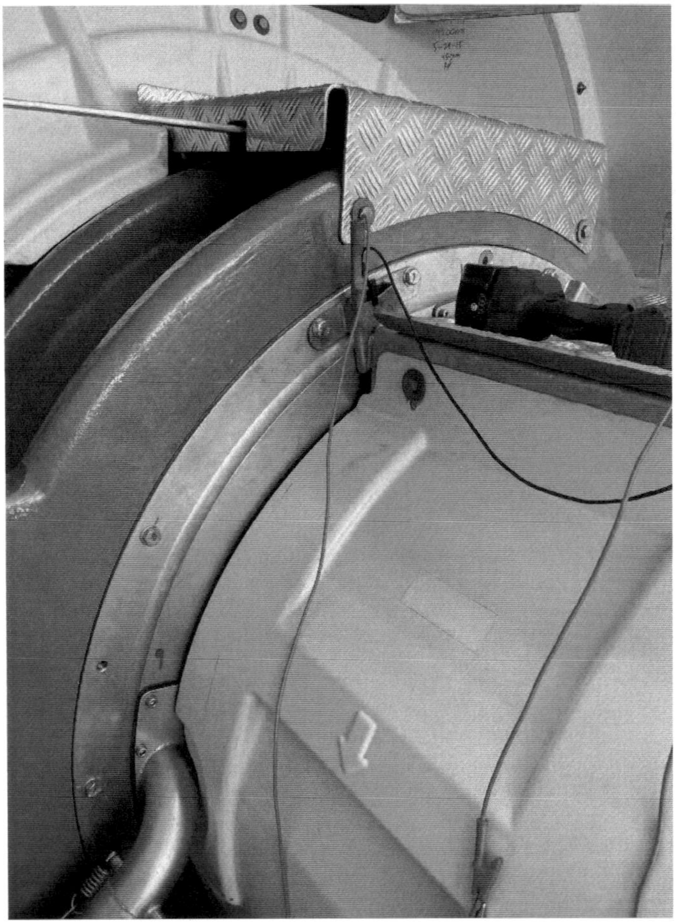

Figure 4: Voltage and ground taken on the main bearing grounding brush and ground.

In the initial study data was collected across the grounding systems within several designs and multiple wind turbines in addition to the data between the generator and controls. This information was utilized to verify specific findings associated with high frequency data. For instance, ground data was taken at the main bearing and then across the shaft brush for the main bearing. The difference would be an estimated amount of current across the bearing.

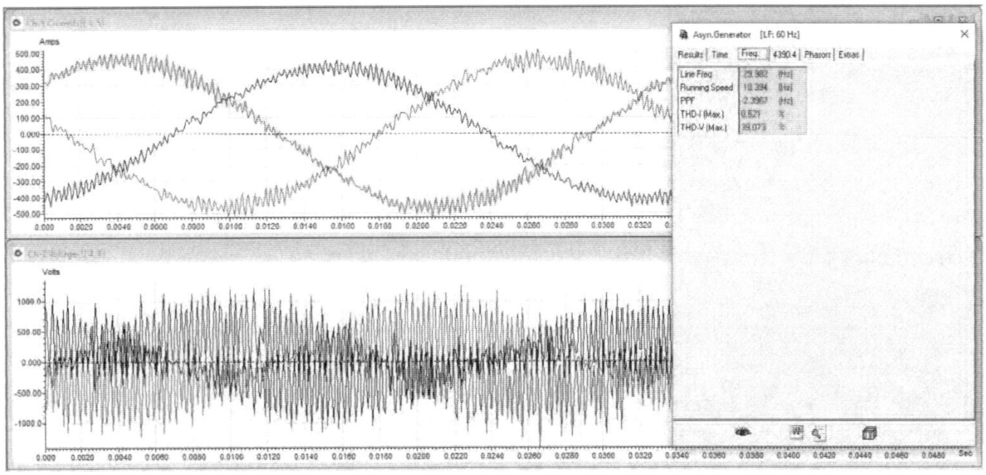

Figure 5: Induction generator voltage, current and frequency.

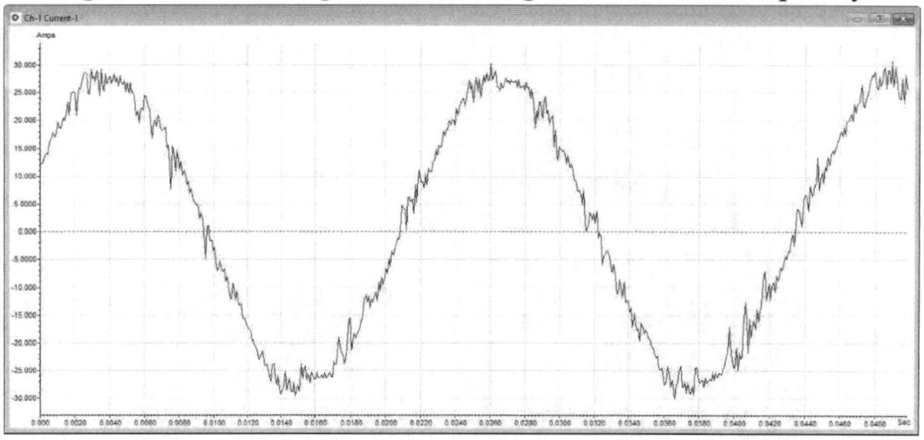

Figure 6: Generator to ground current (nacelle).

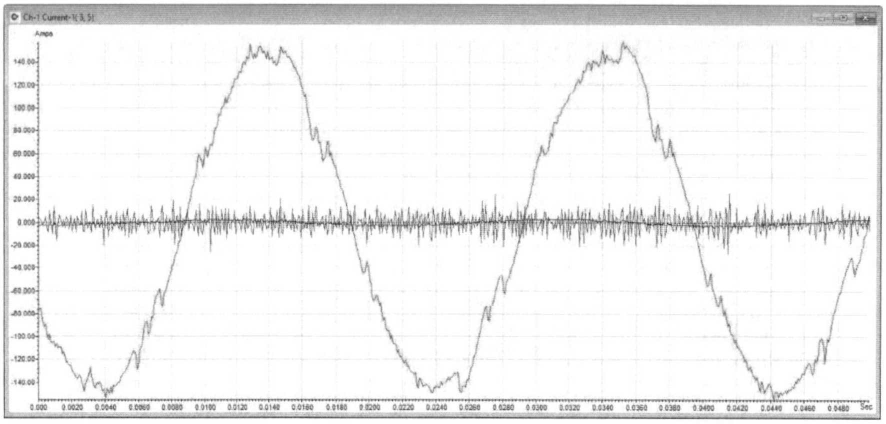

Figure 7: Main bearing ground current and voltage (nacelle).

The traditional evaluation of harmonics has been through the use of percentages of the fundamental, such as total harmonic distortion for voltage and total dynamic distortion for current. No definition for supra-harmonics, or those currents from 2 kHz to 150 kHz, has been provided for in technical literature or standards. A review of higher frequency harmonics using a special CSA device identified that the use of dB values in the same manner as CSA would be more effective. The concept is that the relative energy, or relative 'force,' of higher frequencies as compared to the fundamental current (at line frequency) can be evaluated.

This was then converted to an estimated potential harm versus the conditions if the harmonic current did not exist. These frequencies can be used to determine the impedance (ac resistance) across different components such as the brush and bearing. As shown in Equations 1 and 2, the impedance decreases with the frequency through the bearing and increases with the frequency through the brush.

$$Z_{bearing} = R_b + \frac{1}{j\omega C_b} \quad \text{Equation 1}$$

$$Z_{brush} = R_g + j\omega L_g + \frac{1}{j\omega C_g}$$ Equation 2

Within these equations Z is the impedance, R is basic resistance, jwL is the frequency, L is inductance and C is the capacitance. It is also noted that even with shaft brushes current still flows through the bearing. The idea is to have a low enough impedance to have a majority of the current pass through the brush versus the bearing as shown in Equation 3.

$$I_{bearing} = I_{total} * \left(\frac{Z_{brush}}{Z_{bearing}+Z_{brush}}\right)$$ Equation 3

For the wind turbine above, a chart based upon 100 Amps at the bearing was developed and the amount of current passing through the bearing based upon the calculated bearing and brush (standard and silver) impedances were evaluated and confirmed against actual values.

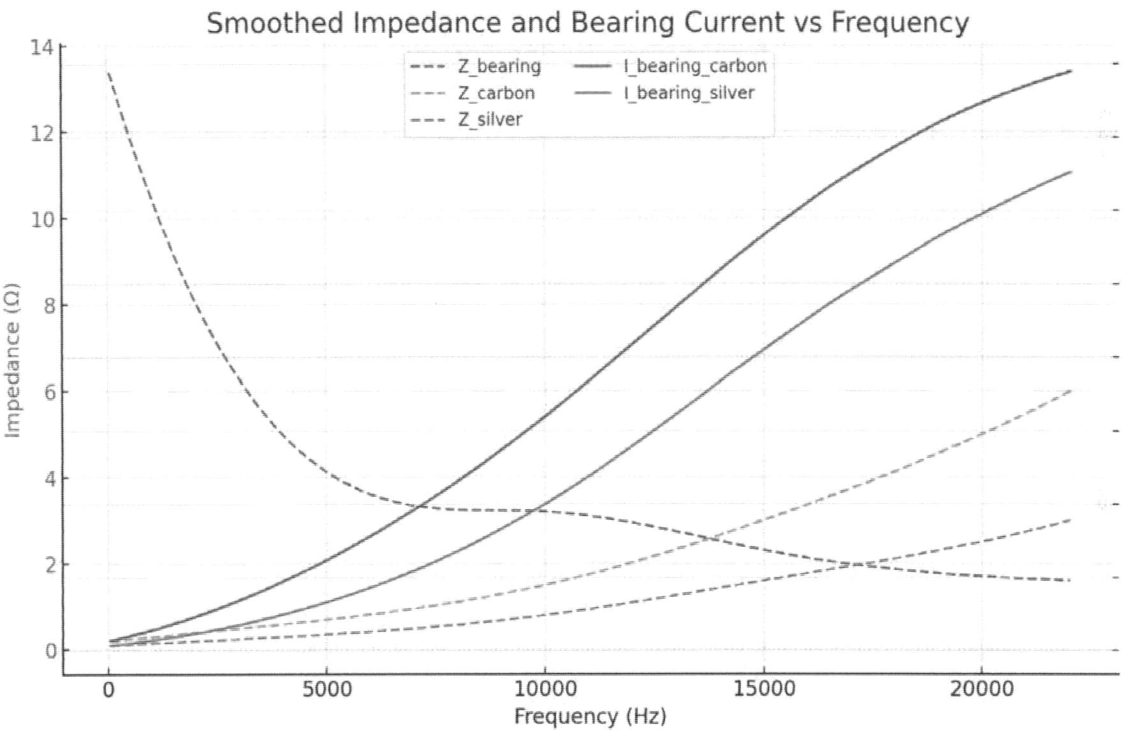

Figure 8: Comparison of impedance and currents versus frequency.

SKF has also published a document[10] that identifies the amount of current and severity for bearings. This can be found as Table 1.

Table 1: The amount of current and potential damage across a bearing

Circulating Current	Severity	Typical Outcome
<10A	Mild	Long-term wear.
10-50 A	Moderate	Premature bearing failures.

50-100 A	Severe	Rapid wear in months.
>100 A	Extreme	Rapid catastrophic failure.

This information was compared to actual values and histories associated with main bearing histories and data collected resulted in a chart based upon the RHI.

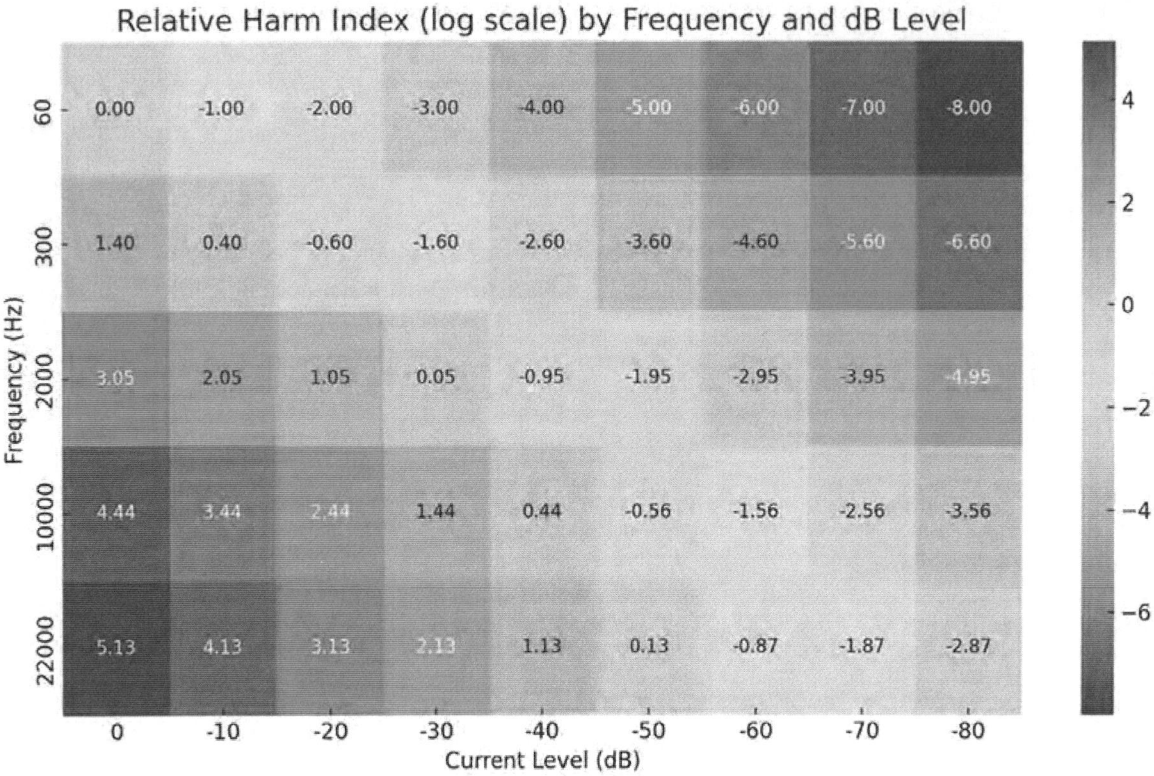

Figure 9: The RHI with 0 being optimal.

The RHI identifies that something is 10^N times more harmful based on frequency and peak even, odd and intra-harmonic. For instance, at 10^1 the condition is 10 times as harmful as if the harmonic condition did not exist. Lower numbers (less than 0) indicate that the equipment should not potentially fail due to bearing currents with the lower numbers indicating a continued reduction in risk.

Additional work identified that this methodology works in industrial and commercial environments to a relative degree of accuracy.

Case Study: Dynamometer

A 1.5 MW, 690 volt dynamometer motor was evaluated from a VFD (Figure 10). The output of the VFD did use inverter rated cabling, but conditions existed that called into question the reliability of the motor. The system was evaluated at the inverter and on the motor ground with the results shown in Figures 11 and 12. Standard electrical and current signature analysis identified both winding and bearing degradation in the equipment.

Figure 10: Case study motor on dynamometer.

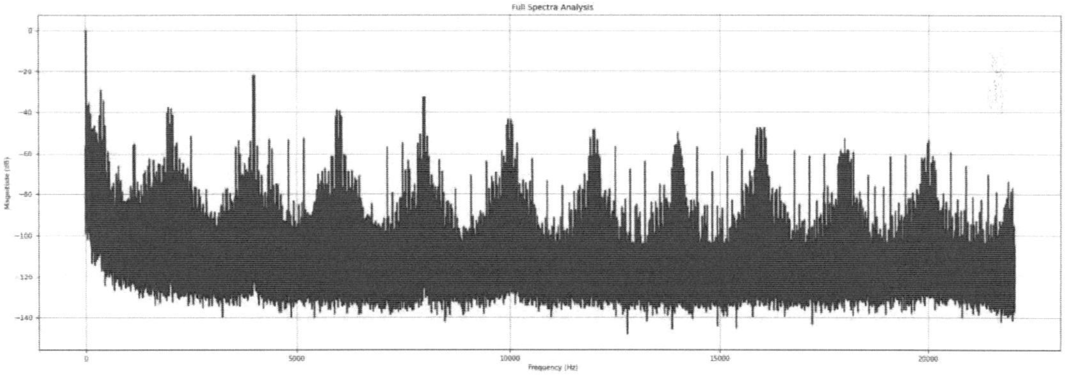

Figure 11: Power harmonics on the power side.

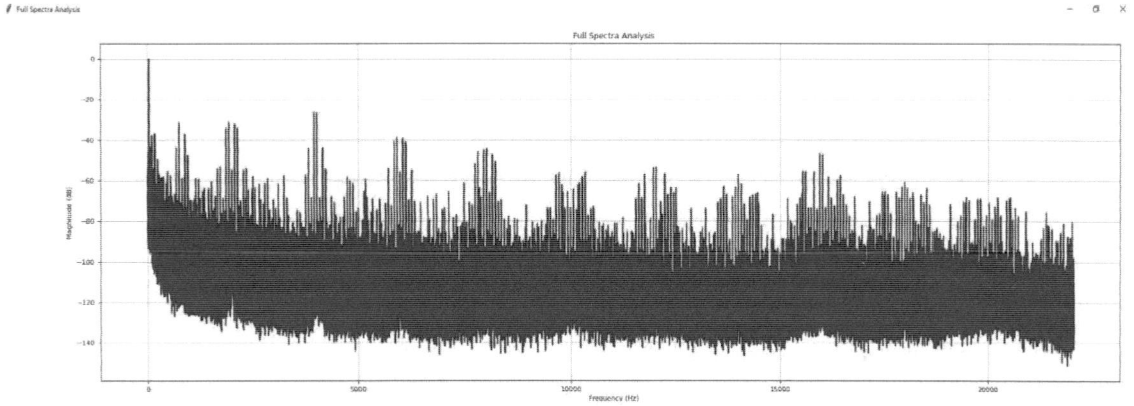

Figure 12: Ground harmonics on the frame of the motor.

The operating current of the machine was 504 Amps during testing and through ground current was 49.6 Amps peak. The combination of the currents can cause possible harm and there is no ground brush. The drive end bearing was found to be in poor shape and the insulation of the motor winding was degrading.

For the RHI we review the highest peaks, which would be the ones over 10 kHz if a brush was present, but is 4 kHz and -20 dB which would result in a value somewhere around 2, which would be 100 times more harmful. The current and severity chart (Table 1) would cite this as Moderate, which would indicate premature bearing failures. In this case, the motor ran an average of 1,000 hours per year for about 6 years before the problems became significant. The high frequency currents also degraded the insulation system, which was expected as well, with motor failing the winding tests during overhaul and the motor requiring a rewind.

Conclusion

The RHI was developed in order to provide a safer and quick way to determine the potential severity of bearing currents and winding degradation in electric machines. While the method was originally created for wind power generators and powertrain components. Further study determined that it is just as effective in commercial and industrial applications. The method can be used for evaluating equipment conditions as well as part of troubleshooting and root cause analysis.

References

[1] ISO 20958, "Electrical Signature Analysis of Three Phase Induction Motors," 2013-08-15, ISO.

[2] IEEE 1068-2015, "IEEE Standard for the Repair and Rewinding of AC Electric Motors in the Petroleum, Chemical and Process Industries," IEEE, Piscataway, USA, 2015.

[4] ACP Gearbox Playbook, "American Clean Power Gearbox Playbook," ACP, Washington, DC, 2023.

[5] CIGRE A1.78, "Guide for the Application of Electrical and Current Signature Analysis of Electric Motors, Generators and Powertrain," CIGRE-USNC.org, 2025.

[6] Haynes and Eissenberg, "Motor Current Signature Analysis Method for Diagnosing Motor Operated Devices," US Patent 4,965,513, Oct 23, 1990.

[7] Penrose, "The Application of Electrical and Current Signature Analysis of DFIG Turbine and Powertrain Defects," IEEE PES, Austin, TX, 2025

[8] IEC 61000-4-30, "Electromagnetic Compatibility Testing and Measurement Techniques. Power Quality Measurement Methods," IEC, 2021.

[9] Penrose, "Evaluation of Wind Turbine Electrical and Mechanical Defects Due to Nacelle Circulating Currents," IEEE Electrical Insulation Conference, South Padre Island, 2025.

[10] “Bearing Damage and Failure Analysis,” SKF Group, June 2017.

TWO-PHASE FLOW PIPING VIBRATION

Author: Nathan Libertowski

VIATC TRAINING CONFERENCE & EXPO
NEWPORT NEWS, VIRGINIA | AUGUST 6-8, 2025

VIATC | TRAINING CONFERENCE & EXPO

NEWPORT NEWS, VIRGINIA | AUGUST 6-8, 2025

TWO-PHASE FLOW PIPING VIBRATION

Nathan Libertowski
Staff Engineer II | Equity Engineering
nlibertowski@e2g.com

Nathan Libertowski, P.E. is an engineer at Equity Engineering with the Rotating Equipment, Vibration, & Dynamics (REVD) team. He has six years of experience as an engineering consultant solving API 579 fitness-for-service (FFS) assessments, with a specialty in piping and structural vibration. Mr. Libertowski holds engineering degrees from Ohio State University (MSME) and Ohio Northern University (MSME). He is a member of ASME and a certified Category III vibration analyst.

Abstract

One common cause of piping vibration in process facilities is two-phase flow. For some processes, it is inevitable to have both liquid and gas phases in the fluid. The gas phase normally has a significantly lower density than the liquid, which results in a much higher gas velocity compared to the liquid. This difference in density and velocity in the flow can cause large dynamic forces that are transferred to the piping, leading to vibration. The fundamentals of two-phase flow are presented, and it is discussed how they relate to real-world piping vibration. For most cases, two-phase flow causes random response vibration. Methods and procedures for evaluating random vibration are discussed. Lastly, a case study that involved evaluating piping vibration caused by two-phase flow is presented.

Key Words: piping vibration, two-phase flow, random vibration

NOMENCLATURE:

CF	crest factor
h_3	coefficient for second-order Hermite model
h_4	coefficient for second-order Hermite model
h_{40}	coefficient for second-order Hermite model
K	coefficient for second-order Hermite model
K_v	stress-to-velocity ratio from FEA model
n	number of data point
$S_{allow,pk}$	allowable peak stress
$S_{allow,rms}$	allowable RMS stress
S_{model}	stress from FEA model
S_{rms}	RMS stress
V_g	Gaussian vibration
V_{ng}	non-Gaussian vibration
u_{model}	displacement from FEA model
v_i	vibration velocity at time i
v_{model}	velocity from FEA model
v_{pk}	peak vibration velocity
v_{RMS}	RMS vibration velocity
Z_g	non-dimensional Gaussian vibration
Z_{ng}	non-dimensional non-Gaussian vibration
γ_3	skewness
γ_4	kurtosis
μ	mean
μ_g	mean of Gaussian distribution
μ_{ng}	mean of non-Gaussian distribution
σ	standard deviation
σ_g	standard deviation of Gaussian distribution
σ_{ng}	standard deviation of non-Gaussian distribution

Two-Phase Flow Fundamentals

For single-phase turbulent flow in a pipe, dynamic forces are transferred to the pipe wall. The turbulent energy is higher at low frequencies and attenuates at higher frequencies. If the piping is not adequately supported (i.e., has a low natural frequency), these dynamic forces can excite the piping natural frequencies, leading to random vibration. This is commonly referred to as flow-induced turbulence (FIT) or flow-induced vibration (FIV) [1].

For two-phase flow, the turbulent forces transferred to the piping can be more severe than single-phase flow. For many process facilities, two-phase flow occurs in piping connected to heat exchangers or fired heaters due to partial vaporization of the fluid when it is heated. To compound this issue, piping that has two-phase flow often contains high-temperature fluids. High temperatures cause significant thermal expansion in the piping, making it difficult to

restrain without inducing excessive stress. In fact, many times piping engineers use spring hangers to support the piping weight, but these provide almost zero stiffness to resist vibration forces. For these reasons, piping with two-phase flow in process facilities is susceptible to high vibration.
Two-phase flow is categorized into different flow regimes, and the flow regime will influence how the piping vibrates. Some examples of flow regimes for vertical and horizontal flow are shown in Figure 1.

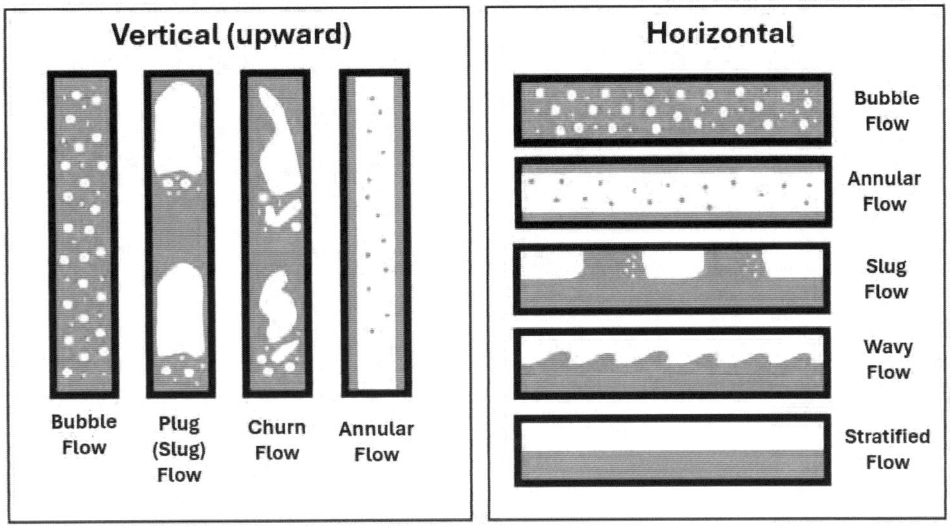

Figure 1: Two-phase flow regimes

The flow regime depends on many factors such as the liquid and gas flow rates, densities, viscosity, surface tension, orientation (horizontal, vertical, or slanted), and direction (upward or downward). It can also depend on the straight length of piping and the upstream conditions. Lastly, flow bends (e.g., flow through a piping elbow) can cause the flow regime to change due to a combination of centrifugal forces, inertia forces, gravity forces, and buoyancy forces [2]. In summation, it can be very difficult to predict the flow regime of two-phase flows.

There have been extensive experimental testing and empirical formulas for predicting flow regimes; however, they are not appliable for all scenarios. Two-phase flow is very well studied for reservoirs and pipelines and is an important aspect to ensure steady flow. Most reservoir and pipeline systems involve very long, straight sections of piping without elbows or other directional changes. In process facilities, most of the flexible piping susceptible to vibration from two-phase flow is constructed with many changes in direction for connection to process vessels and other components. Therefore, care should be taken when applying published flow regime correlations to piping in process facilities. The inertia of the liquid flow around a pipe bend can cause the flow regime to change [3]. Flow regimes are commonly characterized using flow regime maps that are based on empirical testing with long, straight sections of piping. Also, many of these flow regime maps are based on experimental testing of air-water mixtures with small pipe diameters and do not represent the exact physics that are occurring in process facilities. See Figure 2 for an example of a flow regime map for upward vertical flow [4]. Alternatively, there are mechanistic models that can predict the flow pattern for any orientation of piping based on the gas and liquid properties and flow rates [5]. These mechanistic models are more accurate for predicting flow regimes compared to the flow regime maps. The author does not believe that these mechanistic models take into account the effects of fluid inertia due to change in flow direction. To consider the effects of the inertia on the flow regimes due to change in flow direction requires the use of computation fluid dynamic (CFD) simulations. There are numerous articles in the literature that use CFD to predict the flow regime and vibration of piping [2] [6]; however, these simulations are very computationally expensive and are not practical for all applications.

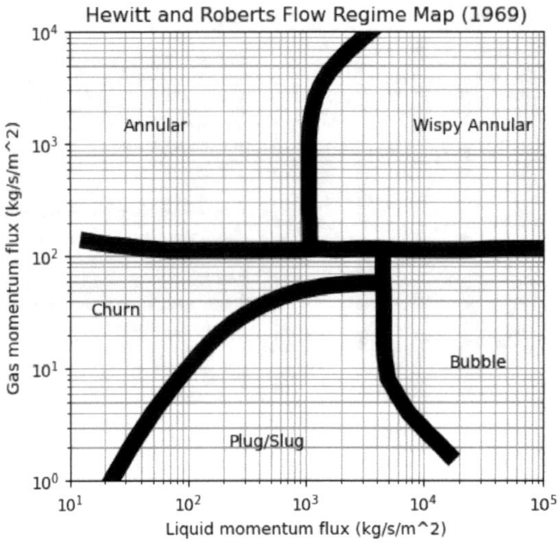

Figure 2: Example of flow regime map for vertical flow [4]

The physics behind the flow regimes for horizontal and vertical flow are different. Flow regimes for horizontal and vertical flow are controlled by buoyancy effects. For horizontal flow, the buoyancy effects cause the liquid and gas to separate to the top and the bottom of the pipe. For vertical (upward) flow, the buoyancy effects cause the gas to rise through the liquid.

- Horizontal flow:
    - For horizontal flow at low velocities, the liquid will flow on the bottom, and the gas will flow on the top of the pipe, which is known as stratified flow.
    - If the velocity is increased, waves start to form at the liquid-gas interface, which is known as wavy flow. If the velocity is increased even more, these waves can become unstable and span the entire cross section of the pipe, leading to intermittent slug flow. Formation of intermittent slug flow typically requires some straight length of pipe. The liquid and gas streams mix where the flow changes direction (e.g., pipe elbows). The mixing of liquid and gas fractions can prevent the liquid wave from covering the entire cross section and can prevent slug flow. Therefore, slug flow could be predicted by an empirical model or flow regime map; however, if there is not a long enough straight section of piping, the slug may not be able to be formed.
    - If the velocity is further increased, the flow will be either bubble, annular, or mist flow depending on the gas and liquid mass fractions. For a high-vapor mass fraction at high-flow velocities, mist flow will occur. For a low-vapor mass fraction at high-flow velocities, bubble flow will occur.
- Vertical (upward) flow:

- For vertical upward flow with low gas fractions (relative to the liquid flow rate), bubble flow will occur. As the name suggests, this consists of gas bubbles in the flow.
- As the gas flow rate increases, small bubbles can combine into larger bubbles and form Taylor bubbles, which take up almost the entire cross section of the pipe. This is referred to as plug or slug flow. The Taylor bubbles will rise though the liquid due to buoyancy effects.
- As the gas velocity further increases, the Taylor bubbles start to become unstable and break apart once they form, which is referred to as churn flow.
- As the gas velocity is further increased, annular flow occurs, where the liquid is on the wall of the pipe and the gas in the center.

Random Vibration Fundamentals

Vibration signals can generally be separated into two categories: periodic and random.

Periodic vibration is repeatable and consistent. In other words, the time waveform repeats itself. Some causes of periodic vibration are mechanical resonance from rotating equipment, acoustic pulsation from reciprocating machinery, and mechanical resonance caused by flow vortex shedding (which is often a problem for flow across thermowells). To characterize a periodic vibration signal for piping vibration analysis, all that is needed is the root-mean-squared (RMS) vibration velocity and the crest factor (CF). As shown in Eq. (1), the CF is the ratio of the peak vibration velocity (v_{pk}) to the RMS vibration velocity (v_{RMS}).

$$CF = \frac{v_{pk}}{v_{RMS}} \tag{1}$$

For a perfect sinusoidal vibration at one frequency, the crest factor of the vibration will be 1.414 $(\sqrt{2})$. If there are other frequencies of vibration, the crest factor will increase. Higher crest factors for periodic vibration commonly occur near reciprocating machinery due to the multiple harmonics of the vibration. Shown in Figure 3 is an example of vibration data collected on piping connected to a reciprocating compressor.

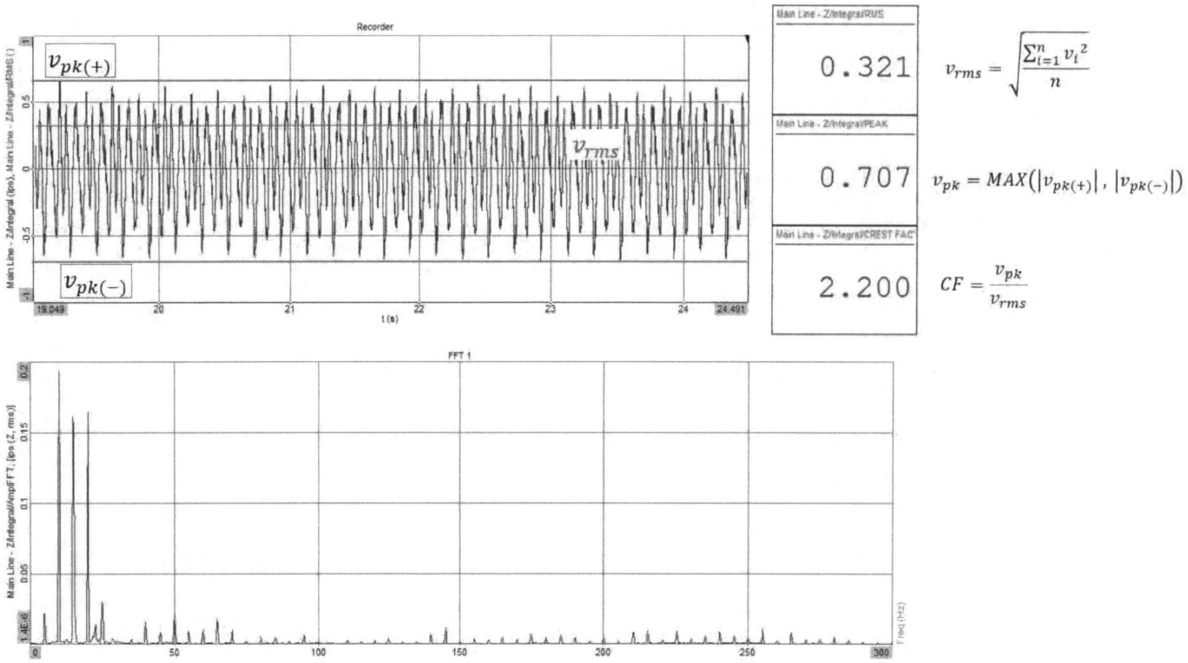

Figure 3: Example of periodic vibration from reciprocating compressor

For periodic vibration, since the time waveform is repeatable, data only needs to be collected for a short period of time to accurately measure the v_{RMS} and the CF. Technically, you only need to measure data for one repeatable section of a time waveform for periodic vibration; however, real-world periodic data is usually not perfectly repeatable, and therefore multiple repeatable sections should be measured.

Random vibration is not repeatable. Some causes of random vibration are turbulent flow in piping, two-phase flow in piping, and non-resonant vortex shedding forces on vessels and stacks. Random vibration can be characterized by v_{RMS} and CF, but because the time waveform is not repeatable, the CF will not be consistent. The value for CF will vary based on the time and duration of the vibration measurement. In general, v_{RMS} for random signals will remain relatively constant, but the peak velocity will vary over time [7]. Figure 4 is an example of random vibration collected on the crossover piping of a fired heater. Figure 4 shows that v_{RMS} remains relatively constant but the peak velocity and crest factor can vary significantly.

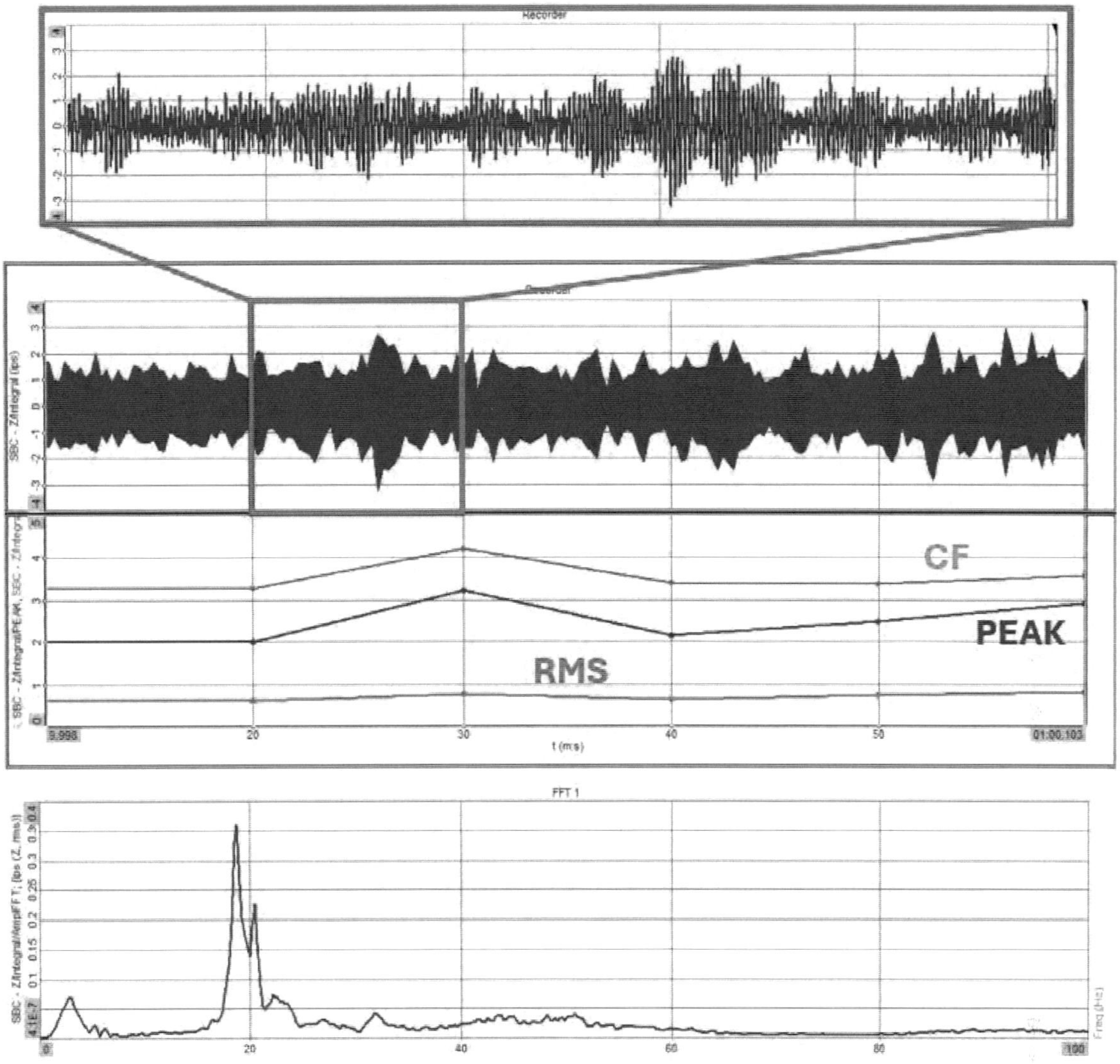

Figure 4: Example of random vibration on furnace crossover piping

Since the crest factor cannot always be reliably measured with random vibration, random vibration is normally characterized by an assumed statistical distribution. It turns out that most random piping vibrations closely match a Gaussian distribution. To define a Gaussian distribution for vibration, one only requires knowing the v_{RMS}, and from there the vibration signal can be inferred from the statistical distribution. Therefore, if the vibration is known to have a Gaussian distribution, then only v_{RMS} needs to be measured, which can be accomplished fairly quickly. The caveat to this is that random piping vibration is not always represented well by a Gaussian distribution. Impacting of the pipe against another surface and intermittent flow are some common reasons why random piping vibration may not be represented well by a Gaussian distribution.

One way to determine if random piping vibration represents a normal Gaussian distribution well is to calculate the kurtosis of the signal. Kurtosis is a measure of how often outliers occur and how large the outliers are. An ideal normal Gaussian distribution will have a calculated kurtosis of 3.0, and therefore if the kurtosis of random piping vibration is approximately 3.0, then it can be assumed that a normal Gaussian distribution will closely represent the vibration data. If the kurtosis is much higher than 3.0, then a normal or Gaussian distribution cannot be assumed, and alternative methods are required to evaluate the vibration. Figure 5 shows examples of real-world random piping vibration data with and without high kurtosis. The upper plot is from vibration caused by high flow turbulence and results in a kurtosis of approximately 3. The amplitude changes with time due to the randomness, but there are generally no sharp transitions. The lower plot is from vibration caused by slug flow and impacting of piping against a support. There are impact events that cause a large spike in the vibration, which causes the kurtosis to be over 11. It should also be noted that for random vibration, it may be difficult to tell if the signal has high kurtosis just by looking at the spectrum; therefore, calculating kurtosis and reviewing the time waveform are important.

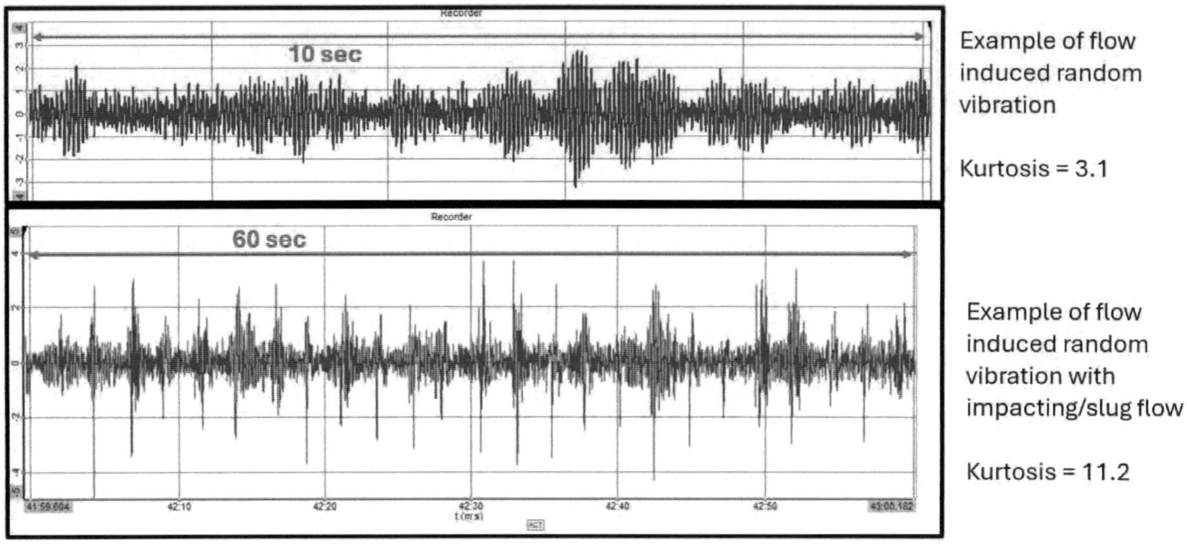

Figure 5: Example of random piping vibration with and without high kurtosis

The next step in piping vibration analysis is to perform a fatigue analysis to determine if the vibration levels cause stresses in the piping to exceed the endurance limit of the material. This can be accomplished multiple ways, with varying levels of complexity. One method for doing this is described in this paper.

The first step is to determine the RMS stress (S_{rms}) in the piping based on the measured vibration using either hand calculations or an FEA model. Next, an allowable RMS stress ($S_{allow,rms}$) is determined by taking the endurance limit of the material and dividing it by the measured CF of the vibration data (Eq. (2)).

$$S_{allow,rms} = \frac{S_{allow,pk}}{CF} \qquad (2)$$

If the calculated S_{rms} is greater than the $S_{allow,rms}$, then the vibration stresses exceed the endurance limit, and remediation actions should be pursued to reduce the vibration to acceptable levels. The equation to calculate the $S_{allow,rms}$ requires knowing the CF. For random vibration, measuring the CF is not reliable unless data is recorded for a long time. To avoid collecting data for long durations, the crest factor can be determined by using the following methods.

As previously mentioned, typical random piping vibration closely matches a Gaussian distribution. If you were to consider a perfect Gaussian distribution, the crest factor is technically unbounded. Albeit very unlikely, there is some very small probability of any amplitude for an ideal Gaussian distribution, and therefore as you accumulate more cycles, the crest factor will continue to increase. For real-world vibration problems, the amplitude of vibration is bounded by the physics of the system, and therefore there is an actual bounding crest factor that can be used for the purposes of the analysis. *For random piping vibration with a kurtosis of approximately 3.0, the CF typically ranges from 4.0 to 5.0. In summation, for random piping vibration with a kurtosis of 3.0, the CF used to calculate the* $S_{allow,rms}$ *can be determined by either measuring the peak vibration for a very long duration or assuming a CF of 4.0-5.0.*

If the kurtosis is not approximately 3.0, then the CF can be approximated by using the following methods. First, to evaluate a non-Gaussian random vibration, there are four parameters that should be considered: the mean (μ), standard deviation (σ), skewness (γ_3), and kurtosis (γ_4). See Eq. (3)–Eq. (6) for how these are calculated. For piping vibration, the mean vibration will most always be approximately zero and can be ignored. If there is a vibration with a non-zero mean, alternative methods may be required to assess the vibration. If the mean vibration is zero, then the standard deviation of the vibration is equal to the RMS vibration. The skewness represents the asymmetry of the distribution, and the skewness of an ideal Gaussian distribution is zero. For most piping vibration, the skewness can be assumed to be zero. As previously mentioned, the kurtosis is a measure of how often outliers occur and how large the outliers are, and an ideal Gaussian distribution will have a kurtosis of 3.0 [8].

$$\mu = \frac{1}{n}\sum_{i=1}^{n} v_i \qquad (3)$$

$$\sigma = \sqrt{\frac{1}{n}\sum_{i=1}^{n} \left(v_i - \mu\right)^2} \qquad (4)$$

$$\gamma_3 = \frac{1}{n\sigma^3}\sum_{i=1}^{n} \left(v_i - \mu\right)^3 \qquad (5)$$

$$\gamma_4 = \frac{1}{n\sigma^4}\sum_{i=1}^{n} \left(v_i - \mu\right)^4 \qquad (6)$$

To evaluate a non-Gaussian vibration (i.e., skewness ≠ 0 and/or kurtosis ≠ 3), a method presented by Winterstein et al [9] can be used. In their paper, they present a method to transform Gaussian distributions to non-Gaussian distributions, based off the skewness and kurtosis, using a second-order Hermite model. Eq. (7) shows the governing equation to convert a Gaussian vibration (V_g) to a non-Gaussian vibration (V_{ng}).

$$V_{ng} = \mu_{ng} + \sigma_{ng} K\left[\frac{V_g-\mu_g}{\sigma_g} + h_3\left(\left(\frac{V_g-\mu_g}{\sigma_g}\right)^2 - 1\right) + h_4\left(\left(\frac{V_g-\mu_g}{\sigma_g}\right)^3 - 3\left(\frac{V_g-\mu_g}{\sigma_g}\right)\right)\right] \quad \textbf{(7)}$$

K, h_3, and h_4 represent factors based on the kurtosis and skewness and will be defined later. As previously mentioned, for almost all piping vibration, the mean can be assumed to be zero, and therefore the equation simplifies to Eq. (8).

$$V_{ng} = \sigma_{ng} K\left[\frac{V_g}{\sigma_g} + h_3\left(\left(\frac{V_g}{\sigma_g}\right)^2 - 1\right) + h_4\left(\left(\frac{V_g}{\sigma_g}\right)^3 - 3\left(\frac{V_g}{\sigma_g}\right)\right)\right] \quad \textbf{(8)}$$

The equation can be further simplified by making the Gaussian and non-Gaussian vibrations non-dimensional by their respective standard deviations (Eq. (9)). The result of this is shown in Eq. (10).

$$Z_{ng} = \frac{V_{ng}}{\sigma_{ng}} \quad , \quad Z_g = \frac{V_g}{\sigma_g} \quad \textbf{(9)}$$

$$Z_{ng} = K\left[Z_g + h_3\left(Z_g^{\ 2} - 1\right) + h_4\left(Z_g^{\ 3} - 3Z_g\right)\right] \quad \textbf{(10)}$$

The equations used to calculate K, h_3, and h_4 are shown in Eq. (11), Eq. (12), Eq. (13), respectively. h_4 is a function of another parameter h_{40}, which is defined in Eq. (14). These equations are only valid for $3 < \gamma_4 < 15$ and $0 \le \gamma_3 < \frac{2(\gamma_4-3)}{3}$ [9].

$$K = \frac{1}{\sqrt{1+2h_3^{\ 3}+6h_4^{\ 2}}} \quad \textbf{(11)}$$

$$h_3 = \frac{\gamma_3}{6}\left(\frac{1-0.15|\gamma_3|+0.3\gamma_3^{\ 2}}{1+0.2(\gamma_4-3)}\right) \quad \textbf{(12)}$$

$$h_4 = h_{40}\left(1 - \frac{1.43\gamma_3^{\ 2}}{\gamma_4-3}\right)^{1-0.1\gamma_4^{\ 0.8}} \quad \textbf{(13)}$$

$$h_{40} = \frac{\left[1+1.25(\gamma_4-3)\right]^{1/3}-1}{10} \quad \textbf{(14)}$$

Therefore, using Eq. (10), a non-Gaussian distribution can be approximated based off the measured kurtosis and skewness. To further simplify these equations the skewness is assumed to be zero, and the resulting equation is shown in Eq. (15).

$$Z_{ng} = \frac{Z_g + h_{40}\left(Z_g^{\ 3} - 3Z_g\right)}{\sqrt{1+6h_{40}^{\ 2}}} \qquad \textbf{\textit{(15)}}$$

Shown in Figure 6 is the result of applying Eq. (15) to a set of random numbers generated by a Gaussian distribution random number generator. For this assessment, 10 million random numbers were generated and input into Eq. (15). The plot compares the probability densities for three different values of kurtosis to the analytical probability density of a Gaussian distribution. For easier interpretation, the results are presented for both a linear and a logarithmic y-axis. It can be seen that using this method with a kurtosis >3, there is actually a higher probability of having a very low amplitude; however, there is a much higher probability of having a very large amplitude compared to a Gaussian distribution.

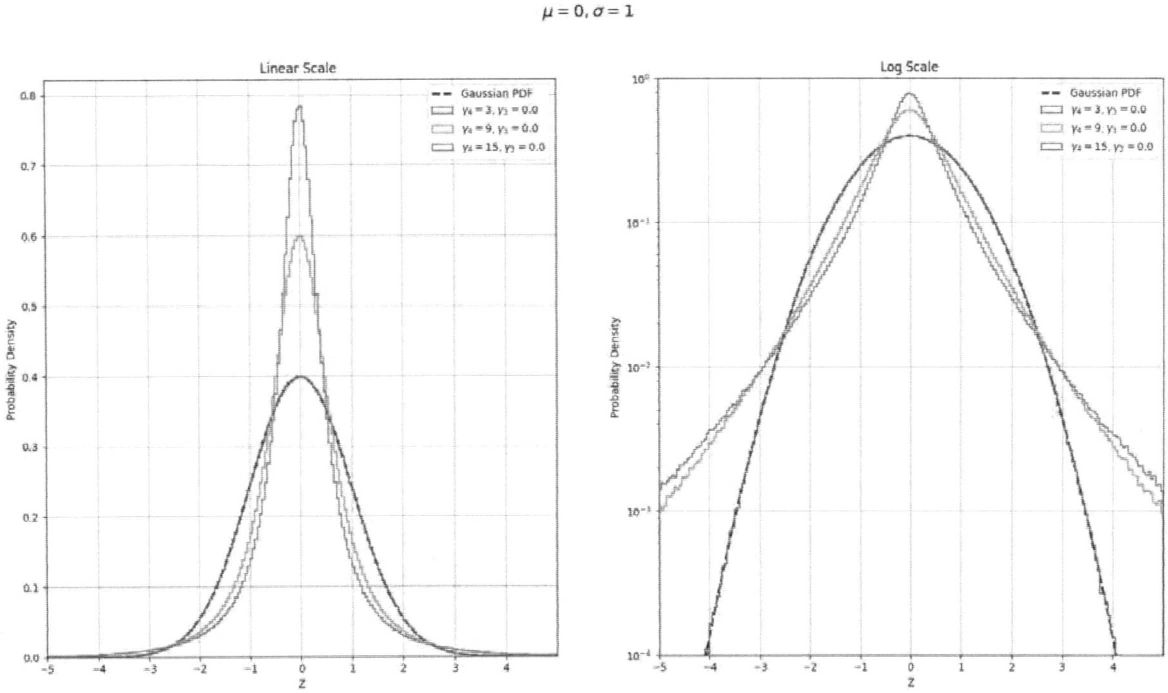

Figure 6: Probability density for Gaussian and non-Gaussian distributions

Eq. (15) represents the response of a non-Gaussian vibration (Z_{ng}) as a function of the Gaussian response (Z_g). This can also be thought of as representing the crest factor of the non-Gaussian vibration (CF_{ng}) as a function of the Gaussian response crest factor (CF_g), as shown in Eq. (16).

$$CF_{ng} = \frac{CF_g + h_{40}\left(CF_g^{\ 3} - 3CF_g\right)}{\sqrt{1+6h_{40}^{\ 2}}} \quad (16)$$

Shown in Table 1 are calculated non-Gaussian crest factors as a function of kurtosis, for different assumed Gaussian crest factors. It can be seen that as the kurtosis increased above 3.0, the crest factor will quickly increase. *In summation, for random piping vibration with a kurtosis of greater than 3.0, the CF used to calculate the allowable RMS stress can be determined by either measuring the peak vibration for a very long duration or calculating the CF using the above method (or Table 1).*

Table 1: Non-Gaussian crest factors

	Non-Gaussian Crest Factor		
Kurtosis	Gaussian Crest Factor: 4.0	Gaussian Crest Factor: 4.5	Gaussian Crest Factor: 5.0
3	4.00	4.50	5.00
4	5.60	6.89	8.39
5	6.64	8.46	10.61
6	7.44	9.65	12.32
7	8.09	10.63	13.72
8	8.64	11.46	14.91
9	9.12	12.19	15.94
10	9.55	12.83	16.86
11	9.93	13.41	17.69
12	10.27	13.94	18.44
13	10.59	14.41	19.13
14	10.88	14.86	19.76
15	11.15	15.27	20.35

Combining Two-Phase Flow with Random Vibration

For two-phase flow vibration, there are two mechanics of vibration: intermittent and turbulent. Intermittent vibration is caused by slug/plug flow and causes large impact of the slugs at pipe bends. Each of these impacts will cause the piping to vibrate. Turbulent vibration is caused by the chaotic and random nature of the two-phase flow. Churn flow results in a large amount of turbulent forces due to the Taylor bubbles breaking apart. The vibration is random in nature and will therefore increase and decrease in time; however, unlike intermittent, there is not one large event that causes the vibration. For some problems, both intermittent and turbulent two-phase flow are present and contributing to the vibration.

From an analysis point of view, vibration that is dominated by turbulence will likely exhibit a nearly Gaussian response (i.e., Kurtosis = 3), while vibration that is dominated by intermittent flow can exhibit a non-Gaussian response (i.e., Kurtosis > 3); however, it is possible for slug flow to cause a Gaussian response (i.e., Kurtosis = 3). As shown in the previous section, non-Gaussian vibrations can result in large crest factors, causing more damage for the same amount of RMS velocity. For this reason, it is common to try to design the piping to avoid intermittent regimes.

Piping with two-phase flow and vibration can mistakenly be attributed to slug flow, when in reality, there is just too much flow turbulence. There are many cases where slug flow is the root cause of the vibration, but there are also cases where the vibration is simply due to the flow turbulence. *This distinction can be very important when deciding on how to remediate the piping vibration*. Vibration from turbulence can be remediated by reducing the turbulence (reducing flow velocity or reducing flow turns/obstructions) or adding additional piping supports to increase stiffness. Vibration from intermittent flow can be mitigated by changing the flow regime or adding additional piping supports to increase stiffness.

For some situations where there is vibration from two-phase flow that is thought to be caused by slug flow, the problem can be solved by decreasing the pipe diameter (i.e., increasing fluid velocity) to change the flow regime so slug flow does not occur. This method does work for some situations; however, decreasing the pipe diameter will lead to more turbulence, which could cause vibration issues. If decreasing the pipe diameter to avoid slug flow is considered, the effect of increasing the velocity should be evaluated for the potential of vibration from flow turbulence. It is important to verify that the vibration is actually caused by slug flow in the first place. If the vibration is not caused by slug flow, decreasing the diameter will likely make the vibration significantly worse.

Case Study for Two-Phase Flow Piping Vibration

At a process facility, inlet piping on one of the heaters had a history of vibration issues. The piping was NPS 8 and was designed to 625 psig at 800°F. As part of an independent study, it was determined to reduce the diameter for a section of pipe to NPS 6 since slug flow was thought to be occurring in the NPS 8 piping. After the piping diameter was changed, two piping vibration-related failures occurred in the same location on a dummy leg support. After these failures, the dummy leg support was replaced with two all-thread hangers, which removed the high stress concentration failure location; however, it made the piping significantly less flexible and visibly increased the vibration of the piping. Vibration measurements were recorded, and a stress analysis was performed to determine the severity of the vibration.

Vibration measurements were recorded at 6 locations on the piping. The locations and directions of the measured vibration are shown in Figure 7. As shown in Figure 8, there were two dominant frequencies of vibration, 3.2 Hz and 5.5 Hz. The 3.2 Hz vibration occurred in the N/S direction while the 5.5 Hz vibration occurred in the E/W direction.

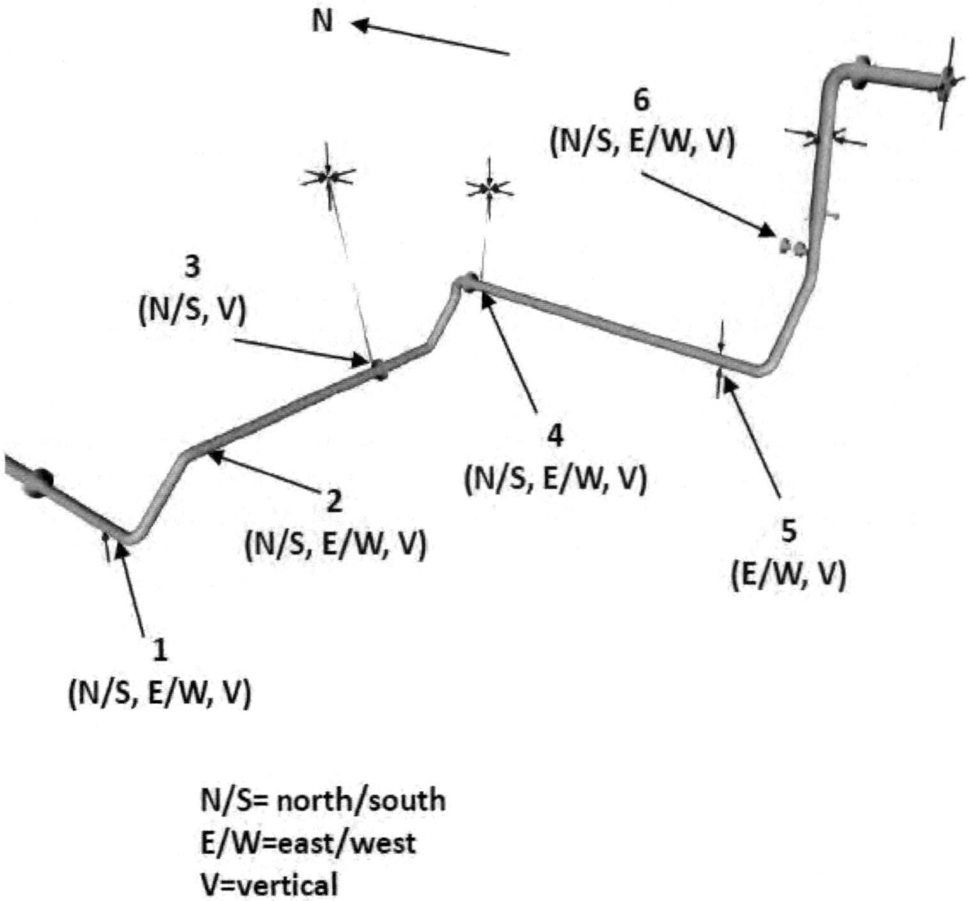

Figure 7: Piping overview with vibration measurement locations and directions

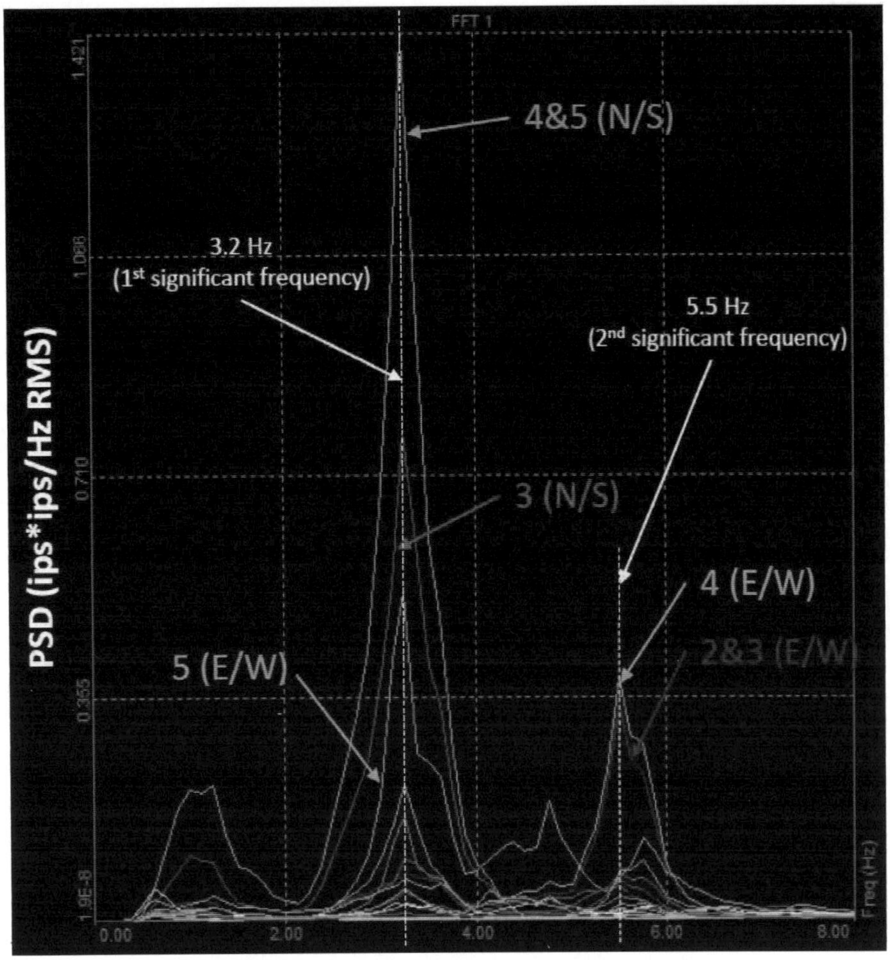

Figure 8: Vibration PSD spectrum

A dynamic model of the piping was created to determine the vibration stresses based on the measured vibration. The support stiffnesses were tuned to match the operating deflection shape (ODS) to the modeled harmonic displacements of the model at the two dominant frequencies. For each frequency, the magnitude and phase of the ODS and the model were compared. Once reasonable agreement was found between the ODS and the model, the maximum stress-to-velocity (K_v) ratio was calculated. K_v is a ratio of a predicted stress in the model to the predicted displacement at a certain location and direction in the model that corresponds to where data was collected (Eq. (17)).

$$K_v = \frac{S_{model}}{v_{model}} = \frac{S_{model}}{2\pi f u_{model}} \tag{17}$$

In order to estimate the stress (S_{rms}), the K_v is multiplied by the measured vibration velocity (v_{RMS}) at the same location and direction that was used to calculate the K_v in the model (Eq. (18)).

$$S_{rms} = K_v v_{RMS} \tag{18}$$

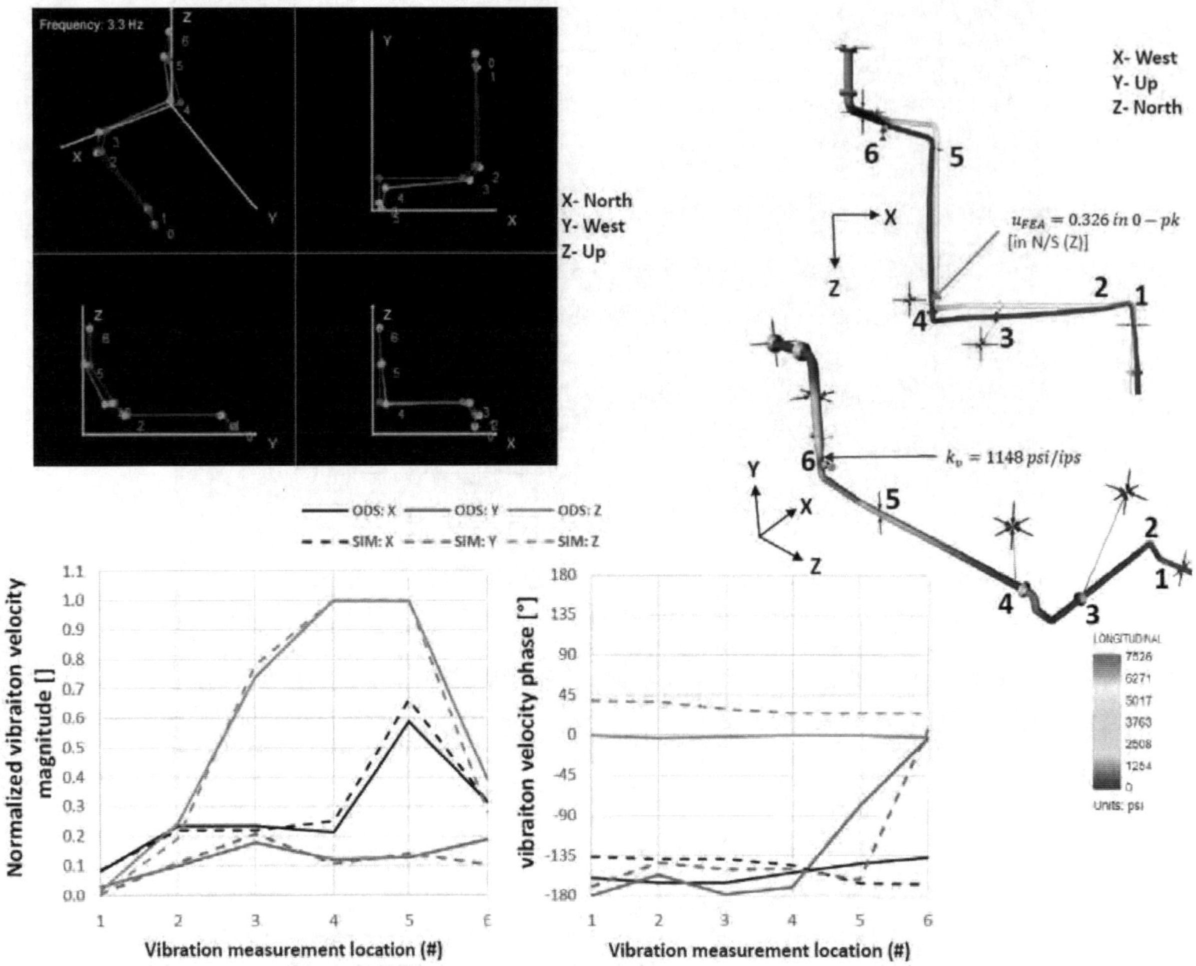

Figure 9: Kv calculation for 3.2 Hz vibration

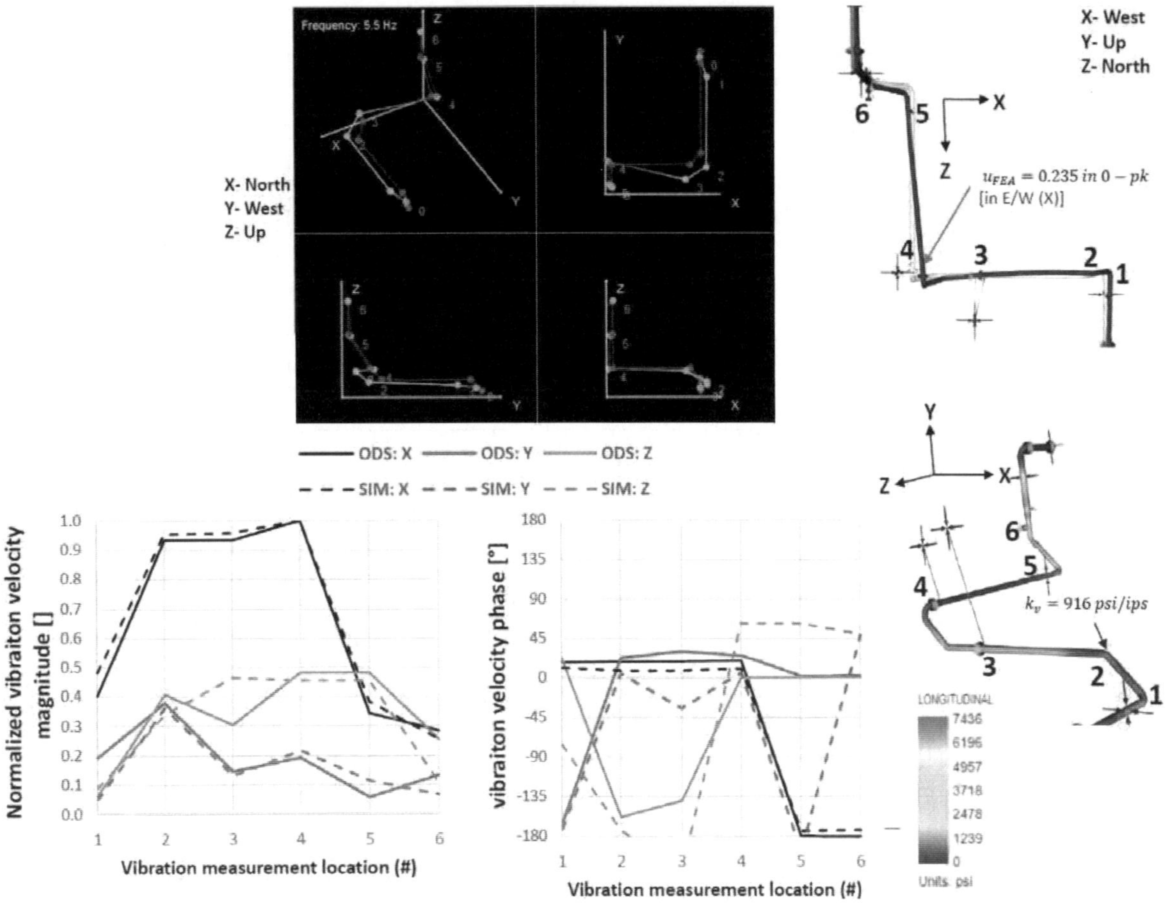

Figure 10: Kv calculation for 5.5 Hz vibration

Figure 9 and Figure 10 show the results of the model tuning and K_v calculation for the two dominant frequencies.

For the 3.2 Hz vibration, location 4 in the N/S direction was used to calculate the K_v, which had an overall vibration velocity of 1.25 ips RMS. For the 5.5 Hz vibration, location 4 in the E/W direction was used to calculate the K_v, which had an overall vibration velocity of 0.70 ips RMS. Since stresses were calculated at every location in the piping model for each frequency, the stresses were combined by taking the RMS of the two calculated stresses. The stress at each location was then compared to the allowable RMS stress ($S_{allow,rms}$). It was determined that 8 locations were predicted to have stresses that exceeded the endurance limit, and therefore it was recommended to pursue adding additional supports to reduce the vibration.

The thermal expansion stresses in the piping were considered when designing the new piping supports. The existing piping already had high thermal stresses, so locations for the new piping supports needed to be analyzed. There were only a few locations that had existing structural steel in close proximity to use for adding new pipe supports. After different support design iterations, it was determined to install two new supports (see Figure 11 for details). This design increased the stiffness of the piping but still kept thermal expansion stresses within allowable limits. Each of the supports consisted of sway struts that connected to the existing structural steel.

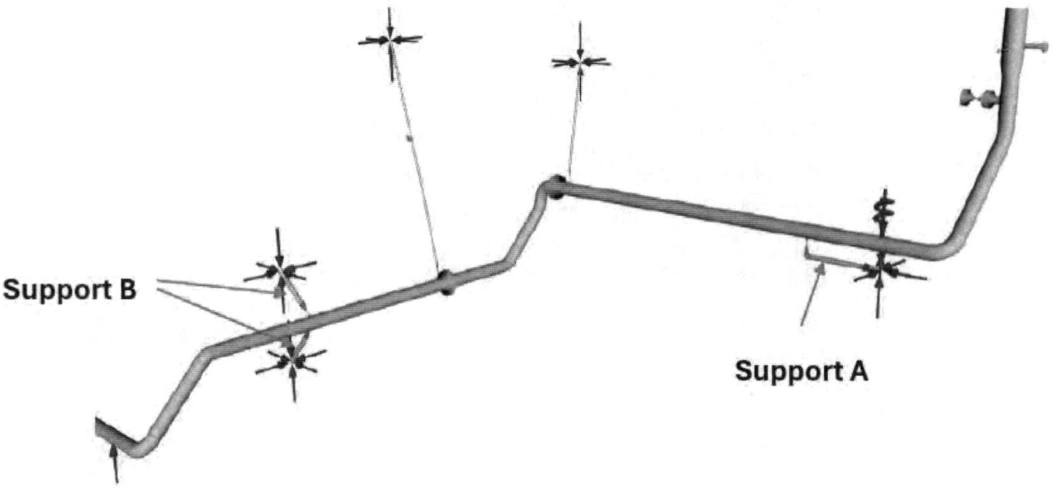

Figure 11: New support recommendations

After the new supports were installed, the vibration was remeasured and found to have significantly decreased and be within acceptable levels.

The conclusions and lessons learned from this case study:

- Decreasing the piping diameter to avoid slug flow can create vibration issues caused by excessive turbulence.
- Designing additional piping supports for high-temperature piping can be difficult to keep thermal expansion stresses to within allowable values
- If sway struts are not installed correctly, they are susceptible to failure.
- The fasteners on sway struts should be double-nutted and/or incorporate thread locker compound to reduce the possibility of the nuts backing off from the vibration.

References

[1] Energy Institute, Guidelines for the avoidance of vibration induced fatigue failure in process pipework, 2nd edition, 2008.

[2] H. Zhu, Y. Hu, T. Tang, C. Ji, and T. Zhou, "Evolution of Gas-Liquid Two-Phase Flow in an M-Shaped Jumper and the Resultant Flow-Induced Vibration Response," *Processes,* vol. 10, no. 10, p. 2133, 2022.

[3] O. Bamidele, W. Ahmed, and M. Hassan, "Characterizing two-phase flow-induced vibration in piping structures with U-bends," *International Journal of Multiphase Flow,* vol. 151, p. 104042, 2022.

[4] G. Hewitt and D. Roberts, "Studies of two-phase flow patterns by simultaneous x-ray and flash photography," Atomic Energy Research Establishment, Chemical Engineering Divison, Harwell, England, 1969.

[5] H. Zhang, Q. Wang, C. Sarica, and J. Brill, "Unified Model for Gas-Liquid Pipe Flow via Slug Dynamics—Part 1: Model Development," *ASME Journal of Energy Resources Technology,* vol. 125, no. 4, p. 266–273, 2003.

[6] T. Firmansyah, M. Rakib, M. Karakaya, M. Musharfy, and M. Suleiman, "Mitigating flow induced vibration in heater radiant coil," *Oil & Gas Science and Technology - Rev. IFP Energies nouvelles,* vol. 73, p. 28, 2018.

[7] L. Breaux, S. McNeill, and G. Szasz, "Fitness-for-service assessment of piping subject to random vibration, PVP2016-63695," in *Proceedings of the Pressure Vessels and Piping Conference*, Vancouver, B.C, Canada, 2016.

[8] D. Benasciutti, Fatigue analysis of random loadings, Saarbrucken, Germany: LAP LAMBERT Academic Publishing, 2012.

[9] S. Winterstein, T. Ude, and G. Kleiven, "SPRINGING AND SLOW-DRIFT RESPONSES: PREDICTED EXTREMES AND FATIGUE VS. SIMULATION," in *7th Intl Conf on the Behaviour of Offshore Structures (BOSS)*, Massachusetts, USA, 1994.

LEAST SQUARES AND MINI-MAX BALANCING USING PYTHON

Author: Steven P. Ciesla, P.E.

VIATC TRAINING CONFERENCE & EXPO
NEWPORT NEWS, VIRGINIA | AUGUST 6-8, 2025

VIATC TRAINING CONFERENCE & EXPO
NEWPORT NEWS, VIRGINIA | AUGUST 6-8, 2025

LEAST SQUARES AND MINI-MAX BALANCING USING PYTHON

Steven P. Ciesla, P.E.
Machinery Diagnostic Services Technical Leader, Bently Nevada
steven.ciesla@bakerhughes.com

Steven P. Ciesla, P.E. is presently a Machinery Diagnostic Services Technical Leader for Bently Nevada. He re-joined the company in 2013. Prior to that he worked as a predictive maintenance engineer at a nuclear power plant where he was responsible for the vibration and lubrication programs. His work history includes being a field engineer for both Bently Nevada and Westinghouse Process Control. He has over 25 years of hands-on vibration analysis experience in the power generation industry.

Steve earned a BS in mechanical engineering from California State Polytechnic University, Pomona and an MSE in mechanical engineering from the University of Nevada, Las Vegas. He is a licensed professional engineer in the states of Nevada and California and certified as a Category IV Vibration Analyst through Mobius Institute.

Abstract

Balancing complex machine trains is often performed using expensive commercial or proprietary software. In both cases, the programs may not be readily available for some users. Regardless of whether the user has access to software, an understanding of the calculation methodology is required to use the computer codes properly. The current paper is an extension of previous papers that reviewed balancing calculation techniques and algorithms used in multiplane balancing and how they can be implemented in a common programming language (Python) and a commercially available spreadsheet program (Excel). The author discusses the implementation of least squares and mini-max (optimization) balancing techniques using Python codes. The results from the programs are compared with data from real machines and compared to the results of commercial multiplane balance program.

Key Words: Balancing, Least Squares, Mini-Max, Multiplane, Optimization, Python

Introduction

Field balancing of complex machine trains may be accomplished using a variety of techniques including single plane, static-couple, or multiplane. Single plane and static-couple techniques provide simple methodologies for performing hand calculations and may be sufficient to achieve a satisfactory overall balance condition for a complex machine train. However, based on the author's experience, full consideration of the machine train is a better approach in many cases.

For the purposes of this paper, a typical complex machine train is a steam or gas turbine generator with rigid couplings, three to twelve bearings, multiple critical speeds, and multiple operating conditions. The techniques discussed, however, are not exclusive for complex machines. The techniques are equally usable on simpler machines with flexible couplings such as motors and fans.

The measured vibration could be shaft vibration relative to the bearings, absolute bearing vibration, or absolute shaft vibration. For this paper, a well-balanced machine is the goal for a balancing program. Well-balanced for this purpose is: shaft vibration at all bearings below about 2.0 mils pp of 1X vibration at normal operating conditions and/or bearing vibration at or below about 0.5 mils pp of 1X vibration. Note that these vibration levels are stated as examples only. Actual vibration amplitudes for a well-balanced machine should be based on OEM and industry standards. The vibration through critical speeds and off-normal operating conditions should be below alarm levels.

A previous paper co-prepared by the author in 2023 [1] reviewed the calculation techniques and algorithms used in multiplane balancing and how they can be implemented in common programming languages and tools, such as Excel, Excel with VBA, and Python. This paper is an extension of that previous work, focusing on the implementation of least squares and mini-max (optimization) balancing techniques using the Python programming language.

Balancing Procedure

The balancing process is well described in a variety of references. The field balancing procedure described above involves the following high-level steps regardless of the number of measurements or balance planes:

- Run 0 – Obtain the reference or as-found data {O}
- Shot 1 – Install a calibration weight in a single balance plane. This "single" plane shot might actually be a static or a couple pair. {Wcal}

- Run 1 – Operate the machine at conditions similar to those when the reference data was collected. {O+C}
- Evaluate the effect of the calibration weight {C} and calculate influence coefficients for all measurement locations and conditions. {H}
- Using Run 1 as a reference, decide whether to optimize Shot 1, whether to install a second shot in a different location, or whether to conclude the balance campaign.

The balancing process described above is typically continued until the as-left vibration is acceptable. Figure 1 provides a flow chart illustrating a typical balancing campaign.

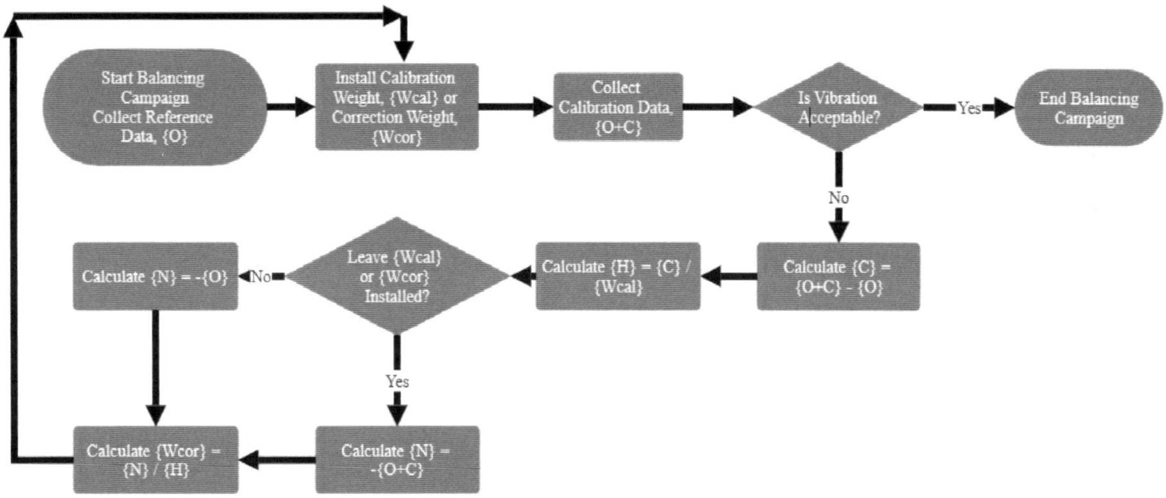

Figure 1: Balancing campaign flow chart

The basis of least squares balancing was presented by Goodman in 1964 [2]. The balancing process presented by Goodman is summarized as:

> "Given N balance planes, M = K x L vibration readings, find the optimum magnitudes and angular locations of correction masses in the N balance places."
> M = Vibration readings
> K = Different operating conditions (speed or load)
> L = Different pickup locations (vibration probes)

When M is less than or equal to N, then an exact solution is possible where the resultant M vibration readings will be zero. When M is greater than N (which is normally the case for real machines) then the N correction masses must be chosen to make the M vibration amplitudes all simultaneously as small as possible. This is accomplished by the least squares method.

The balancing process requires conventions and notation. The author used the convention adopted by their employer, Bently Nevada. Goodman [2] used different conventions. Table 1 provides a comparison of the nomenclature used by Bently Nevada and in the Goodman paper.

Table 1: Nomenclature comparison

Variable description	Goodman	Bently Nevada
Original vibration vector (Zero rotor data)	A	O
Vibration after adding calibration (trial) weight	B	O+C
Vibration due to calibration weight	B-A	C
Influence coefficients (response coefficients)	α	H
Number of measurements	M	M
Number of balance planes	N	N
Number of operating conditions	K	
Probe (pickup) locations	L	
Residual vibration	ε	R

A convention for displaying influence coefficient angles is also required. The calculation process does not depend on the adoption of any specific convention for influence coefficient angles. The process does require that the influence coefficient convention is consistent when sharing influence coefficients from one tool to another. Similarly, comparing results between the tools used for the analysis requires an understanding of the conventions used by the tools. When field balancing, the angular location of the balance weights are typically provided in reference to the Keyphasor target on the applicable rotor(s). This is a physically identifiable feature, such as a notch or projection on the shaft. As such, the examples provided in this paper utilize the balance weight angle relative to the Keyphasor target on the applicable rotor.

The Bently Nevada convention is that all calculations are performed with phase angles based on the probes. The location of the Keyphasor is not required for these calculations. Zero is located on the rotor at the point in time when the Keyphasor fires for the probe of interest. The zero positions for X and Y probes are therefore inherently different physical locations on the rotor.

Another convention could be to convert all phase angles such that zero degrees would be the Keyphasor notch. This convention requires that the Keyphasor probe location is known and input accurately.

The author's advice on this subject is to always try to obtain the data required to calculate influence coefficients. Then the influence coefficients can be calculated in the user's standard convention.

Python Implementation – Single Plane

Basic information about the Python programming language and the Spyder environment was discussed in a previously co-authored paper [1].

Three Python libraries were utilized in the balancing programs described in this paper. The NumPy (Numerical Python) library is the fundamental package for scientific computing in Python and allows the use of matrices, linear algebra, and complex numbers. This library was used for performing balancing calculations within Python. The Pandas library was installed and utilized to import applicable vibration data from an Excel spreadsheet. The CVXPY library was used to calculate the mini-max balancing optimization solution.

Balancing vibration data including the original vibration (O), original vibration plus effects of the calibration weight(s) (O+C), and the calibration weight(s) (Wcal) are documented in Sheet1 of an Excel spreadsheet using polar coordinates, (r,φ) as shown in Figure 2. The data is transformed to cartesian coordinates (x,y) in Sheet2 of the spreadsheet (Figure 3). The data is then transformed to complex coordinates (x+yj) in Sheet 3 of the spreadsheet (Figure 4).

	A	B	C	D	E	F
1	O(r)	O(ϕ)	O+C(r)	O+C(ϕ)	Wcal(r)	Wcal(ϕ)
2	0.2	350	0.2	200	322	304
3	4.7	285	2.0	20		
4	1.4	100	0.5	100		
5	2.9	289	1.2	40		

Figure 2: Excel worksheet Sheet1 showing O, O+C, and Wcal data in polar coordinates

	A	B	C	D	E	F
1	O(x)	O(y)	O+C(x)	O+C(y)	Wcal(x)	Wcal(y)
2	0.196961551	-0.034729636	-0.187938524	-0.068404029	180.0601149	-266.9500984
3	1.216449512	-4.539851384	1.879385242	0.684040287		
4	-0.243107449	1.378730854	-0.086824089	0.492403877		
5	0.944147648	-2.742003869	0.919253332	0.771345132		

Figure 3: Excel worksheet Sheet2 showing O, O+C, and Wcal data in cartesian coordinates

	A	B	C
1	**O(x+yj)**	**O+C(x+yj)**	**Wcal(x+yj)**
2	0.196961550602442-0.0347296355333863j	-0.187938524157182-0.0684040286651337j	180.06011491758-266.950098362724j
3	1.21644951198185-4.53985138355862j	1.87938524157182+0.684040286651337j	
4	-0.243107448733702+1.37873085421709j	-0.0868240888334652+0.492403876506104j	
5	0.944147647925753-2.74200386923802j	0.919253331742774+0.771345131623847j	

Figure 4: Excel worksheet Sheet3 showing O, O+C, and Wcal data in complex coordinates

The manipulation of the vector (amplitude and phase) measurements to x- and y- coordinates and then to complex number notation in Excel is not mandatory. Python has the capability of performing calculations and transformations that would mimic the work performed in Excel. However, since vibration data is commonly tabulated in Excel for presentation and reporting purposes, it felt natural to the author to manipulate the data in Excel into complex number format and have Python read the data directly.

In Python, the hashtag (#) is used for comments. The beginning of the following Python code contains multiple comments about the libraries installed and descriptions of the variables used in the single plane balancing code (Figure 5). Additionally, the code shows how the O, O+C, and Wcal data is read from an Excel worksheet using the Pandas library.

Prior to determining the correction weight (Wcor) via the Least squares method, it is necessary to calculate the Null (N) vectors and influence vectors (H). The Null vectors are the same length of the Original (O) vectors, but their phase measurements are 180 degrees opposite. The crux of balancing is for the balance weight(s) to create a vibration response that is opposite the original vibration (which corresponds to the Null vector). When the original vibration response is added to the response from the balance weight(s), the residual vibration is theoretically zero (exact solution) or minimized. The RMS residual is minimized via the least squares methodology. Calculation of the Null (N) vectors, influence vectors (H), and correction weight (Wcor) in Python is shown in Figure 6. Note that the calculation of the Null vectors is dependent on whether the calibration weight is left installed or removed after the calibration run. The Least square calculation is performed using NumPy's linear algebra, least squares algorithm, np.linalg.lstsq(H,N). Python code for formatting of the solution and the calculation of the RMS Residual and Mini-max Amplitude is shown in Figure 7.

```
#Installed numpy, from command prompt, pip install numpy
#Installed pandas, from command prompt, pip install pandas
#Installed cvxpy, from command prompt, pip install cvxpy
# O: Original Vibration vectors
# O_C: O+C Vibration vectors
# Wcal: Calibration (trial) weight
# N: Null Vector, equals -O
# H: Influence vectors
# Eps: Residual vibration
# S: Sum of squared residuals
# R: RMS Residual
# Wcor: Correction Weight based on Least Squares (minimize RMS Residual)
# W_Mini: Correction Weight based on Mini-Max optimization

import numpy as np
import pandas as pd
import cvxpy as cp

#Read in O, O+C, & Wcal data from Excel data file

#file = "Turb_Pump_Data.xlsx"
#file = "ID_Fan_Data.xlsx"
#file = "Steam_Turbine.xlsx"
#file = "Vertical_Pump.xlsx"
file = "CT_Generator.xlsx"
#file = "Steam_Turbine_Static.xlsx"

xl = pd.ExcelFile(file)
df = xl.parse("Sheet3")
Temp = df.to_numpy()

O = Temp[:,0]
O = O.astype(complex)
O = O.reshape(-1,1)

O_C = Temp[:,1]
O_C = O_C.astype(complex)
O_C = O_C.reshape(-1,1)

Wcal = Temp[0,2]
Wcal = complex(Wcal)
```

Figure 5: Portion of Python code for single plane balancing showing the libraries being imported and the O, O+C, and Wcal data being read from the Excel worksheet (Sheet3)

```
N = -1. * O #Calculates the Null Vector
#N = -1. * O_C #Calculates the Null Vector if Wcal is left installed

H = (O_C - O) / Wcal #Cacluates the influence vectors

Wcor = np.linalg.lstsq(H, N)[0] #Solves for the Correction Weight via Least Squares
Wcor = Wcor[0]
```

Figure 6: Portion of Python code for single plane balancing showing the calculation of the Null (N) and influence (H) vectors and solving for the correction weight (Wcor) via Least squares

```
Wr = np.round(np.abs(Wcor),2) #The amplitude/weight of the correction weight, Wcor
Wphi = np.mod(np.angle(Wcor, deg=True),360) #The angle of the correction weight, Wcor
Wphi = np.rint(Wphi).astype(int)
print(" ")
print ("RMS Correction Weight =",*Wr, "at", *Wphi, "degrees")
print(" ")

M = len(O)
O = -1. * N #Accounts for Wcal remaining installed or removed
Eps = H * Wcor + O #Calculate the residual vectors, Epsilon
Eps_Amp = np.abs(Eps) #Calculate the amplitude of the residual vectors
Eps_Max = np.max(Eps_Amp)
S = np.sum(np.square(Eps_Amp)) #Calculate S "Sum of Squared Residuals"
R = np.sqrt(S/M) #Calculate R "RMS Residual"

print("RMS Residual = ", np.round(R,6))
print("Mini-Max Amplitude =", np.round(Eps_Max,3))
print(" ")
```

Figure 7: Portion of Python code for single plane balancing showing the formatting of the correction weight solution and calculation of the RMS Residual and Mini-max Amplitude

Goodman [2] states:

> "When the sum of the squares of the M residual vibration amplitudes is minimized, it is possible that one or two of the residual amplitudes may still be considerably larger than the others. If the goal of the balancing procedure is to minimize the maximum residual amplitude, then the plain least squares solution is not the optimum."

These statements are then followed by the description of an iterative procedure (weighted least squares) using the residuals from the least squares results to "weight" the influence vectors that leads to a solution which in general reduces the larger residual amplitudes at the expense of increasing the smaller residual amplitudes. Machine owners and operators are typically interested in achieving an acceptable balance condition in the shortest amount of time and fewest number of runs. An acceptable balance condition is usually determined from the maximum vibration amplitude observed on the rotor being balanced. As such, minimizing the maximum amplitude (mini-max) is a preferable balance solution to the least squares solution.

The mini-max solution is calculated using the CVXPY library's Minimize function (Figure 8). In CVXPY, the @ character is used for matrix multiplication. Earlier in the program, the Null (N) vector was calculated based upon whether the calibration weight (Wcal) was removed or left installed. The implementation of CVXPY's Minimize function utilizes the original (O) vibration data, but this starting point changes if Wcal is left installed. As such, the O vector is calculated from the Null (N) vector that was previously modified to account for the status of the calibration weight (removed/installed). A dummy variable for the solution weight (W_Mini) is created and the residual calculation (H @ W_Mini + O) is being minimized using the infinity "inf" norm. The "inf" norm corresponds to the largest amplitude in the residual vector calculation, thereby minimizing the maximum amplitude (mini-max). The least squares and mini-max output results from the single plane balancing program are displayed in Figure 9. Note that the least squares (RMS) solution minimizes the RMS calculation while the mini-max solution minimizes the mini-max amplitude.

```
#Mini-Max with CVXPY
O = -1. * N #Accounts for Wcal remaining installed or removed
WM=cp.Variable((1,1),complex=True)
objective=cp.Minimize(cp.norm((H @ WM + O),"inf"))
prob=cp.Problem(objective)
prob.solve()
W_Mini = WM.value
W_Mini = W_Mini[0]
Wr = np.round(np.abs(W_Mini),1) #The amplitude/weight of the correction weight, WMini
Wphi = np.mod(np.angle(W_Mini, deg=True),360) #The angle of the correction weight, WMini
Wphi = np.rint(Wphi).astype(int)

Eps = H * W_Mini + O #Calculate the residual vectors, Epsilon
Eps_Amp = np.abs(Eps) #Calculate the amplitude of the residual vectors
Eps_Max = np.max(Eps_Amp)
S = np.sum(np.square(Eps_Amp)) #Calculate S "Sum of Squared Residuals"
R = np.sqrt(S/M) #Calculate R "RMS Residual"
print ("Mini-Max Correction Weight =",*Wr, "at", *Wphi, "degrees")
print(" ")
print("RMS Residual = ", np.round(R,6))
print("Mini-Max Amplitude =", np.round(Eps_Max,3))
print(" ")

print("The program has completed.")
```

Figure 8: Portion of Python code for single plane balancing showing the mini-max optimization, solution formatting and calculation of the RMS residual and mini-max amplitude

```
RMS Correction Weight = 283.8 at 324 degrees

RMS Residual =  0.395525
Mini-Max Amplitude = 0.712

Mini-Max Correction Weight = 314.4 at 321 degrees

RMS Residual =  0.525554
Mini-Max Amplitude = 0.618

The program has completed.
```

Figure 9: Single plane results from the least squares and mini-max solutions

Python Implementation – Two Plane

Extending the program to handle two balance planes requires some modification regarding the data input and tracking the balance weights. For simplicity, independent single plane and two plane balance programs were created, using the single plane program as a template for the two plane program. Python source code for both single plane and two plane balancing programs is included in the Appendix of this paper. For the two plane program, the data entry spreadsheet is modified to include amplitude and phase data for the second balance weight plane and is denoted as O+C2 as shown in Figure 10. Also note that a second calibration weight is included.

The steps to read in the O, O+C, and Wcal data to read from an Excel worksheet using the pandas library are slightly modified to account for the extra column of O+C2 data and the extra calibration weight data. When performing a two plane balance, the balancing engineer must decide on whether to leave the first calibration (trial) weight installed or remove it prior to adding the second calibration (trial) weight. There are technical and practical reasons to pursue either of these options, but many times the decision boils down to the following question, did the trial weight improve the vibration response or was the negative impact of the trial weight manageable? The decisions concerning the removal or keeping of calibration weights affects both the Null (N) and Calibration (C) vector calculations as shown in the code (Figure 11). The equations associated with the non-applicable conditions are commented out prior to executing the program.

	A	B	C	D	E	F	G	H
1	O(r)	O(ϕ)	O+C1(r)	O+C1(ϕ)	O+C2(r)	O+C2(ϕ)	Wcal(r)	Wcal(ϕ)
2	1.040	94	0.43	96	0.29	35	28.0	255
3	0.990	298	0.34	308	1.08	197	18.0	135
4	1.460	53	0.91	64	1.03	47		
5	1.490	182	0.93	196	0.96	182		
6	1.310	132	1.61	138	0.48	110		
7	2.200	308	2.71	317	0.94	230		
8	2.47	37	1.49	39	0.91	41		
9	1.650	179	1.1	190	0.78	191		

Figure 10: Excel worksheet Sheet1 showing O, O+C1, O+C2, and Wcal data in polar coordinates

```

N = -1. * O #Calculates the Null Vector
#N = -1. * O_C1 #Calculates the Null Vector if Wcal1 is left installed
#N = -1. * O_C2 #Calculates the Null Vector if Wcal2 is left installed

C1 = O_C1 - O #Use if removing Wcal1 after O_C1 run
C2 = O_C2 - O #Use if removing Wcal2 after O_C2 run
#C2 = O_C2 - O_C1 #Use if leaving Wcal1 installed after O_C1 run

```

Figure 11: Two plane balancing code for determining the N and C vectors depending on if the calibration weight (TW #1) is left installed or removed after the run

The influence vectors (H) are determined independently from the Calibration (C) and Calibration Weight (Wcal) vectors and then concatenated in a matrix (Figure 12). The least squares and mini-max solutions are calculated identically to the single plane program. The two plane program can be further extended to three, four, or more planes with similar modifications that were performed when extending the program from a single plane to two planes.

```

H1 = C1 / Wcal[0,0]
H2 = C2 / Wcal[1,0]

#Need to put the H1 and H2 vectors together into a single matrix first.
H = np.concatenate((H1,H2)).reshape((-1,2), order="F")
H = np.asmatrix(H)

```

Figure 12: Two plane balancing code for determining the influence vectors (H1 and H2) and then concatenating them into a matrix

Test Case #1

A generator driven by a gas turbine was exhibiting elevated direct and 1X vibration amplitudes above the OEM's alert limit at the NDE bearing during full speed no load (FSNL) and baseload conditions. The generator is directly coupled to the gas turbine's power turbine and operates at 3600 rpm. Vibration data is collected and monitored by pairs of orthogonally mounted proximity probes on the DE and NDE bearings. Single plane balancing using the balance plane on the exciter's fan wheel was performed to reduce the vibration amplitudes. Vibration measurements collected prior to and following the installation of a net 33-gram weight centered at 161 degrees against rotation from the Keyphasor target are summarized in Table 2.

The least squares and mini-max solutions are compared in Figure 13. Both solutions recommend a larger balance weight, 56 grams (least squares) and 51 grams (mini-max), but the angles are significantly different, 187 degrees and 156 degrees respectively. Also note that the least square solution minimizes the RMS Residual (1.85 versus 2.00) while the mini-max solution minimizes the maximum vibration amplitude (3.14 versus 2.57). Note that the results from the 33-gram balance weight reduced the vibration amplitudes below the alarm limits and were deemed acceptable, so the balance program was terminated.

Table 2: Summary of pre-balance and post-balance vibration measurements for the generator bearings in Test Case #1, TW#1 (33 grams at 161 degrees against rotation from the Keyphasor notch on the exciter fan ring)

Location	Pre-Balance		Post-Balance	
	Direct Amplitude (mil pp)	1X Vibration (mil pp @ deg)	Direct Amplitude (mil pp)	1X Vibration (mil pp @ deg)
	FULL SPEED NO LOAD CONDITIONS			
GENERATOR DE BRG Y	1.45	1.26 @ 182	2.32	2.02 @ 158
GENERATOR DE BRG X	0.67	0.42 @ 322	0.63	0.42 @ 297
GENERATOR NDE BRG Y	3.99	3.87 @ 350	2.92	2.81 @ 6
GENERATOR NDE BRG X	3.57	3.38 @ 129	2.58	2.40 @ 147
	BASELOAD CONDITIONS			
GENERATOR DE BRG Y	2.61	2.44 @ 190	2.78	2.53 @ 174
GENERATOR DE BRG X	0.89	0.68 @ 313	0.99	0.69 @ 306
GENERATOR NDE BRG Y	2.72	2.56 @ 351	2.32	2.13 @ 1
GENERATOR NDE BRG X	3.42	1.76 @ 143	2.04	1.72 @ 158

```
RMS Correction Weight = 56.1 at 187 degrees

RMS Residual =  1.850731
Mini-Max Amplitude = 3.139

Mini-Max Correction Weight = 51.4 at 156 degrees

RMS Residual =  2.00174
Mini-Max Amplitude = 2.565

The program has completed.
```

Figure 13: Single plane results (1X vibration) from the least squares and mini-max solutions for Test Case #1

Test Case #2

A generator driven by a gas turbine was exhibiting elevated direct and 1X vibration amplitudes near the OEM's alert limit at both the DE and NDE bearings loaded operation. The generator is directly coupled to the gas turbine's power turbine and operates at 3600 rpm. Vibration data is collected and monitored by pairs of orthogonally mounted proximity probes on the DE and NDE bearings. Two plane balancing using the balance planes on the exciter's fan wheel and the power turbine to generator coupling was performed to reduce the vibration amplitudes. Vibration measurements collected prior to and following the installation of a net 28-gram balance weight on the exciter fan ring centered at 255 degrees against rotation from the Keyphasor target are summarized in Table 3. The results indicated improvement at the generator NDE bearing at the expense of the vibration amplitudes at the generator DE bearing.

Table 3: Summary of vibration measurements for the generator bearings in Test Case #2 collected before balancing and after the installation of TW#1 (28 grams at 255 degrees against rotation from the Keyphasor notch on the exciter fan ring)

Location	Pre-Balance		TW #1	
	Direct Amplitude (mil pp)	1X Vibration (mil pp @ deg)	Direct Amplitude (mil pp)	1X Vibration (mil pp @ deg)
	FULL SPEED NO LOAD CONDITIONS			
GENERATOR DE BRG Y	1.40	1.04 @ 94	0.87	0.43 @ 96
GENERATOR DE BRG X	1.41	0.99 @ 298	0.95	0.34 @ 308
GENERATOR NDE BRG Y	1.81	1.46 @ 53	1.27	0.91 @ 64
GENERATOR NDE BRG X	1.80	1.49 @ 182	1.34	0.93 @ 196
	BASELOAD CONDITIONS			
GENERATOR DE BRG Y	1.76	1.31 @ 132	1.86	1.61 @ 138
GENERATOR DE BRG X	2.55	2.20 @ 308	2.98	2.71 @ 317
GENERATOR NDE BRG Y	2.73	2.47 @ 37	1.84	1.49 @ 39
GENERATOR NDE BRG X	2.00	1.65 @ 179	1.40	1.10 @ 190

To address the generator DE bearing vibration amplitudes, a balance weight (TW #2) of 18 grams was placed at 135 degrees against rotation from the Keyphasor notch on the power turbine to generator coupling. The results from the balance weight on the coupling are summarized in Table 4.

The least squares and mini-max solutions are presented in Figure 14. Both solutions indicate similar angular locations (258/246 degrees for the exciter weight plane versus 255 degrees where the balance weight was installed) and (122 degrees for the coupling weight plane versus the 135 degrees where the balance weight was installed). The mini-max solution predicts an increase of 0.1 for the RMS residual and a decrease of approximately 0.5 mil pp for the maximum vibration amplitude. Like Test Case #1, the balance weights installed in the field were sufficient to reduce the vibration amplitudes to acceptable values, so confirmation of the least squares and mini-max solutions were not realized.

Table 4: Comparison of vibration measurements for the generator bearings in Test Case #2 collected with TW#1 installed and after the installation of TW#2 (18 grams at 135 degrees against rotation from the Keyphasor notch on the power turbine to generator coupling)

Location	TW #1		TW #2	
	Direct Amplitude (mil pp)	1X Vibration (mil pp @ deg)	Direct Amplitude (mil pp)	1X Vibration (mil pp @ deg)
	FULL SPEED NO LOAD CONDITIONS			
GENERATOR DE BRG Y	0.87	0.43 @ 96	0.68	0.29 @ 35
GENERATOR DE BRG X	0.95	0.34 @ 308	1.63	1.08 @ 197
GENERATOR NDE BRG Y	1.27	0.91 @ 64	1.36	1.03 @ 47
GENERATOR NDE BRG X	1.34	0.93 @ 196	1.27	0.96 @ 182
	BASELOAD CONDITIONS			
GENERATOR DE BRG Y	1.86	1.61 @ 138	1.04	0.48 @ 110
GENERATOR DE BRG X	2.98	2.71 @ 317	1.60	0.94 @ 230
GENERATOR NDE BRG Y	1.84	1.49 @ 39	1.36	0.91 @ 41
GENERATOR NDE BRG X	1.40	1.10 @ 190	1.13	0.78 @ 191

```
RMS Correction Weight Plane #1 46.49 at 258 degrees
RMS Correction Weight Plane #2 17.7 at 122 degrees

RMS Residual =  0.625758
Mini-Max Amplitude = 1.26

Mini-Max Correction Weight Plane #1 38.98 at 246 degrees
Mini-Max Correction Weight Plane #2 14.58 at 122 degrees

RMS Residual =  0.723581
Mini-Max Amplitude = 0.777

The program has completed.
```

Figure 14: Two plane results from the least squares and mini-max solutions for Test Case #2 assuming that both TW#1 and TW#2 are removed

Test Case #3

A rotor kit was used to collect vibration data and compare balancing results using both the least squares and mini-max solution methods. The rotor kit was configured with three masses (disks) supported between two bearings spaced at 20 inches on center. The masses were installed at distances of 5 inches, 9 inches, and 12 inches from the inboard bearing, resulting in an asymmetric configuration (Figure 15). Orthogonal proximity probe pairs (X/Y) were installed at both ends of the rotor with the Y probes mounted at top-dead-center and the X probes mounted at 90° R when viewing from the driver to the driven. Unbalance weights were introduced into each of the disks, but only the 1st and 3rd disks were used in the balancing exercise, mimicking real machine conditions wherein the location of the balance planes do not always match the location of the imbalance.

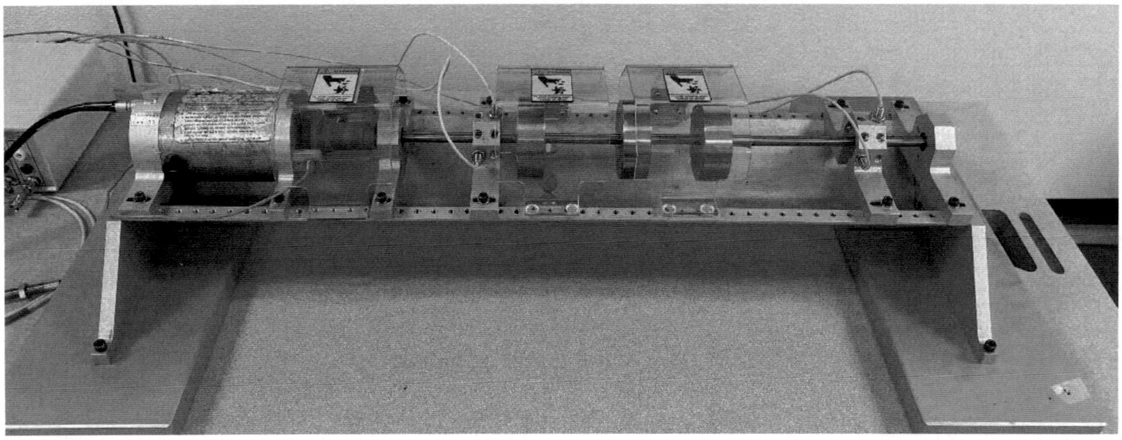

Figure 15: Rotor kit with a 20" bearing span and three masses (disks) spaced at 5", 9" and 12" from the inboard bearing

The original (Pre-Balance) 1X vibration collected at five speed points (1150, 1404, 5300, 6400, and 9990 rpm) is compared to the calibration runs (TW #1 and TW #2) in Table 5. Both trial weights were 0.4 grams and were installed on the first mass (disk) for TW #1 and the last mass (disk) for TW #2. The trial weights were installed at 180 degrees (TW #1) and 270 degrees (TW #2) against rotation from the Y probe when the Keyphasor notch was located under the Keyphasor probe. Note that the trial weights were installed and removed for each specific run.

Least squares and mini-max solutions were calculated (Figure 16). The results show meaningful differences in the weights and angles and an approximately 0.8 mil pp improvement in the maximum amplitude for the mini-max solution relative to the least squares solution. The rotor kit balance weights are available in 0.1, 0.2, 0.4, 0.8, 1.0, 1.2, and 1.4 gram sizes. The rotor kit masses (disks) have 16 holes spaced at 22.5 degrees. In order to best match the calculated least squares and mini-max solutions, weight splitting was required. The calculated correction weights for each solution are compared to the installed weights in Table 6. The maximum deviations from the calculated and installed weights were 0.02 grams and 4 degrees.

Table 5: Comparison of vibration measurements for the original (Pre-Balance), TW#1, and TW#2 runs in Test Case #3 collected at five different speed conditions

Location	Pre-Balance		TW #1		TW #2	
	1X Amplitude (mil pp)	1X Phase (deg)	1X Amplitude (mil pp)	1X Phase (deg)	1X Amplitude (mil pp)	1X Phase (deg)
	1150 rpm					
IBY	5.20	354	5.46	358	6.03	351
OBY	3.46	356	3.78	1	4.17	355
	1404 rpm					
IBY	8.06	153	7.07	151	7.83	145
OBY	5.38	155	4.88	153	5.58	148
	5300 rpm					
IBY	6.77	187	13.43	211	6.51	166
OBY	3.57	45	12.00	55	2.57	23
	6400 rpm					
IBY	5.08	16	6.56	0	3.34	341
OBY	6.96	181	9.54	176	6.55	157
	9990 rpm					
IBY	2.25	349	2.14	68	1.48	101
OBY	3.82	163	5.39	170	3.99	154

```
In [1]: %runfile 'C:/Users/204028477/OneDrive - Baker Hu
2025/LeastSquares_Minimax_2Plane.py' --wdir
RMS Correction Weight Plane #1 0.12 at 319 degrees
RMS Correction Weight Plane #2 0.61 at 209 degrees

RMS Residual =  3.956251
Mini-Max Amplitude = 6.669

Mini-Max Correction Weight Plane #1 0.2 at 234 degrees
Mini-Max Correction Weight Plane #2 0.75 at 203 degrees

RMS Residual =  4.50993
Mini-Max Amplitude = 5.821

The program has completed.
```

Figure 16: Two plane results from the least squares and mini-max solutions for Test Case #2 assuming that both TW#1 and TW#2 are removed

Table 6: Comparison of calculated and installed correction weights based upon the least squares and mini-max solutions

Location	Least Squares - Calculated		Least Squares - Installed		Mini-Max - Calculated		Mini-Max - Installed	
	Weight (gram)	Angle (deg)	Weight (gram)	Angle (deg)	Weight (gram)	Angle (deg)	Weight (gram)	Angle (deg)
BP1	0.12	319	0.10	315	0.20	234	0.20	236
BP2	0.61	209	0.59	210	0.75	203	0.74	203

Correction weights for both the least squares and mini-max solutions were installed independently and the results are compared to the Pre-Balance condition in Table 7. The Bode plots of the compensated1X vibration from the IB Y probe are shown in Figure 17 to compare the Pre-Balance, Least Squares, and Mini-Max runs. The calculated RMS values were 5.34 (Pre-Balance), 3.75 (Least Squares), and 3.79 (Mini-Max). The calculated mini-max values were 8.06 (Pre-Balance), 6.61 (Least Squares) and 6.28 (Mini-Max). Both least squares and mini-max solutions resulted in improved vibration responses at the five speed conditions used for the calculations. The solutions responded as expected with Least Squares producing the smallest RMS value and Mini-Max producing the smallest maximum vibration amplitude.

Table 7: Comparison of vibration measurements for the original (Pre-Balance), Least Squares, and Mini-Max runs in Test Case #3 collected at five different speed conditions

Location	Pre-Balance		Least Squares		Mini-Max	
	1X Amplitude (mil pp)	1X Phase (deg)	1X Amplitude (mil pp)	1X Phase (deg)	1X Amplitude (mil pp)	1X Phase (deg)
	1150 rpm					
IBY	5.20	354	5.00	330	4.51	343
OBY	3.46	356	3.25	332	3.00	345
	1404 rpm					
IBY	8.06	153	6.61	144	6.28	138
OBY	5.38	155	4.60	146	4.42	140
	5300 rpm					
IBY	6.77	187	2.33	181	2.75	195
OBY	3.57	45	1.91	106	2.44	90
	6400 rpm					
IBY	5.08	16	3.03	82	3.98	70
OBY	6.96	181	2.66	192	3.32	201
	9990 rpm					
IBY	2.25	349	2.63	123	2.68	114
OBY	3.82	163	2.88	153	2.84	159

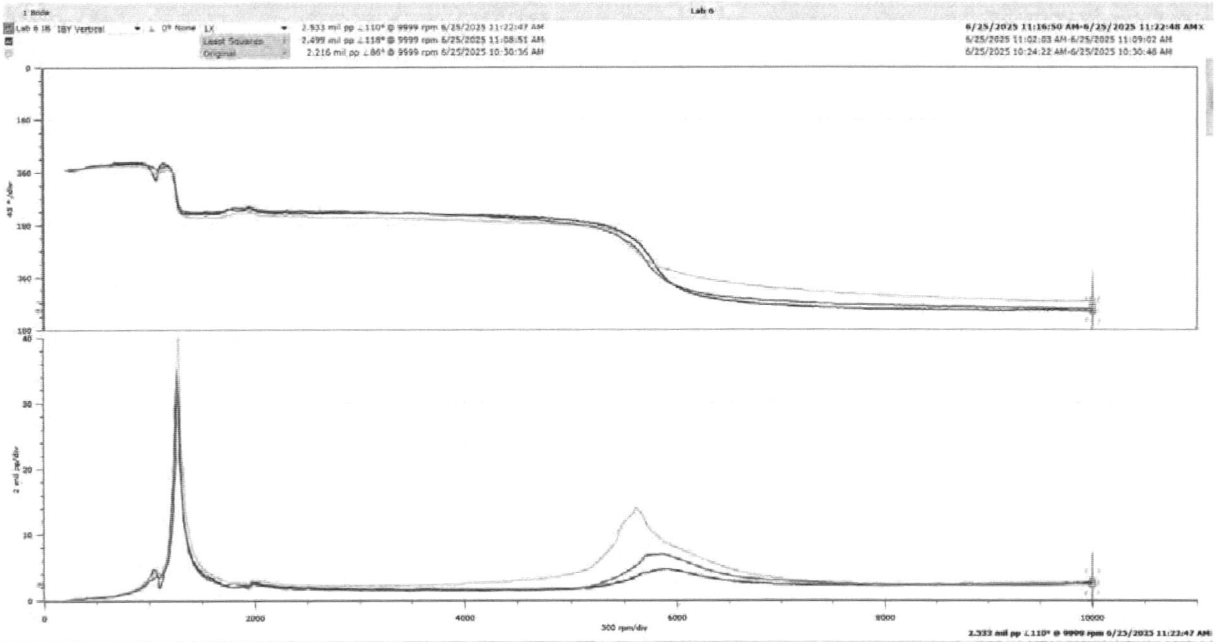

Figure 17: Comparison of compensated 1X vibration Bode plots the IB Y probe for the Pre-Balance (green), Least Squares (red) and Mini-Max (blue)

Conclusions and Future Work

Based on the work performed for this paper, the author concluded the following:

- Implementing least squares and mini-max balancing techniques in Python is simplified when using the NumPy and CVXPY libraries.
- Using the Pandas library in Python allows data to be input and read from a standard Excel spreadsheet.
- Extending the balance program from a single plane to two planes was demonstrated and requires minimal changes to the source code. Further extension to three or more planes is readily achievable if required.
- The mini-max solution offers an improvement over the least squares solution in the context of the maximum vibration amplitude observed. This characteristic is typical of the desired deliverable for most machine owners/operators.

The author is planning to continue this work with the intent of adding a simple graphical user interface for data entry, making the code more general and adding plots to compare the balancing results.

Nomenclature

C = vibration caused solely by the calibration weight(s)
H = influence vector(s)
Mini-max = maximum value from a group of values
MP = multiplane
N = null vector(s)
M = number of vibration measurements which is a combination of probe locations, speed conditions, and/or load conditions
O = original vibration
O + C = vectorial combination of original and calibration response vibration
R = RMS (root-mean square) residual
S = Sum of squared residuals
Wcal = Calibration weight
Wcor = Correction weight
W_Mini = Mini-max correction weight

References

[1] M. J. Titone and S. P. Ciesla, "Multiplane Balancing Using Commonly Available Modern Tools," in *Vibration Institute Annual Training Conference*, Niagara Falls, 2023.

[2] T. P. Goodman, "A Least squares Method for Computing Balance Corrections," *Journal of Engineering for Industry,* pp. 273-279, August 1964.

Appendix - Python Source Code

Single Plane Least Squares and Mini-max

```
#Installed numpy, from command prompt, pip install numpy
#Installed pandas, from command prompt, pip install pandas
#Installed cvxpy, from command prompt, pip install cvxpy
# O: Original Vibration vectors
# O_C: O+C Vibration vectors
# Wcal: Calibration (trial) weight
# N: Null Vector, equals -O
# H: Influence vectors
# Eps: Residual vibration
# S: Sum of squared residuals
# R: RMS Residual
# Wcor: Correction Weight based on Least squares (minimize RMS Residual)
# W_Mini: Correction Weight based on Mini-max optimization

import numpy as np
import pandas as pd
import cvxpy as cp

#Read in O, O+C, & Wcal data from Excel data file

file = "CT_Generator.xlsx"

xl = pd.ExcelFile(file)
df = xl.parse("Sheet3")
Temp = df.to_numpy()

O = Temp[:,0]
O = O.astype(complex)
O = O.reshape(-1,1)

O_C = Temp[:,1]
O_C = O_C.astype(complex)
O_C = O_C.reshape(-1,1)

Wcal = Temp[0,2]
Wcal = complex(Wcal)

N = -1. * O #Calculates the Null Vector
#N = -1. * O_C #Calculates the Null Vector if Wcal is left installed

H = (O_C - O) / Wcal #Cacluates the influence vectors

Wcor = np.linalg.lstsq(H, N)[0] #Solves for the Correction Weight via Least
squares
Wcor = Wcor[0]
```

```
Wr = np.round(np.abs(Wcor),2) #The amplitude/weight of the correction weight,
Wcor
Wphi = np.mod(np.angle(Wcor, deg=True),360) #The angle of the correction
weight, Wcor
Wphi = np.rint(Wphi).astype(int)
print(" ")
print ("RMS Correction Weight =",*Wr, "at", *Wphi, "degrees")
print(" ")

M = len(O) #Determines the total number of vibration measurements
O = -1. * N #Accounts for Wcal remaining installed or removed
Eps = H * Wcor + O #Calculate the residual vectors, Epsilon
Eps_Amp = np.abs(Eps) #Calculate the amplitude of the residual vectors
Eps_Max = np.max(Eps_Amp)
S = np.sum(np.square(Eps_Amp)) #Calculate S "Sum of Squared Residuals"
R = np.sqrt(S/M) #Calculate R "RMS Residual"

print("RMS Residual = ", np.round(R,6))
print("Mini-max Amplitude =", np.round(Eps_Max,3))
print(" ")

#Mini-max with CVXPY
O = -1. * N #Accounts for Wcal remaining installed or removed
WM=cp.Variable((1,1),complex=True)
objective=cp.Minimize(cp.norm((H @ WM + O),"inf"))
prob=cp.Problem(objective)
prob.solve()
W_Mini = WM.value
W_Mini = W_Mini[0]
Wr = np.round(np.abs(W_Mini),1) #The amplitude/weight of the correction weight,
WMini
Wphi = np.mod(np.angle(W_Mini, deg=True),360) #The angle of the correction
weight, WMini
Wphi = np.rint(Wphi).astype(int)

Eps = H * W_Mini + O #Calculate the residual vectors, Epsilon
Eps_Amp = np.abs(Eps) #Calculate the amplitude of the residual vectors
Eps_Max = np.max(Eps_Amp)
S = np.sum(np.square(Eps_Amp)) #Calculate S "Sum of Squared Residuals"
R = np.sqrt(S/M) #Calculate R "RMS Residual"
print ("Mini-max Correction Weight =",*Wr, "at", *Wphi, "degrees")
print(" ")
print("RMS Residual = ", np.round(R,6))
print("Mini-max Amplitude =", np.round(Eps_Max,3))
print(" ")

print("The program has completed.")
```

Two Plane Least squares and Mini-max

```
#Installed numpy, from command prompt, pip install numpy
#Installed pandas, from command prompt, pip install pandas
#Installed cvxpy, from command prompt, pip install cvxpy
# O: Original Vibration vectors
# O_C: O+C Vibration vectors
# Wcal: Calibration (trial) weight
# N: Null Vector, equals -A
# H: Influence vectors
# Eps: Residual vibration
# S: Sum of squared residuals
# R: RMS Residual
# Wcor: Correction Weight based on WLS (minimize RMS Residual)
# W_Mini: Correction Weight based on Mini-max optimization

import numpy as np
import pandas as pd
import cvxpy as cp

file = "Gen_2Plane.xlsx"

xl = pd.ExcelFile(file)
df = xl.parse("Sheet3")
Temp = df.to_numpy()

O = Temp[:,0]
O = O.astype(complex)
O = O.reshape(-1,1)

O_C1 = Temp[:,1]
O_C1 = O_C1.astype(complex)
O_C1 = O_C1.reshape(-1,1)

O_C2 = Temp[:,2]
O_C2 = O_C2.astype(complex)
O_C2 = O_C2.reshape(-1,1)

#Wcal = Temp[0:2,3]
Wcal = Temp[:,3]
Wcal = Wcal.astype(complex)
Wcal = Wcal.reshape(-1,1)

N = -1. * O #Calculates the Null Vector
#N = -1. * O_C1 #Calculates the Null Vector if Wcal1 is left installed
#N = -1. * O_C2 #Calculates the Null Vector if Wcal2 is left installed

C1 = O_C1 - O #Use if removing Wcal1 after O_C1 run
C2 = O_C2 - O #Use if removing Wcal2 after O_C2 run
#C2 = O_C2 - O_C1 #Use if leaving Wcal1 installed after O_C1 run

H1 = C1 / Wcal[0,0]
```

```
H2 = C2 / Wcal[1,0]

#Need to put the H1 and H2 vectors together into a single matrix first.
H = np.concatenate((H1,H2)).reshape((-1,2), order="F")
H = np.asmatrix(H)

Wcor = np.linalg.lstsq(H, N)[0] #Solves for the Correction Weight via Least
Squares
W1 = Wcor[0,0]
W2 = Wcor[1,0]

Wr1 = np.round(np.abs(W1),2) #The amplitude/weight of the correction weight,
Wcor1
Wphi1 = np.mod(np.angle(W1, deg=True),360) #The angle of the correction weight,
Wcor1
Wphi1 = np.rint(Wphi1).astype(int)
Wr2 = np.round(np.abs(W2),2) #The amplitude/weight of the correction weight,
Wcor2
Wphi2 = np.mod(np.angle(W2, deg=True),360) #The angle of the correction weight,
Wcor2
Wphi2 = np.rint(Wphi2).astype(int)
print ("RMS Correction Weight Plane #1",Wr1, "at", Wphi1, "degrees")
print ("RMS Correction Weight Plane #2",Wr2, "at", Wphi2, "degrees")
print(" ")

M = len(O) #Determines the total number of vibration measurements
Wcor = np.asmatrix(Wcor)
O = np.asmatrix(O)
Eps = H * Wcor + O #Calculate the residual vectors, Epsilon
Eps_Amp = np.abs(Eps) #Calculate the amplitude of the residual vectors
S = np.sum(np.square(Eps_Amp)) #Calculate S "Sum of Squared Residuals"
R = np.sqrt(S/M) #Calculate R "RMS Residual"
Eps_Max = np.max(Eps_Amp)
print("RMS Residual = ", np.round(R,6))
print("Mini-max Amplitude =", np.round(Eps_Max,3))
print(" ")

# Mini-max with CVXPY
O = -1. * N #Accounts for Wcals remaining installed or removed
WM=cp.Variable((2,1),complex=True)
objective=cp.Minimize(cp.norm((H @ WM + O),"inf"))
prob=cp.Problem(objective)
prob.solve()
W_Mini = WM.value
W_Mini1 = WM.value[0,0]
W_Mini2 = WM.value[1,0]
Wr1 = np.round(np.abs(W_Mini1),2) #The amplitude/weight of the correction
weight, WMini1
Wphi1 = np.mod(np.angle(W_Mini1, deg=True),360) #The angle of the correction
weight, WMini1
Wphi1 = np.rint(Wphi1).astype(int)
Wr2 = np.round(np.abs(W_Mini2),2) #The amplitude/weight of the correction
weight, WMini2
```

```
Wphi2 = np.mod(np.angle(W_Mini2, deg=True),360) #The angle of the correction
weight, WMini2
Wphi2 = np.rint(Wphi2).astype(int)

Eps = H * W_Mini + O #Calculate the residual vectors, Epsilon
Eps_Amp = np.abs(Eps) #Calculate the amplitude of the residual vectors
Eps_Max = np.max(Eps_Amp)
S = np.sum(np.square(Eps_Amp)) #Calculate S "Sum of Squared Residuals"
R = np.sqrt(S/M) #Calculate R "RMS Residual"
print ("Mini-max Correction Weight Plane #1",Wr1, "at", Wphi1, "degrees")
print ("Mini-max Correction Weight Plane #2",Wr2, "at", Wphi2, "degrees")

print(" ")
print("RMS Residual = ", np.round(R,6))
print("Mini-max Amplitude =", np.round(Eps_Max,3))
print(" ")

print("The program has completed.")
```